EXPLORING ANTENNAS AND TRANSMISSION LINES BY PERSONAL COMPUTER

EXPLORING ANTENNAS AND TRANSMISSION LINES BY PERSONAL COMPUTER

John A. Kuecken

VNR VAN NOSTRAND REINHOLD COMPANY
─────────────────────────── New York

Copyright © 1986 by Van Nostrand Reinhold Company Inc.

Library of Congress Catalog Card Number: 85-9226
ISBN: 0-442-24714-1

All rights reserved. No part of this work covered by the copyright hereon may be reproduced or used in any form or by any means—graphic, electronic, or mechanical, including photocopying, recording, taping, or information storage and retrieval systems—without permission of the publisher.

Manufactured in the United States of America

Published by Van Nostrand Reinhold Company Inc.
115 Fifth Avenue
New York, N.Y. 10003

Van Nostrand Reinhold Company Limited
Molly Millars Lane
Wokingham, Berkshire RG11 2PY, England

Van Nostrand Reinhold
480 Latrobe Street
Melbourne, Victoria 3000, Australia

Macmillan of Canada
Division of Gage Publishing Limited
164 Commander Boulevard
Agincourt, Ontario M1S 3C7, Canada

15 14 13 12 11 10 9 8 7 6 5 4 3 2 1

Library of Congress Cataloging-in-Publication Data

Kuecken, John A.
  Exploring antennas and transmission lines by personal computer.

  Includes index.
  1. Antennas (Electronics)—Data processing.
2. Microcomputers. I. Title.
TK7871.6.K794 1985 621.38'.028'30285 85-9226
ISBN 0-442-24714-1

Preface

In the not too distant past, the student intent upon exploring some of the mysteries of antennas and transmission lines was faced with the study of some very sophisticated mathematical techniques. The direct solution of a number of the principal problems of the art involved using some very long-winded and laborious arithmetic. In tackling real antenna problems with a slide rule and log table, he might face weeks or months of tedious computation. On the other hand, the solution to a few very special-case problems could be obtained by very sophisticated alalytical techniques. These permitted one to obtain a "feel" for the way antenna problems worked but did not permit a detailed grasp of very many practical problems.

When the computer first made its appearance, it held the promise of a fantastic expansion of capabilities and a great widening of understanding. As a practical matter, however, it was sometimes a help and sometimes a snare and delusion. Some of the early machines that had to be programmed with jumper wires in a frame that looked a good deal like a furnace filter could, under the guidance of gifted programmers, produce answers to questions the antenna man had only been able to speculate about. On the other hand, on more than one occasion this writer has seen his allocation dissappear into a welter of charges while the resulting data was nothing but garbage. All too often the money alloted was gone long before a reasonable answer had been derived.

As matters progressed, the machines were programmed with punched cards that one handed to a clerk, and the results arrived later. If something was wrong with the program, the process of debugging was tediously slow since it always involved a queue for the run of cards. An attempt at correction could proceed at about the rate of one try per day.

At least for this writer, the first really satisfactory solution came with the introduction of the Hewlett-Packard 9800 desk-top computer. When equipped with a plotter and printer, this machine opened up whole avenues of investigation that had been too time-consuming to be practical. The machine itself could chew on antenna problems for considerable periods of time, but it was tireless and one eventually received the answers. Furthermore, it was right in the office, and one did not have stand in line to try something on it. If the program had bugs, you became aware right away and could work to correct them rather than having to pick up a pile of sheets of garbage answers the next day. Compared to current personal computers these machines were relatively inflexibly devoted to math and had very restricted data memory. Moreover, the output data was not easily labeled in plain English. Nevertheless, one learned to work around these restrictions. The convenience of having the machine at one's immediate disposal was so great that any complaint about its shortcomings seemed like quibbling.

The collection of programs in this book was started at about this time and grew by accretion. They cover a gamut of practical problems as well as a number of purely instructive manipulations and investigations. Since many of them have been found instructive or useful by a number of associates, it seemed worthwhile to assemble them into a comprehensive antenna and transmission line book.

As originally written, many of the programs were in the "calculator keyboard" language of the machines used. This text, on the other hand, is written in either BASIC or Applesoft. Translation is seldom difficult, and other languages can be readily accommodated. BASIC has the advantage of being easily understandable and fairly self-documenting. You will find a number of comments included in some of the programs for documentation purposes. Removal of these can sometimes result in faster execution.

This text is intended to cover much the same ground as the earlier ANTENNAS AND TRANSMISSION LINES by the same writer. An effort has been made to use the computer to facilitate understanding of antenna problems, and therefore a number of exercises have been included. In addition, a chapter on the use of a digital computer to solve numerical problems has been added. The intention throughout the text has been to make the subject matter as clear and understandable as possible.

This text is not intended as a first book on personal computer programming, nor, for that matter, is it intended as a first book on algebra. It is intended as an aid to understanding antennas for those already possessing these skills.

JOHN A. KUECKEN

Contents

Preface/v
1. **A Brief History/1**

    The Radiomen/7

2. **Antenna Math with a Personal Computer/11**

    Complex Notion/11
    The Operator "j"/14
    Integration/18
    Coordinate Transformations/20
    Great Circle Calculations/22

3. **Gain and Huygens Principle/25**

    Friis Transmission Formula/26
    Pattern Integration/27
    Huygens Principle/30

4. **Arrays of Sources/38**

    Printing with the Silentype/46
    Odd-Order Arrays/47
    Young's Wave Interference Demonstration/48

5. **Aperture Distributions/52**

    Aperture Taper/52
    Phasing and the End-fire Condition/58
    End-Fire Arrays/61
    Ring Arrays/65

6. **Geometric Collimators/70**

    The Spherical Mirror/71
    The Newtonian System (The Parabola)/73
    The Ellipse and the Hyperbola/76
    Aperture Integration/79
    Aperture Blocking/85

7. Off-Axis Feeds and Phase Errors/89

Moving Off Focus/90
Focusing/99
Phase Errors/101
Stacked-Beam Antennas/106

8. Tracking Antennas/109

The Conical Scan/109
Monopulse/112
Amplitude Monopulse/113
Feed Crowding/114
Noise and Tracking/117
Feed Arrangements/118
Phase Monopulse/119
Amplitude Versus Phase Monopulse/124
Combination Monopulse/124
Cassagranian Systems/125
Arrays/125

9. Shaped Beam Antennas/126

The Cardioid Antenna/126
The Cosecant-Squared Pattern/130
Physical Realization/132
The FFT/IFT/134

10. Unfilled Arrays/147

The Simple Interferometer/147
The Two-Element Interferometer/149
The Unfilled Array/151
The Correlation Antenna/157

11. The Telegraphers Equation/161

Voltage Reflection Coefficient/173
Time Domain Reflectometers/173
Attenuation/174

12. The Steady-State AC Case/176

Line Terminated in a Short Circuit/181
The Open-Circuited Line/182

13. The Smith Chart/187

Using a Smith Chart/194
The Computer Smith Chart/196
Finding Line Length and Attenuation/206

14. Elementary Radiators/209

The Elementary Dipole/213
Radiation Resistance/217

15. Wire Antennas/220

The Dipole Antenna/223
Radiation Patterns/227
Reviewing the Patterns/231
The Traveling-Wave Antenna/231

16. Mutual Impedance/236

Sine, Cosine, and Exponential Integral Relations/242

17. Electrically Small Antennas/247

The Hairpin Monopole/263
Impedance Calculation/264
Fundamental Limitations/266

18. Mobile and Vehicular Antennas/270

Land Mobile Antennas/270
The Shipboard Antenna/275
Antenna Couplers/276
Broadband Shipboard Antennas/287
Airborne Antennas/287

19. Reflections in Space and Antennas above Earth/291

Transmission Line Reflection Analogy/291
Finite Thickness/294
Oblique Incidence/296
Brewster's Angle/299
Electric Field Parallel to Interface/301
Imperfect Dielectrics/304
Horizontal Dipole Patterns/306
The Vertical Dipole/311

Index/317

EXPLORING ANTENNAS AND TRANSMISSION LINES BY PERSONAL COMPUTER

1
A Brief History

It has always seemed to this writer that a brief historic background can help us understand the basics of a particular art by placing things in the perspective of time and providing a concrete "jumping-off" place for our studies. Virtually nothing we do in the arts and sciences springs into life full-grown clothed and armored like Athene from the brow of Zeus. Since each of us living today stands upon the shoulders of giants, let us spend a few moments looking at the past before we launch into a pursuit of our present subject.

Looking back to the most ancient records, we can tabulate the various events that would eventually lead to a mastery of electromagnetic propagation:

600 BC   Thales of Miletus records the ability of rubbed amber to attract lint, dust, and light particles.

77 AD   Pliny describes the use of water-filled globes as burning glasses and magnifiers.

1280   Meissner describes spectacles as "beneficial to older people."

1296   Marco Polo brings a magnetic compass to Venice from the court of the Kublai Khan. (He also brought pasta!)

1600   William Gilbert shows that a magnetized sphere effects a compass in the same way as the earth does and describes the difference between electrostatic and electromagnetic attraction.

1614   John Napier invents logarithms.

1630   Richard Delamain invents the slide rule.

1646   Sir Thomas Browne uses the word *electricity* for the first time.

2 EXPLORING ANTENNAS AND TRANSMISSION LINES

1650   Otto Von Guericke builds the first electrostatic machine.

1665   Sir Isaac Newton invents the calculus.

1668   Newton invents the reflecting telescope.

1687   Newton publishes *Principia*.

1704   Newton demonstrates the spectrum of white light and publishes *Opticks*.

1733   Charles Du Fay distinguishes two kinds of electricity, "resinous" and "vitreous," and notes that articles charged with like charges repel and articles charged with opposite charges attract.

1735   Stephen Gray discovers that electricity can be transmitted by conductors such as metals but cannot be transmitted by "electrics" (dielectrics) such as silk and amber.

1745   E. G. Von Kleist and Pieter van Musschenbroek independently invent the Leyden Jar.

1746   Benjamin Franklin misnames the two kinds of electricity: negative for an excess of electrons and positive for a deficiency of electrons!

1752   Benjamin Franklin demonstrates the electrical nature of lightning with his kite-and-key experiment.

1769   James Watt patents the steam engine.

1780   Luigi Galvani notices that a nearby electrical spark excites the leg muscles in a disected frog.

1800   Alessandro Volta develops the battery, or voltaic cell. For the first time experimenters were provided with a source of continuously flowing electric current.

1805   Oliver Evans invents the high-pressure (250 psi) steam engine to propel an amphibious dredge—the Oruckter Amphibolos.

A BRIEF HISTORY

1820	Hans Christian Oersted discovers electromagnetism.
1822	André Marie Ampère formulates the law quantitatively describing the force between wires carrying electric currents. Electrical measurements are placed on a quantitative basis for the first time.
1826	Georg Simon Ohm formulates the law relating voltage, current and resistance.
1831	Michael Faraday in England and Joseph Henry in the US simultaneously discover electromagnetic induction. Faraday invents the dynamo-electric machine.
1832	Hypolite Pixii invents the first practical generator to provide the means for coupling the power of the steam engine to electricity.
1841	Arc lights are demonstrated in Paris.
1844	Samuel F. B. Morse and Alfred Vail exchange the first long-distance telegram between Washington and Baltimore.
1861	The transcontinental line of Western Union is completed, linking the coasts.
1969	Thomas Alva Edison invents the improved stock Ticker, or automatic printing telegraph receiver.
1876	Alexander Graham Bell applies for a patent on the telephone.
1877	Edison invents the phonograph.
1879	Edison invents the incandescent lamp.

Any antenna and transmission line book owes considerable debt to James Clerk Maxwell (1831–1879), among others. Our listing has glossed over certain pivotal figures in the drama, but we shall attempt to do justice to some of these as we proceed in an orderly fashion leading up to Maxwell and his contributions.

Following the pioneering work of Ampère, Dr. James Prescott Joule developed an experimental technique for determining the mechanical equivalent of heat. Joule used a falling weight to stir a paddle in water and measured the temperature rise caused by the work. He also determined the electrical equivalent of work. For the first time it became possible to equate a given amount of electrical work to an amount of mechanical work.

The contribution of Volta cannot be underestimated. Before the invention of the voltaic cell, electricity was a matter of brief sparks and discharges too swift to be subject to experimentation. The voltaic cell made it possible to quantify electricity, allowing Charles Augustin de Coulomb to establish a value for the unit of electricity that would plate out a gram molecular weight of an element. In investigating the nature of work, however, Joule was able to demonstrate that a grain of zinc in a battery was able to deliver only about one-twentieth of the mechanically equivalent work provided by a grain of coal. And the zinc was 20 times more expensive. Joule stated in a letter that "I can scarcely doubt that electromagnetism will eventually be substituted for steam in propelling machinery." The hard fact remained, however that electricity continued to be expensive as long as its generation depended upon the consumption of zinc. Thus, although a number of experimenters produced electric locomotives and canal boats and although the Paris Opera was lighted with arc lamps at a very early date, the first really practical use of electricity had to await the Morse Telegraph. Since the telegraph used very little electricity compared to the service it offered, the high cost of the batteries required could be easily supported.

Although the science of electrical engineering continued to advance, the early state of knowledge about electricity can be gauged from the fact that it took some free advice from a student of Joseph Henry for Samuel Morse to learn that the wires wound on his electromagnets had to be insulated! Henry personally showed Morse how to use a relay "repeater" so that his telegraph could cover appreciable distances.

The quantifying work of Ohm, Coulomb, Ampère, Faraday, and Henry helped to transform electricity from a mysterious art into a science. In 1858, the first transatlantic telegraph cable was completed by Cyrus Field. The unit failed electrically after several weeks, but even if it had not, it would have been a financial failure because of its dreadfully slow rate of operation. At any speed faster than a few words per minute, the dots and dashes became hopelessly scrambled. Sensitivity was not the problem because the mirror galvanometer invented by William Thomson provided the unit with a large and solid response. There seemed to be something wrong with the cable itself.

Thomson's success in taking over the engineering supervision of the revitalized cable company was to bring him fame and fortune. He was knighted Lord Kelvin by Queen Victoria. The new cable was both a technical and a business success. News, business information, and personal communications began a steady flow across the Atlantic. The Telegraphers Equations formulated by Kelvin represented the first time that an electrical engineering problem of major proportions had been solved in advance by a mathematical proceedure.

James Clerk Maxwell was born in Edinburgh, Scotland and educated at Edinburgh College and Trinity College, Cambridge. He came under the influence of Michael Faraday, and while teaching at Kings College, London, in 1864, he predicted the existence of electromagnetic waves. His *Treatise on Electricity and Magnetism*, published in 1873, provided a completely unified theory of electricity and magnetism.

The existence of the electromagnetic wave was not as generally accepted then as one might think today. There were still serious objections to it, and the arguments were fairly well-founded, given the state of science at the time.

As early as 1675, Ole Roemer had measured the velocity of light by astronomical means. Roemer was compiling an ephemeris of the eclipses of the moons of Jupiter, tables that could presumably be used to synchronize chronometers for navigation anywhere in the world. Roemer discovered that his data was superimposed by an annual sinewave of 1000-seconds peak-to-peak amplitude. He reasoned that the delay resulted from the 186-million-mile difference in distance between the earth and Jupiter caused by earth's orbit around the sun and that the 1000-second difference meant that light propagated at the enormous velocity of 186,000-miles per second.

A second factor influenced the considerations. In 1645 Blaise Pascal had carried a mercury tube barometer up Mont Blanc. He noted that the mercury column fell from its 30-inch sealevel height at the rate of about one inch per thousand-feet. By linear extrapolation, the atmosphere would then be only 30,000-feet deep, and the space between the planets and the sun would be a vacuum—filled with nothing at all!

Before Newton, Robert Hooke, who developed the spring law and invented the Gregorian telescope, had suggested a wave nature for light. Newton was not convinced. If light was a wave, then a wave in what medium? Space had been shown to be a vacuum, and it had been proven that sound could not propagate through a vacuum. Christian Huygens proposed a wave theory in which the waves were seen as propagating through the *Aether*.

The strongest of Newton's arguments against this theory was based on the fact that light propagates in straight lines. If light were of a wave nature, then why did it not fill in behind obstacles, like waves behind a piling? As we shall see shortly, Huygens came up with an explanation for this paradox, but Newton's view prevailed at the time and it was accepted that light consisted of *corpuscles*.

There are several other considerations that argue against the Aether theory. The velocity of sound in gases is proportional to the square root of the bulk modulus divided by the mass per per unit volume. Since the velocity of light is roughly a million times the velocity of sound in air, if the Aether were a million times less dense than air, it would also have to be a million times less compressible. Furthermore, a gas will not sustain a polarized wave, only a pressure wave.

On the other hand, if the Aether were a solid, the velocity would be determined by the square root of Young's modulus divided by the mass. Although a solid will sustain a transversely polarized wave, the mass and the modulus of most solids are related; toolsteel and aluminum, for example, exhibit just about the same sound velocity. Thus, if the Aether were a hundred thousand times less dense than toolsteel, it would have to be three million times stiffer!

The thought of the planets having to move through a substance three million times stiffer than toolsteel is not all that convincing!

From the time of Newton in 1727 until about 1800, the corpuscular theory of light held sway, resting largely upon the reputation of Newton himself. In 1800, Thomas Young, a professor at the Royal Institute in London, proposed a theory that explained the phenomenon of the colors both in soap films and Newton's Rings on the basis of wave interference. In 1808, he set up an experiment using light passing through two pinholes that clearly demonstrated the presence of wave interference. The experiment had another advantage in that it permitted the calculation of the wavelength of light.

In 1816, Sir David Brewster experimentally demonstrated the transverse polarization of light, thereby proving that if the Aether existed, it had to be a solid!

In 1845, Faraday demonstrated that the plane of polarization of light could be rotated by an electromagnet, implying that light and magnetism were somehow related.

In 1856, Weber and Kohlrauch demonstrated that the relationship between the electrostatic and electromagnetic units was 3×10^{10}, or nearly identical to the velocity of light in-cm/sec.

It was the mathematical genius of Maxwell that assembled the pieces into a unified picture. The Aether was dispensed with, and electromagnetic fields, which could be demonstrated to propagate freely though a vacuum, were substituted. The fact that light was electromagnetic had not been proven, but the theoretical evidence was there.

The man who provided the practical proof was Heinrich Rudolf Hertz (1857–1894). Hertz was born in Hamburg and studied under Heinrich Helmholtz. Helmholtz had extensively studied sound and hearing and had invented a number of resonators that could be used to isolate individual tones. Hertz set about to create electromagnetic waves from purely electrical sources. His generators and detectors drew heavily upon the concepts developed by Helmholtz concerning resonance and resonant reinforcement. Working at wavelengths of 10 to 20-cm, Hertz was able to demonstrate wave interference, polarization, refraction, reflection, and focusing. Between 1886 and 1888, Hertz was able to measure the refractive indices of more than 100 substances at microwave frequencies. He demonstrated conclusively that microwave signals produced by electrical apparatus behaved exactly as light waves.

The success of the Hertz experiments sparked a flurry of activity among physicists around the world. All manner of quasi-optical experiments were performed, the pinnacle, perhaps, being achieved by J. C. Bose in India. Bose measured wave properties using wavelengths as short as 3-cm using the equivalent of a waveguide horn antenna!

By the turn of the century, the interest in quasi-optical waves had waned, and there was little experimentation in this wavelength region until the mid-1930's when the interest was reawakened by the attempt to develop radar.

THE RADIOMEN

After the first big burst of experimentation, the interest in quasi-optical waves waned because investigators could see little practical use for them. The original apparatus was pretty crude and quite insensitive, such as the typical Hertz transmitter and ring receiver shown in Fig. 1.1. In the transmitter, a large voltage, usually intermittent, was applied, and the arc at the ball gap excited the dipole made up of paddles "d." The receiver worked by displaying a much smaller arc across its gap. Hertz used a parabolic cylinder reflector to concentrate the energy. Some of his receivers looked just like his transmitters but with a smaller gap. One had to have enough signal to make the receiver spark on sympathy.

In 1890, E. Branley developed the *Coherer*, which made detection considerably more sensitive. In Marconi's version, a glass tube 4-cm long was fitted with two silver pole pieces separated by about half a millimeter. This gap was filled with a mixture of fine nickel and silver filings and a trace of mercury and then evacuated to a pressure of about 4-mm Hg. In its normal condition, this device was nearly an open circuit to the driving battery. When a high-frequency signal passed through it (by displacement current—capacitance), the particles would "cohere," and the dc resistance would drop dramatically. Marconi used an interruptor apparatus similar to a doorbell to "de-cohere" the device. The clapper would beat on the coherer, thereby vibrating the powder and breaking the circuit. If the RF signal was still present, the cycle would repeat. The buzz of the tapper gave an audible signal of the RF.

When Marconi began his researches, he used apparatus similar to that of Hertz. By 1897, however, he had increased to a wavelength of 120-cm, or 250-MHz. He was discovering that he could signal over greater ranges with longer waves. In 1900, he was able to report successful communications between U.S. and British battleships over a distance of 38-miles. By 1902, he was able to pick up strong signals over 1551-miles and decipherable ones over 2099-miles, but only at night. During daylight hours, signals failed at distances over 700-miles because of the Kennely-Heaviside effect (more about that later).

The change in wavelengths is explained by the sketch of the transmitter and receiver Fig. 1.2. (The schematics for the Marconi apparatus are copied from his July 13, 1897 U.S. Patent, with some liberties taken to make the circuits conform to modern circuit conventions.) A large overhead aerial tuned the system to a wavelength on the order of 200 to 300-meters. When the key was depressed, the circuit primary was energized. After the magnetic field had risen to a suitable level, interrupter B pulled open, causing the magnetic field to collapse. The spark across the dual spark gap excited the aerial into oscillation. A similar aerial tuned coherer J, which operated as previously described. It is noteworthy that Marconi used RF chokes K to isolate the de-coherer circuitry.

Some idea of the transmitter can be had from the description in 1897 that Mr. Marconi had gone from a coil that made 6-in. sparks to one that produced 20-in. sparks!

On December 12, 1901, Marconi received the first transatlantic signal, which originated from his transmitter in Poldhu, England, at his receiving station in St John's, Newfoundland. In 1905, Marconi received a patent on a horizontal directional aerial for long waves, and in 1909 he shared the Nobel Prize in Physics with Ferdinand Brown, the inventor of the cathode ray tube.

The Edison effect discovered in 1888 was put to use in the Flemming Valve of 1904 as a detector. In 1906, Lee De Forest placed a grid between the filament and plate to create the amplifier. In the same year,

A BRIEF HISTORY 9

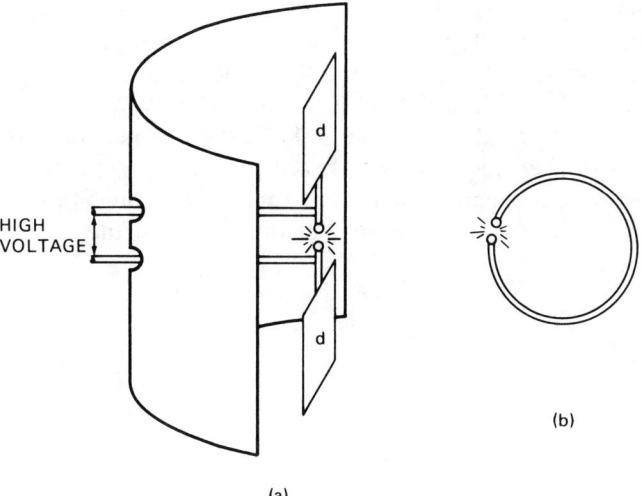

Fig. 1.1 The Hertz transmitter (a) and receiver (b).

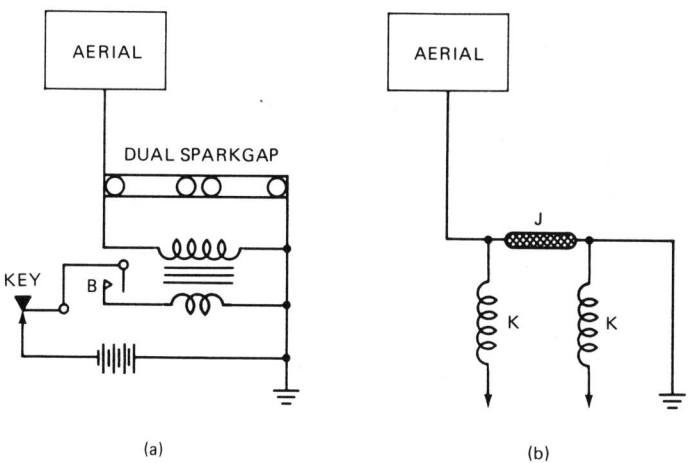

Fig. 1.2 The Marconi transmitter (a) and receiver (b).

Prof. Reginald Fessenden broadcast phonograph music using a continuous wave signal developed by a mechanical rotating alternator.

In 1912, Edwin Armstrong invented the Superheterodyne receiver.

In 1920, station KDKA in Pittsburg and station WWJ in Detroit began regular commercial broadcasting, and in 1923, the first network broadcast from what was to become NBC was heard.

The group of investigators that were to follow wrought tremendous accomplishments on this foundation, but the names of the pioneers still ring the loudest.

2
Antenna Math with A Personal Computer

Antenna work is prone to be quite mathematical in nature. Not only is the design phase mathematical but the usage and testing as well. When one measures the impedance of an antenna, what one usually measures is actually the impedance as seen at the near end of a cable of an appreciable length and probably with significant losses. Only after mathematical "rotation" can one obtain the information of interest, namely the impedance at the antenna feedpoint. Similarly, when it comes time to use the antenna, one may have to calculate the great circle path to the target, the bearing angle to the satellite, or the MUF over a path in order to determine the Optimum Working Frequency (OWF).

In the design process, one frequently must perform computations for impedance matching on lower frequency antennas and distribution and pattern calculations on higher frequency and microwave units. In this chapter, we shall attempt to explain some of the fundamentals and their application to antenna computations involving personal computers. Some generally useful techniques are treated in this chapter and will be referred to in later chapters.

COMPLEX NOTATION

An overwhelming part of the math found in this text will use "complex" notation, or vector algebra. The use of this form of math for alternating-current problems was introduced by Charles Proteus Steinmetz at the General Electric Co. shortly after the turn of the century. Its usefulness in solving circuit problems remains in force to this day.

This math is termed "complex" because it makes use of both "real" and "imaginary" numbers. The term "real" has a different meaning in this context from what is generally meant by it in computer parlance. The Apple II + manual recognizes three types of numbers: scientific notation, integer, and "real," the latter referring to floating-point numbers. This usage came from Fortran originally. In this text, we will be using the Electrical Engineers' definition of "real," where it is taken to mean resistive, in-phase, or the like.

12 EXPLORING ANTENNAS AND TRANSMISSION LINES

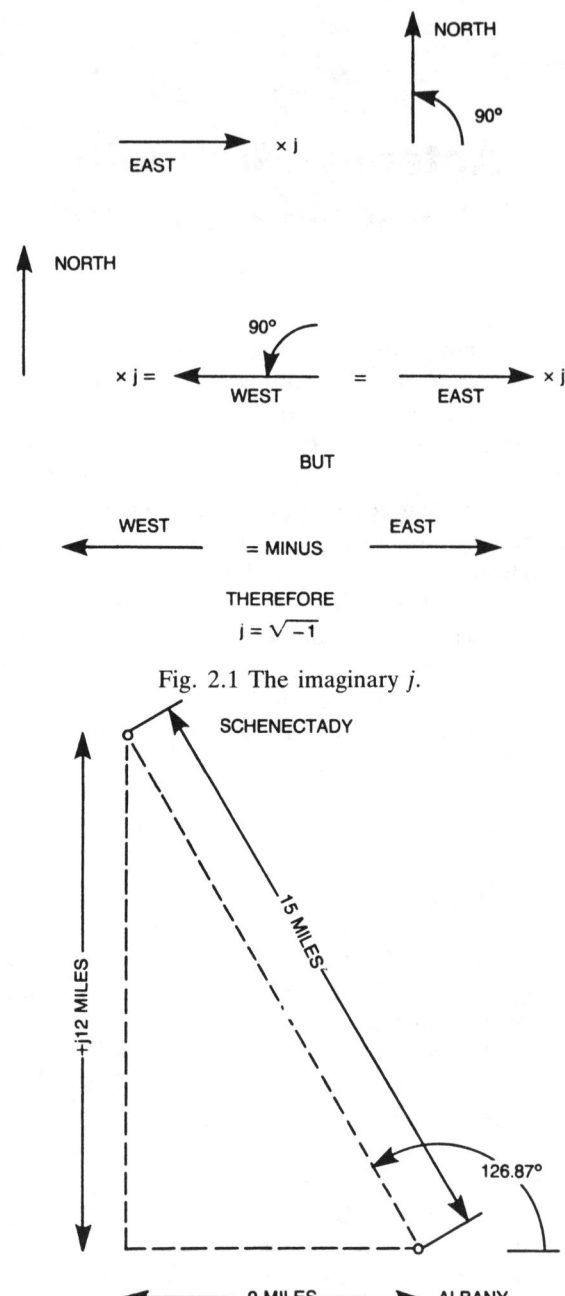

Fig. 2.1 The imaginary j.

Fig. 2.2 Steinmetz illustration of vector quantities.

There is no problem associated with this change of definition, and the computer does not care about it in the least, but the reader should be aware of the difference. We shall be expanding upon this topic in a moment. Along with our "real" numbers, we shall also be dealing with "imaginary numbers," which in truth are no less real than the "real" numbers.

For example, if someone tells you he is departing at a rate of 25-mph, he has given you little hint of where to find him in the future. On the other hand, the pilot who tells En-Route Traffic Control that he is passing checkpoint B with a ground speed of 575-mph on course 272-degrees magnetic has given a complete description of where he may be found in the immediate future. He has provided a vector description of his flight path that indicates where the vector starts, where it is pointed, and its length. In this example, the vector consists of a velocity, but it could have been any of a number of other quantities such as acceleration, distance, force, voltage, etc. The only important characteristic is the fact that a vector always has a point of origin, points in some direction, and has a certain magnitude. The direction need not be a physical direction. It can just as well be a phase angle that does not directly apply to space but is rather a mathematical concept.

THE OPERATOR "j"

Now let us suppose that we would like to define an "operator" that will rotate a vector through an angle of 90-degrees. As noted, the angles in question do not have to be physical angles in space, but it sometimes helps to think of them in that way. For our present purpose, let us consider that our prime, or starting, direction is east and that our "operator j" will rotate the vector from east to north. The vector has a length of 1 and begins at the geographic location of the nonpointed end of the vector labeled East (see Fig. 2.1).

We see from Fig. 2.1 that if one application of "j" rotates the vector from east to north, then a second will rotate it from north to west. Since west is equal to minus east, however, $(j*j) = -1$, and j must be equal to the square root of minus one. This is what is called an imaginary number since no number multiplied by itself yields a negative result.

Steinmetz was fond of saying that there is nothing more imaginary about an imaginary number than there is about the distance between Albany and Schenectady (where he worked for General Electric). The illustration of Fig. 2.2 was used to make his point. It also serves to show the two principal mechanisms for describing vector quantities. The description that gives the distance as $-9 + j12$-miles is the Cartesian

14 EXPLORING ANTENNAS AND TRANSMISSION LINES

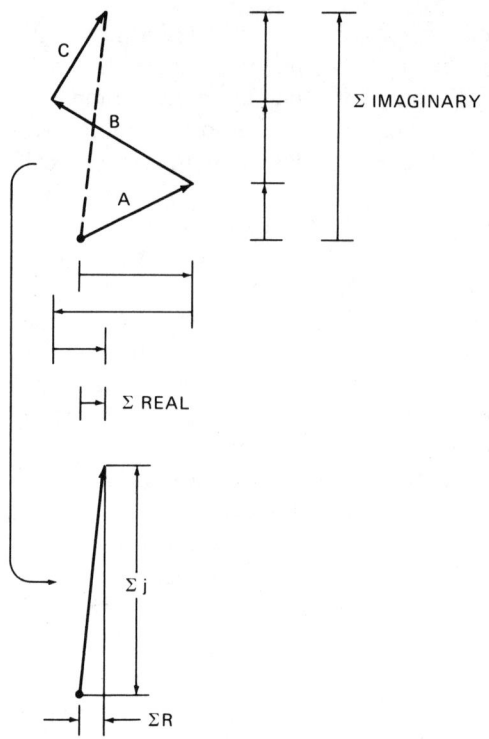

Fig. 2.3 Vector addition, in which the resultant is formed by separately reals and imaginaries.

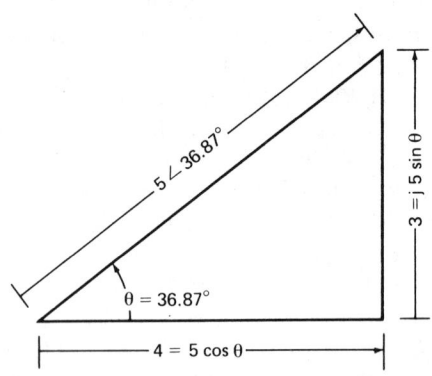

Fig. 2.4 Graphic illustration of Euler's equation.

ANTENNA MATH WITH A PERSONAL COMPUTER   15

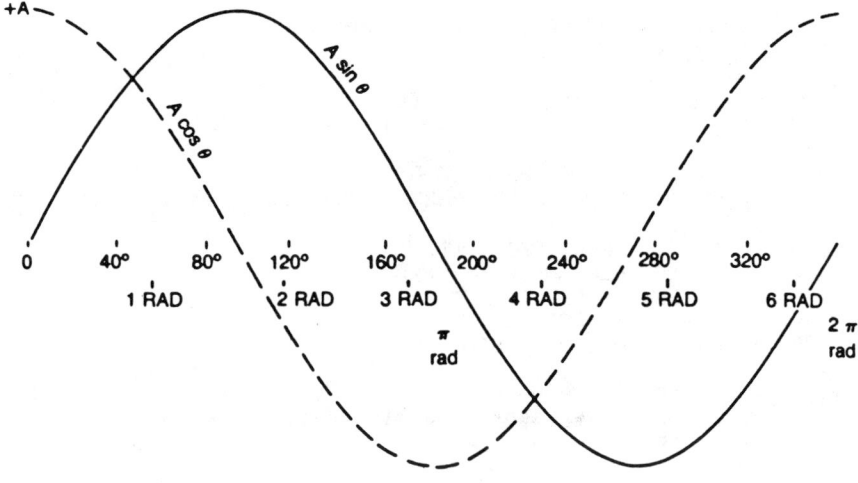

Fig. 2.4 (*continued*)

(after René Descartes) or complex (meaning real-plus-imaginary) description, whereas 15-miles ∠126.87-degrees is the polar description.

Most ac calculations switch between the concepts. For example, addition and subraction are much more easily performed in complex notation. In the example shown in Fig. 2.3, you can see that the resultant is formed quite simply by separately adding (algebraically) the reals and the imaginaries. The two sums are then used to construct the resultant.

Vector multiplication and division are accomplished best using the polar form To multiply two polar vectors, one simply multiplies the magnitudes and adds the angles. Division is accomplished by dividing the magnitudes and subtracting the angle of the denominator from the angle of the numerator. Calculation of powers and roots follows from this procedure. For example, to find the cube root of a vector, one takes the cube root of the magnitude and divides the angle by three.

Also related to these matters is Eulers equation, sometimes referred to as *exponential notation*. This is given by:

$$A\varepsilon^{j\theta} = A\cos\theta + jA\sin\theta \tag{2.1}$$

$$A\varepsilon^{-j\theta} = A\cos\theta - jA\sin\theta \tag{2.2}$$

where ε = natural log base (= 2.718) and $j = \sqrt{-1}$.

The term $A\varepsilon^{j\theta}$ describes the location of the tip of a vector rotating through an angle θ. Note that the two equations follow from the fact that reversing the sense of either phase in a two-phase system reverses the direction of rotation of the phase sequence and therefore of the motor

16 EXPLORING ANTENNAS AND TRANSMISSION LINES

```
2000   PRINT "ENTER DEGREES OR RADIANS-D/R"
2005 PI = 3.14159
2010   INPUT DR$
2020   IF DR$ = "D" THEN D = 1: GOTO 2100
2030   IF DR$ = "R" THEN D = 0: GOTO 2100
2038   HOME
2040   PRINT "IT HAS TO BE D OR R"
2045   PRINT "%%%%%%%%%%%%%%%%%%%%%%%%%%%%%%%%%%%%%%%%"
2050   GOTO 2000
2100   PRINT "IS INPUT DATA POLAR"
2110   PRINT "OR RECTANGULAR?--P/R"
2120   INPUT PR$
2130   IF PR$ = "P" GOTO 2200
2140   IF PR$ = "R" GOTO 2500
2145   HOME
2150   PRINT "EITHER P OR R"
2160   PRINT "########################################"
2200   PRINT "ENTER MAGNITUDE"
2210   INPUT M
2220   PRINT "ENTER ANGLE"
2230   INPUT TH
2240   IF D = 1 THEN TH = TH * PI / 180
2250 R = M *  COS (TH)
2260 J = M *  SIN (TH)
2270   PRINT R,J
2280   STOP
2500   REM %%%%%%%%%%%%%%%%%%%%%%%%%%%%%%%%%%%%%
2510   REM   RECTANGULAR TO POLAR
2520   REM ####################################
2530   PRINT "INPUT REAL"
2540   INPUT R
2550   PRINT "INPUT IMAGINARY"
2560   INPUT J
2570 M =  SQR ((R * R) + (J * J))
2580 TH =  ATN (J / R)
2590   IF D = 1 THEN TH = 180 * TH / PI
3000   PRINT M,TH
5000   STOP


ENTER DEGREES OR RADIANS-D/R
?D
IS INPUT DATA POLAR
OR RECTANGULAR?--P/R
?P
ENTER MAGNITUDE
?5
ENTER ANGLE
?36.869898
4.00000161      2.99999785

BREAK IN 2280
```

Program 2.1

connected to the system. Figure 2.4 illustrates the derivation of the terms.

It is obvious that good polar-to-rectangular and rectangular-to-polar conversion routines would be worthwhile aids. Many engineering-grade pocket calculators offer them, but few BASIC program computers do. The routine given in Program 2.1 performs these operations. The example shows its solution for a 4-3-5 triangle.

There are a few points worthy of interest in this program. Let us suppose that we want to convert perfectly valid data with real = 0.00 and imaginary = 10 to polar. The machine will respond with:

DIVISION BY ZERO ERROR IN 2580

To avoid this, we could add:

2545 IF R = 0 THEN R = .0000001

If the same problem is now attempted, the machine will return a magnitude of 10 and an angle of 90.0000755-degrees, which are values close enough for most work.

A second point is more subtle. Since most BASICS offer only the inverse tangent function, running the program for real = −1 and imaginary = 1000 will print the angle as −89.9427802-degrees. The ATN function will return the same angle if *either* the real or the imaginary part is negative.

If the latter angle and magnitude are run through the machine in reverse, one obtains the answer that the original components were 0.999999098 and −1000! The sine and cosine functions are thus not properly adjusted. If we wish to place the angle in the correct quadrant, we can do so by adding the following statements:

2581 IF R > 0 AND J > 0 GOTO 2590
2582 IF R < 0 THEN TH = TH + PI : GOTO 2590
2583 IF R > 0 AND J < 0 THEN TH = (2*PI) − TH

This alteration will locate the result in the proper quadrant and with all angles positive. Although this information is not particularly necessary if one uses only the tangent function, it may be necessary when one uses other trig functions, when angles must be added or divided, or the like.

The reason for stressing these points is the fact that the majority of the bugs in programs created by people experienced in the art consist of just this sort of thing. One writes program and after it starts to run, the computer beeps and one sees the DIVIDE BY ZERO ERROR or a similar annoying flag. Either that, or the program seems to run well and then spits out ridiculous results. As the values progress, the vector rolls around, and the machine swaps quadrants on you.

The author estimates that a good 75-percent of the debugging time on the programs in this book where for items of this ilk. They can be insidious. A program can be run on a proof problem that doesn't expose the bug because it does not get out of the first quadrant and on so.

INTEGRATION

A great deal of the math needed for antennas involves integration. Although a digital computer cannot actually do integration, it can do summation with progressively finer increments and thus approach integration. As an example, Program 2.2 shows integration over the surface of a sphere (see Fig. 2.5). The example was selected because the

```
135  HOME
140  PI = 3.14159
150  PRINT "ENTER STEP SIZE IN DEGREES"
160  INPUT S
170  S = (S / 180) * PI
180  A = S / 2
190  REM  THE AREA OF THE ANNULUS IS
200  DA = 2 * PI * SIN (A) * S
210  AS = AS + DA: REM  THE SUMMATION
220  A = A + S: REM  INCREMENT THE ANGLE
225  IF A < PI GOTO 200
230  PRINT "AREA EQUALS,",AS
240  PRINT "COMPARED TO TRUE  ";AS / (4 * PI)
1000  STOP

]RUN
ENTER STEP SIZE IN DEGREES
?20
AREA EQUALS,    12.6303862
COMPARED TO TRUE  1.00509505

BREAK IN 1000
]RUN
ENTER STEP SIZE IN DEGREES
?1
AREA EQUALS,    12.5665195
COMPARED TO TRUE  1.00001269

BREAK IN 1000
```

Program 2.2

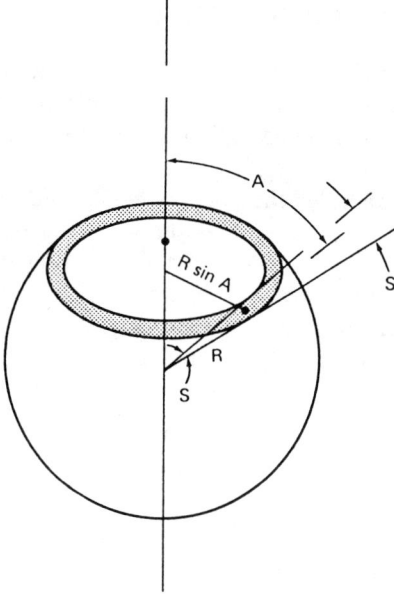

Fig. 2.5 Integration over the surface of a sphere.

area of a sphere is known to be 4*PI*R*R. We can thus compare the attained answer with the true answer. For the example, we assumed that $R = 1$. The program first of all calculates the area of the annulus, which is taken as the product of the circumference times the width, as shown in line 200. The areas of the annuli are then summed in line 210. It can be seen that even for a step as large as 20-degrees, the accuracy is not too bad. This form of integration is used in a great number of antenna problems.

A few other points are worthy of note. First of all, not all BASICs offer PI; the Apple, for example, does not. Also, most of them will run trig functions only in radians. If you wish to use degrees, you must perform your own conversion, as shown in line 170. Of course, this formula uses the step angle as the width of the strip, which is inherently in radians anyway.

Since there are quite a few antennas that lend themselves to description in spherical coordinates, as in this example, the use of spherical coordinates is fairly common in antenna problems.

COORDINATE TRANSFORMATIONS

One of the common features of antenna work is the frequent need to transform coordinates. An antenna pattern described very simply in one coordinate system may have to be evaluated in another coordinate system for various reasons.

Suppose, for example, that we are concerned with a horizontal dipole antenna lying along the Y axis, as shown in Fig. 2.6. This antenna in free space can be described entirely in terms of angles c and d. We need to describe the response, however, in terms of the departure from the zenith angle (W) and the azimuth angle (TH). The latter requirement could arise from a desire to evaluate the performancee of the antenna with steeply descending signals.

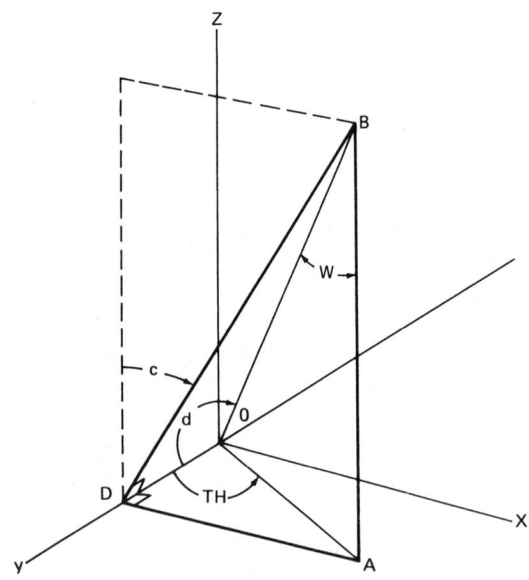

Fig. 2.6 Three-dimensional coordinate transformations.

Although the length of the radius vector is often not required, it can be calculated by using the Pythagorean Theorem, as follows:

$$OA = OB\sin(W)$$

$$OD = OA\cos(TH) = OB\sin(W)\cos(TH)$$

$$DA = OB\sin(W)\sin(TH)$$

$$AB = OB\cos(W)$$

Then,

$$DB = \sqrt{((DA)^2 + (AB)^2)}$$

$$\angle d = \text{Arctan}((DB)/(OD))$$

Angle d is calculated as the arc tangent of DB/OD. It could also be calculated as the arc sine of DB/OB, but the Apple (and most other BASICs for that matter) doesn't provide this function directly. The same cautions regarding the use of the tangent noted earlier apply here.

The inclusion of line 545 in Program 2.3 prevents the DIVIDE BY ZERO ERROR from stopping the program. The problem here does not so much concern the program proper as it does a routine that is incorporated as a portion of a larger program. It is annoying to have a program trip in its middle because it comes across a divide by zero error or some other flaw.

In this particular example, angle c was not calculated since the pattern of the dipole is symmetrical about the axis of the dipole and is therefore not a function of c. In cases where it might be required, it may be seen that it is equal to arctan $((DA)/(AB))$.

```
490  REM  0000000000000000000000000000000000000000
495  REM  COORDINATE TRANSFORMATION
500  PI = 3.14159
510  DA =   SIN (W) *  SIN (TH)
520  AB =   COS (W)
530  DB =   SQR ((DA * DA) + (AB * AB))
540  OD =   SIN (W) *  COS (TH)
545  IF  ABS (OD) < .001 THEN D = PI / 2: GOTO 570
550  D =   ATN (DB / OD)
560  REM  ****************************************
570  STOP
```

Program 2.3

22 EXPLORING ANTENNAS AND TRANSMISSION LINES

Another point is worthy of note. In the coordinate system we have been using for spherical problems, angle d begins with value 0 at the axis and increases to value PI at the axis in the other direction. These evaluations are different from those used in geography where latitudes are given as $+\text{PI}/2$ at the North Pole, $-\text{PI}/2$ at the South Pole, and zero at the equator. In fact any self-consistent system of coordinates can be used, but the key word here is "self-consistent." It is relatively easy to get caught up in an inconsistent coordinate system and waste a good deal of time trying to figure out why the answers are crazy.

GREAT CIRCLE CALCULATIONS

Program 2.4 provides a routine that will produce most of the information needed for long-haul shortwave work. Up to line 490, the program

```
130 PI = 3.141592654
140   HOME
150   PRINT "ENTER FIRST LATITUDE"
160   PRINT "DEGREES"
165   INPUT Z
170   PRINT "MINUTES"
180   INPUT Y
190 Z = Z + Y / 60
200   PRINT "SECONDS"
210   INPUT Y
220 Z = Z + Y / 3600
230   PRINT "ENTER FIRST LONGITUDE"
240   PRINT "DEGREES"
250   INPUT X
260   PRINT "MINUTES"
270   INPUT W
280 X = X + W / 60
290   PRINT "SECONDS"
300   INPUT W
310 X = X + W / 3600
320   PRINT "ENTER SECOND LATITUDE"
330   PRINT "DEGREES"
340   INPUT V
350   PRINT "MINUTES"
360   INPUT Y
370 V = V + Y / 60
380   PRINT "SECONDS"
390   INPUT Y
```

Program 2.4

```
400 U = U + Y / 3600
410  PRINT "ENTER SECOND LONGITUDE"
420  PRINT "DEGREES"
430  INPUT U
440  PRINT "MINUTES"
450  INPUT W
460 U = U + W / 60
470  PRINT "SECONDS"
480  INPUT W
490 U = U + W / 3600
491  HOME : PR# 1
492  PRINT "FIRST LATITUDE    ";Z;"  DEGREES"
493  PRINT "FIRST LONGITUDE   ";X;"  DEGREES"
494  PRINT "SECOND LATITUDE   ";V;"  DEGREES"
495  PRINT "SECOND LONGITUDE ";U;"  DEGREES"
520 C =  ABS (X - U)
522  PRINT "TIME DIFERENCE    ";C / 15;"  HOURS"
525 C = C * PI / 360
530  IF  ABS (Z) >  ABS (V) THEN LB = Z:LA = V: GOTO 550
540 LB = V:LA = Z
550 NN = LB - LA
560 ND = LB + LA
570 NN = NN * PI / 360
580 ND = ND * PI / 360
585  IF ND = PI / 2 THEN ND = ND + .000001
586  IF ND = 0 THEN ND = ND + .000001
590 SA =   ATN ((1 /  TAN (C)) *  COS (NN) /  SIN (ND))
595 SB =   ATN ((1 /  TAN (C)) *  SIN (NN) /  COS (ND))
600 A1 = SB - SA
610 A2 = SB + SA
612  PRINT ""
620  PRINT "BEARING FROM LOCATION CLOSEST"
625  PRINT "TO EQUATOR    ",A1 * 180 / PI
630  PRINT " DEGREES EAST OF NORTH"
640  PRINT "BEARING FROM OTHER LOCATION"
650  PRINT A2 * 180 / PI;" DEGREES EAST OF NORTH"
660  PRINT ""
670  REM *****************************************
680  REM     THE DISTANCE CALCULATION
690  REM )))))))))))))))))))))))))))))))))))))))))
700 ZI = (90 - Z) * PI / 180
710 VI = (90 - V) * PI / 180
720 C =  ABS ((X - U) * PI / 180)
730 ZW = (  COS (ZI) *  COS (VI)) + ( SIN (ZI) *  SIN (VI) *  COS (C))
740 ZD =  -  ATN (ZW /  SQR ( - ZW * ZW + 1)) + 1.5708
750  PRINT "ZD=  ";ZD * 180 / PI;" DEGREES"
760  PRINT "GREAT CIRCLE DISTANCE=";ZD * 180 * 111.12 / PI;"  KM"
1990  PR# 0
2000  STOP
```

Program 2.4 (*continued*)

```
FIRST LATITUDE      40.8111111  DEGREES
FIRST LONGITUDE     73.2527778  DEGREES
SECOND LATITUDE    -22.9525     DEGREES
SECOND LONGITUDE   43.3686111   DEGREES
TIME DIFERENCE      1.99227778  HOURS

BEARING FROM LOCATION CLOSEST,
TO EQUATOR         -23.7330377
 DEGREES EAST OF NORTH                        EXAMPLE
BEARING FROM OTHER LOCATION
150.681509 DEGREES EAST OF NORTH

ZD=  69.5492328 DEGREES
GREAT CIRCLE DISTANCE=7728.31074   KM
```

Program 2.4 (*continued*)

consists simply of the data entry and unit conversion. Map data is usually given in degrees, minutes and seconds, but the Apple requires this information in radians. The conversion proceeds first to decimal degrees and then to radians. Time differences are calculated since it is helpful to know the time along the path and the time at the destination when the frequency selection is being made.

Lines 585 and 586 are included to prevent the DIVIDE-BY-ZERO message. In line 730, the term ZW is actually the cosine of the great circle angle between the points. Since the Apple does not have an arc cosine function, the routine of line 740 is substituted. (This relationship can be found on page 103 of the Applesoft manual.) The last printed line gives the great circle distance in kilometers. If the readout is desired in other units, substitute 69.05 for the factor 111.12 on line 760 for an answer in statute miles and substitute 60 for an answer in nautical miles.

The example at the end of the program represents the path from Brentwood, New Jersey, to Rio De Janeiro. This problem has been used as an example in the *ITT Handbook for Radio Engineers* for the past 35-years.

A particular note of caution is in order when using the program. *If an entry is negative, then all its parts, meaning the minutes and seconds as well, must be given a negative sign when they are entered.*

The programs selected for this chapter were chosen to illustrate typical antenna arithmetic and to indicate some of the uses to which a personal computer may be put. In subsequent chapters, we shall be developing more sophisticated concepts.

3
Gain and Huygens Principle

To begin with, it seems appropriate to question the use of the term *gain*. With respect to an amplifier, the use of the term is fairly evident. If an amplifier input impedance and output impedance are both 50-Ω and we put 1-W in and get 10-W out, we obtain a gain of 10-dB. But how can an antenna be said to have gain? A simple piece of wire has no mechanism for deriving additional power from a power supply or some other source of energy. How could it possibly have "gain"?

The answer is to be found in the definition of *antenna gain*. It has a great deal to do with the Inverse Square Law. Referring to Fig. 3.1, we see a sphere of radius R and therefore an area of 4∗PI∗R∗R. At the

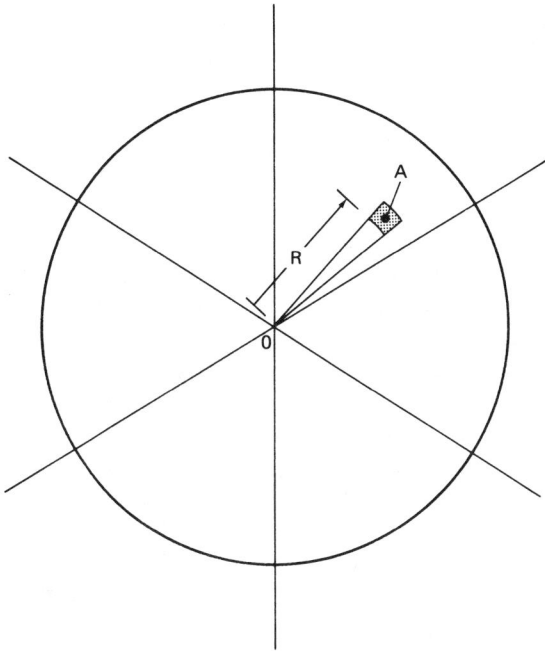

Fig. 3.1 Plotting radiation on the unit sphere the sphere has an area of $4\pi R^2$.

center of the sphere, we have a radiator that is radiating energy into space. If the radiator disperses the energy equally in all directions, then the amount of energy collected over an area A is simply the ratio of area A to the area of the entire sphere. Since the area of the sphere increases as the square of the radius of the sphere, the fraction of the energy collected over area A is inversely proportional to the square of the radius of the sphere.

There is nothing particularly electromagnetic about this reasoning. If we were talking about painting the entire sphere with a fixed amount of paint, the amount to be found on a given square meter of the sphere would also be proportional to the reciprocal (the inverse) of the square of the radius. The property, in other words, is simply a function of the principles of solid geometry.

FRIIS TRANSMISSION FORMULA

Now let us suppose that at the center of the sphere we have a transmitter that is radiating a power, PT, through an antenna that scatters this power *isotropically* (meaning equally in all directions). Let us suppose, moreover, that we are collecting the power at the receiving end with a device of area A. The power collected we will call PR. The device used for the collection is not important at this point; it could be a bucket as well as an antenna. The amount of power being collected would be as follows:

$$PR = ((PT)*A)/(4*PI*R*R)$$

This expression is simply the ratio of the collecting aperture to the area of the sphere.

It is a natural property of antennas that they do not radiate equally in all directions. As a matter of fact, it is fairly easy to show that it is physically impossible to make an antenna do so. Referring to Fig. 3.2, we see a "tribble," a perfectly spherical animal covered with fur. No matter how hard we try to comb the hair of the tribble so that it lays flat, he will always wind up with a cowlick or two (usually two). Applying this state of affairs to the radiation from an antenna, it indicates that at some point in the sphere we will have to look at the polarization vector end-on. No antenna can radiate equally in all directions in a single polarization!

Since power cannot be radiated equally in all directions, it follows that it must be stronger in one direction than in others. It may have several strong points and several weak points, but it cannot be equal all over the sphere. This being the case, a fortunate receiver in one of the "strong" directions will receive more *PR* than it would if the power were equally—

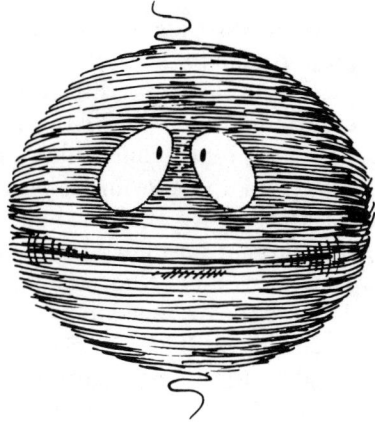

Fig. 3.2 The "tribble."

that is, isotropically—distributed. The ratio of the power it actually receives and what it would have received isotropically is termed the antenna "gain."

Our version of the Friis Transmission Formula is thus modified as follows:

$$(PR) = (PT)*(GT)*A/(4*PI*R*R)$$

where GT is the ratio between the power received in the strongest direction and the power that would have been received if the antenna had distributed the power equally. This "gain" of the antenna functions in the formula as a numeric ratio with no units, although it can also be stated in decibels, or dB. It is important to note that the gain represents the actual increase in signal strength received by a distant receiver and therefore includes all losses. These may or may not be significant. The apparent increase in signal strength resulting from the geometry of the antenna pattern is called the *directivity*. In an absolutely lossless antenna, the directivity and the gain would be equal, but for all practical purposes the directivity will always be greater than the gain. In a large microwave dish, it may be only a few tenths of a decibel greater, but in something like a Rhombic or a Yagi, the difference can amount to several decibels.

PATTERN INTEGRATION

As an example of a pattern integration, let us suppose that we have a pattern that is a figure of revolution about the coordinate axis having the spherical coordinates used in Program 2.2. Each annulus would receive an amount of energy proportional to the square of the voltage in direction H (see Fig. 2.5) and the area of the annulus. Note that this means that an antenna with little energy about the "waist" ($A = PI/2$) will have a higher gain than an antenna with its energy concentrated there.

Program 3.1 was written for the case in which the printed pattern is given in voltage versus direction B. The program first asks for the step size and then proceeds to prompt for the voltage at a sequence of angles. Once the sequence has been completed, the energy will have been integrated over the sphere, and it remains only to divide the result into 4∗PI. Note from line 130 that a 2∗PI has been cancelled out of both numerator and denominator; the summation is therefore the divisor of 2 in line 250.

Program 3.2 shows this routine altered to self-input a pattern in which a single lobe extends along the axis in one direction only. The pattern varies as the cosine of $3A$. It can be seen that the directivity is quite high, as would be expected since only a fraction of the sphere receives any energy at all. The pattern is a slender teardrop shape and not unlike the radiation pattern from a waveguide horn.

In some cases, the radiation pattern is a function of the circumferential angle that was labeled "c" in Fig. 2.5. Often the pattern occurs in two perpendicular planes. It is reasonably accurate to do a pattern integration for each plane and then calculate the directivity as the geometric mean of the two. Alternatively, the integration routine could be rewritten to integrate around each annulus, assuming that the voltage varies sinusoidally between the values for the two planes.

A third rule-of-thumb type of gain estimate observes that there are 4∗PI square radians on a sphere and 41,253 square degrees. For a narrow-beam pattern a good estimate of the directivity can be obtained from the following equation:

$$G\theta = 41,253/(\theta E * \theta H)$$

where θE and θH represent the two half-power beamwidths, in degrees.

The more astute will recognize that a PI/4 has been deleted from the denominator, but this omission actually improves the accuracy of the estimate in most cases. The estimate is usually good to within a decibel or so provided the antenna pattern does nothing too fancy below the −3-dB point.

```
130  REM   G0=(4*PI)/((2*PI)*SIGMA:(P0*SIN(0)*D0))
140  PI = 3.14159
150  PRINT "ENTER STEP SIZE IN DEGREES"
160  INPUT S
170  S = (S / 180) * PI
180  A = S / 2
190  B = (A / PI) * 180
200  PRINT "ENTER VOLTAGE AT  ";B;"  DEGREES"
210  INPUT V
220  P = P + (V * V) *  SIN (A) * S
230  A = A + S
240  IF A < PI THEN   GOTO 190
250  G0 = 2 / P
260  PRINT "DIRECTIVITY= ";G0
265  PRINT "WHICH EQUALS   ";(4.3429 *  LOG (G0));"  DB"
1000  STOP
```

Program 3.1

```
130  REM   G0=(4*PI)/((2*PI)*SIGMA:(P0*SIN(0)*D0))
140  PI = 3.14159
160  S = 1
170  S = (S / 180) * PI
180  A = S / 2
190  B = (A / PI) * 180
200  VA = 3 * A
205  IF VA > PI THEN  GOTO 250
210  V =  COS (VA)
220  P = P + (V * V) *  SIN (A) * S
230  A = A + S
240  IF A < PI THEN   GOTO 190
250  G0 = 2 / P
255  PR# 1
260  PRINT "DIRECTIVITY= ";G0
265  PRINT "WHICH EQUALS   ";(4.3429 *  LOG (G0));"  DB"
270  PR# 0
1000  STOP
```

```
DIRECTIVITY= 8.2351048
WHICH EQUALS  9.15659681  DB
```

Program 3.2

HUYGENS PRINCIPLE

Christian Huygens (1629–1695) was a Dutch astronomer and physicist who contributed a great deal to our knowledge of nature. He was 13-years older than Newton and, aside from the latter, probably the most respected scientist of his day. In the course of building a telescope in 1655, Huygens developed an improved technique for polishing the objective lens. This so dramatically improved the performance of the telescope that it allowed him to discover the true nature of the rings of Saturn. Earlier telescopes had showed only a pair of blurred extensions. Requiring a more accurate clock for astronomical observations, he invented the horologium, or grandfather clock. Both his concise exposition of the true relationship between the length of the pendulum and its period and his theorem of centrifugal force greatly assisted Newton's formulation of the law of gravitation.

Following the work of Robert, Hooke, Huygens developed the wave theory of light and published the property we shall be studying here in his *traite de la lumiere* in 1690. As noted earlier, this gained only limited acceptance because of the awesome reputation of Newton.

One of the principal objections of Newton was the fact that light travels in straight lines whereas waves fill in behind obstacles. If light consists of waves, then why did it not fill in behind obstacles as waves fill in behind a piling or a rock?

Huygens was faced with explaining two seemingly disparate properties about waves. On the one hand, when a pebble is tossed into a pond, it produces a set of evenly spreading symmetric circles. On the other hand, the straight-fronted waves of the North Sea impinging upon a harbor mouth gave rise to a set of waves that marched straight-fronted across the harbor in a well-defined column. Huygens argued that both of these phenomena can be explained if we assume that each small segment of a wavefront can be considered to be the source of a new disturbance—a *Wavelet*.

Figure 3.3 shows how this assumption affects the circular and the straight-fronted wave. Each point on both wavefronts is considered to be the source of a new disturbance. The wave spreads in circular fashion from these disturbances, but regardless of whether the original wavefront was straight or curved, the only place where the wavelets consistently reinforce one another is along the outermost tangent. It is by this mechanism that we see the shape of both the circular wave and the straight wave conserved. Each of them propagates in a direction normal to the wavefront. It can be seen that only a straight front and a circular front will preserve their shape by this mechanism. For the circle, the

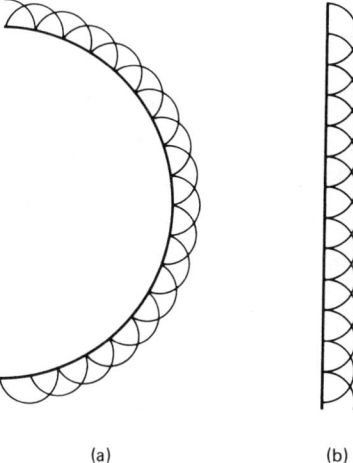

Fig. 3.3 The Huygens wavelet: (a) circular wave and (b) straight-fronted wave.

normal to any point is a point radially outward. For the straight line, all of the normals align, and the line will propagate perpendicularly to the wavefront. A wave with a front that is convex to the direction of travel will eventually cross over itself.

Another interesting property can be deduced from the Huygens wavelet argument, namely, that spreading on a straight wavefront takes place only at the ends. Figure 3.3 makes it clear that in the center of the wavefront there is as much activity from left to right as from right to left; as a result, all activity cancels. Only at the edges does cancellation fail to take place. It would therefore seem that the rate of spreading would be related to the length of the wavefront compared to the wavelength. We shall shortly show this relationship to be true.

Consider the schema of Fig. 3.4. Here we see that straight wavefront EF is the source of a number of wavelets that will eventually reach the parallel locus containing CD. A and B are simply two of the many wavelet sources on the wavefront. We see that the distance in wavelengths from A to C and from B to C in is simply a function of the geometry of the placement of A, B, and C. If B had been chosen to be just slightly closer to A, then A and B would have cancelled at point C. On the other hand, if A and B were just slightly farther apart, they would have completely reinforced each other. Although A and B represent only two of a large number of wavelets, we see that all of the cases of reenforcement and cancellation will be represented in the second wavefront, and obviously the geometry will favor some points over others.

32 EXPLORING ANTENNAS AND TRANSMISSION LINES

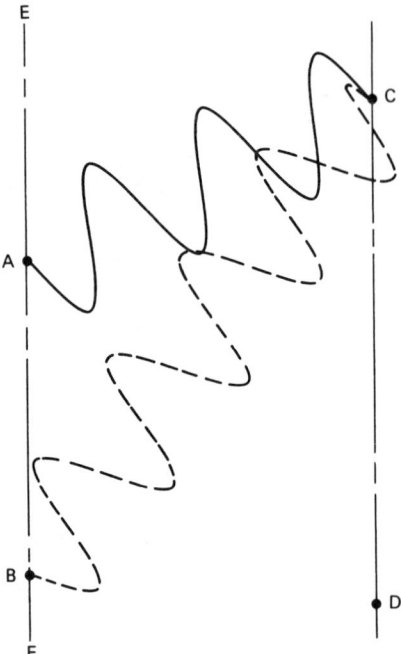

Fig. 3.4 Additive and cancellation effects in a straight wavefront.

The Inverse Square Law also influences the matter. Although we have been examining spreading in only two dimensions, the problem in electromagnetic radiation usually involves square-law spreading in three dimensions. If power varies as the reciprocal of the distance (radius) squared, then the voltage will vary as the reciprocal of range since the power is proportional to voltage squared for a given resistance. From this we can see that the radiation to the second wavefront is influenced more by the closest wavelets than by more distant ones.

A wavefront infinite in length would be completely inform since each segment of the new wave would see exactly the same picture as every other segment. On a finite wavefront, however, the segments near the ends will see an assymetric distribution and therefore need not be identical to their neighbors. For example, consider a point on the second wavefront opposite the left end point. This point is excited from wavelets on the left but not from wavelets on the right, and its situation is thus different from a point in the center, which experiences symmetrical excitation.

At this point in such a discussion, it is usual to present an integral showing that the excitation of each point on the second wavefront is equal to the vector sum of the contributions from each individual wavelet. Since we are concerned in this text with the use of digital computers, however, we will take a slightly different tack and consider the summation of a series of closely spaced points on the wavefront rather than a proper integral.

This summation is specified by lines 250 through 300 of Program 3.3. Line 250 simply calculates the distance, using the Pythagorean Theorem, from a point Q on the wavefront to a point P on the new wavefront (see Fig. 3.5). The potential energy is assumed to vary as the reciprocal of this distance, as shown in line 270, and the phase angle is calculated in line 282. Note that the distances for H and EF and initially given in radians [(2*PI) radians = one wavelength].

Initially, we set aperture size EF at 5 wavelengths and distance H at three wavelengths. In the course of the program, however, distance QP becomes very large, and therefore line 282 subtracts the largest integral number of wavelengths.

The real and imaginary parts of the contribution from the wavelet are added into the sums at 290 and 300 by the conventional vector addition described in Chap. 2. Distance A is incremented and tested in lines 310 and 320 to see whether point F has been attained.

As we see from line 195, since A is incremented by four steps per wavelength, a five-wavelength aperture will require 20 circuits of the loop from 250 to 320 to calculate the data for a given pixel. This program runs very s−l−o−w−l−y! The plot of the output requires about 4.5-hours of runtime.

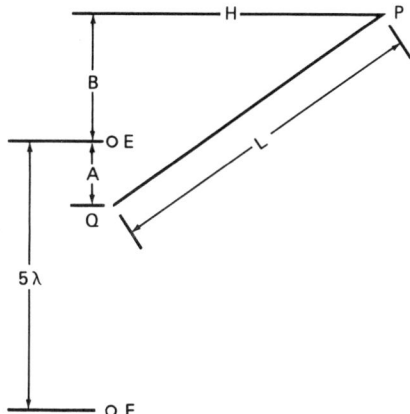

Fig. 3.5 Distance between points on an advancing wavefront.

```
150 PI = 3.14159
160 D = 10 * PI: REM   THE APERTURE SIZE
170 H = 6 * PI: REM   THE DISTANCE TO THE NEW WAVEFRONT
190 A = 0:B = 15 * PI
195 LA = PI / 2
197 S = 2 * PI
200 REM  ::::::::::::::::::::::::::::::::::::::::::::
202 REM         THE PLOT SETUP
204 REM  ============================================
205 HGR2 : HCOLOR= 3
210 POKE  - 12524,0
215 POKE  - 12525,64
217 POKE  - 12529,255
240 REM  >>>>>>>>>>>>>>>>>>>>>>>>>>>>>>>>>>>>>>>>>>>>
242 REM         THE CALCULATION
248 REM  &&&&&&&&&&&&&&&&&&&&&&&&&&&&&&&&&&&&&&&&&&&&
250 L =  SQR ((H * H) + ((A + B) * (A + B)))
270 V = 1 / L
280 PH = L
282  IF PH > S THEN PH = PH - S: GOTO 282
290 VR = VR + V *  COS (PH)
300 VI = VI + V *  SIN (PH)
310 A = A + LA
320  IF A < (10 * PI) GOTO 250
330 A = 0
340 B = B - .330699
350 VP =   SQR ((VR * VR) + (VI * VI))
360 X = XL + 3 * H * VP
365  IF X > 279 THEN  STOP
370 Y = ((15 * PI) - B) * 3.0239
375 VR = 0:VI = 0
377  IF Y > 190 GOTO 410
380  HPLOT X0,Y0 TO X,Y
390 X0 = X:Y0 = Y
400  IF B > ( - 5 * PI) GOTO 250
410 XL = XL + 15
420  IF XL > 250 THEN  STOP
430 A = 0:B = 15 * PI
440 X0 = XL:Y0 = 0
450 H = H + 2 * PI
460  GOTO 250
1000  STOP
```

Program 3.3

Line 350 simply obtains the magnitude of the vector. We could have calculated the phase angle as

$$\text{ANGLE} = \text{ATN}(VI/VR)$$

but would have had to apply the quadrant correction scheme discussed in Chap. 2. Since the phase angle is not usually of interest in antenna pattern calculations, however, this step was omitted.

The size of the increment of B in line 340 and the calculation for the vertical pixel position in line 370 are such that the program produces a pixel per calculation.

The test in line 400 confines the plot to the upper half of the wavefront since the wavefront will be symmetrical. Note that it is necessary to sum the wavelets across the entire aperture.

Figure 3.6 shows the output of this program. The other half of the wavefront was obtained by a photocopy reversal. The interpretation of the figure is easiest if one looks at it from a grazing angle at a point to the left of the aperture. A quasi-three-dimensional effect will appear that is proportional to the amplitude of the wave at that point. The wavefront is actually straight in each case. It is easy to see the sideways spread caused by the edge effect. In the initial front at three wavelengths, the amplitude distribution is not very much like the uniform wave that first hit the aperture nor is it like the antenna pattern that it will become.

Harking back to Fig. 3.4, we see that, when the distance between the wavefronts gets large enough, angles EAC and EBC will become nearly identical and AC and BC will be essentially parallel. The region from the aperture to this point is termed the *near-field zone* and the region beyond is termed the *far-field zone*. These are sometimes called the Fresnel and the Fraunhofer regions, respectively. The latter terminology helps to distinguish the regions from the near- and far-field regions of an individual radiator that are caused by induction, which is desirable since these represent a different physical interpretation. Unfortunately, the Fresnel/Fraunhofer nomenclature is not widely used in American antenna practice.

It can be shown that at a distance measured in wavelengths and equal to twice the square of the aperture dimension, there will be no more than a quarter wavelength difference in the distance from the two aperture edges to any point. In the Fraunhofer region, the antenna pattern can be described in terms of angle alone, that is, without regard to distance.

The pattern at the top of Fig. 3.6 was calculated for the aperture in the Fraunhofer region. It looks like the familiar $(\text{SIN}(X))/X$ function, and that is what it is. We will get a little deeper into this subject later.

36 EXPLORING ANTENNAS AND TRANSMISSION LINES

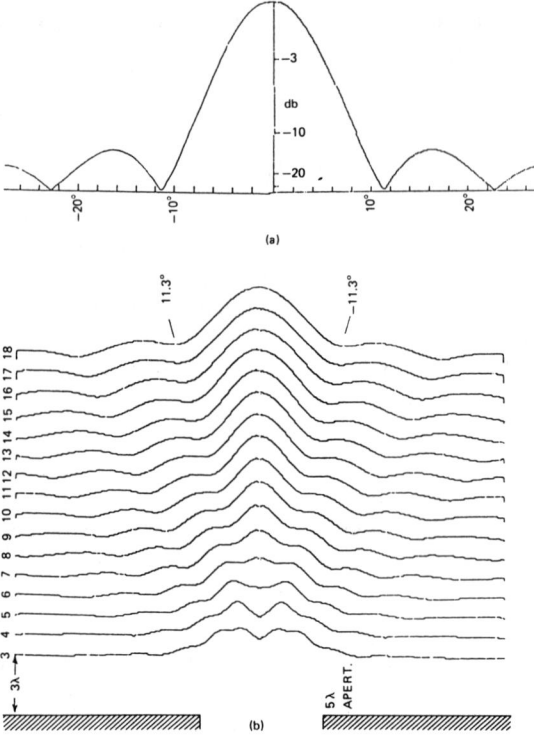

Fig. 3.6 Huygens diffraction: (a) Pattern in the far field where paths AC and BC are essentially parallel and $R > 2D^2/\lambda$ ($R > 50\lambda$ in this case) and (b) pattern in the near field.

It is significant that the near field is not fully formed even beyond 18-wavelengths, a distance that represents 3.6 aperture widths. Note that the dips at 11.3-degrees are not yet fully formed nulls as they are in the Fraunhofer pattern, although the angle is correct. In this connection, note that the scales in the X and Y planes are not equal but rather have been adjusted to enhance the proportions of the figure.

In the past, antenna measurements and calculations were confined almost entirely to the Fraunhofer region because of the long computations required to make any sense of measurements made in the Fresnel Zone. In recent years, however, the availability of computing power has started to change this situation for reasons that will be discussed more fully later.

Returning to our earlier rhetorical question about why waves of light don't spread behind obstacles, the answer is that, as shown by Fresnel

Zone calculation, they do! The thing that neither Newton nor Huygens had any way of knowing was the extremely small wavelength of light. The yellow line from sodium has a wavelength of 0.589-meters. Our five-wavelength aperture would have a width of 0.000116-inches for sodium D light, or about one twenty-sixth the diameter of a blonde hair! Nearly any device that was used in their days had very large apertures, and the diffraction was very closely spaced in angle.

Huygens developed a micrometer for measuring the angle between double stars, and when its very fine hair obscured a star, he thought he could see a bright diffraction line inside the shadow. This was difficult to see, however, and hard to get observers to agree upon. In the next chapter we shall see how the matter was eventually resolved.

4
Arrays of Sources

In this chapter we shall examine the conditions of wave interference in some detail. This issue is central to the problems of diffraction and to most forms of antenna pattern shaping. For the moment, we can leave behind the arguments about the nature of waves and treat them on a purely phenomenological basis. For our purposes, waves need only propagate at a fixed velocity and be capable of interference. We shall also confine our investigations to the Fraunhofer Zone.

Consider the diagrams of Fig. 4.1. We see two sources of waves, $L1$ and $R1$, arrayed symmetrically about the center of a coordinate system. When measuring point P is sufficiently distant so that rays from the two sources to P are essentially parallel, it is not difficult to see that waves from $R1$ will reach point P ahead of waves from $L1$. We could define the phase reference with respect to any location, but we have elected to define it with respect to the center of the array for reasons that will shortly become evident.

If D is a small fraction of a wavelength, it can be seen that the contributions from the two sources will never be much separated in phase and therefore that the radiation will be pretty much the same in all directions. Note that we are here talking about the "steady state" condition of waves.

Each of the sources radiates with a certain frequency, and the contributions from the two radiators possess a certain phase jWt (where $W = 2\pi \times$ frequency) with respect to their source. It is rare, however, that we are interested in the actual phase of the radiation. (The exception is the Omega Navigation System, in which the received phase carries the position information.) More generally, we are concerned only with the relative phases of the arriving signals since these determine the way in which the signals combine.

Note in Fig. 4.1 that the phase shifts are identical except for their direction. That is, as $A1$ increases, the phase of $R1$ grows increasingly early, or leads, and the phase of $L1$ grows increasingly late, or lags, until we reach $A1 = PI/2$.

For a spacing in which D is a quarter-wavelength long in the direction in which $A1 = PI/2$, we see that the radiation from $R1$ will arrive a full half-wavelength ahead of the radiation from $L1$. Under such circum-

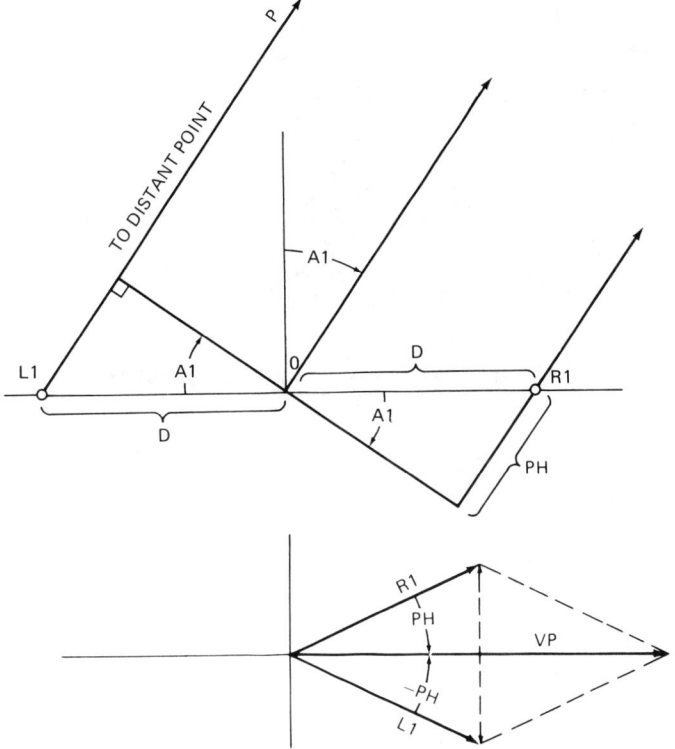

Fig. 4.1 An array of radiators.

stances, if the sources are identical and in phase at the start, the radiation will cancel. Program 4.1 illustrates this fact. The plot at the bottom of the program shows that the radiation is cancelled along the axis of the two sources. If DL (D expressed in wavelengths) is shrunk to a tenth of a wavelength, the pattern becomes nearly circular. Conversely, a wider spacing tends to produce some very interesting results: patterns that are very "flower petaled" and detailed.

Figure 4.2 shows some of the flowerlike patterns that may be obtained. Figure 4.2(a) is for $D = 3\lambda/4$ and Fig. 4.2(b) for $D = 5\lambda/4$. The fact that Fig. 4.2(a) has three petals and Fig. 4.2(b) five petals on each side of the axis denotes that DL has grown to the point that it can make multiple revolutions while $A1$ is going around just once.

The selections for DL were made in the sequence that they were so that the radiators would be equally spaced if the three arrays were all put together. The spacing in the original set was $2*D$; therefore, the spacing

40 EXPLORING ANTENNAS AND TRANSMISSION LINES

```
125 X0 = 140:Y0 = 96
130 PI = 3.14159
140 DL = 2 * PI / 4
150  REM  DL=2*PI*D/WAVELENGTH
160  HGR2
170  HCOLOR= 3
180  POKE  - 12524,0
190  POKE  - 12525,64
200  POKE  - 12529,255
205  REM  %%%%%%%%%%%%%%%%%%%%%%%%%%%%%%%%%%%%%%%%%%
210  REM  THAT COMPLETES THE GRAPHICS SETUP
215  REM  %%%%%%%%%%%%%%%%%%%%%%%%%%%%%%%%%%%%%%%%%%
220 PH = DL *  SIN (A1)
230 VP =  ABS (96 *  COS (PH))
240  REM ##########################################
250  REM  THAT DOES THE CALCULATION
255  REM  FOR VP
260  REM ##########################################
270 X1 = X0 + (VP *  SIN (A1))
280 Y1 = Y0 + (VP *  COS (A1))
290  IF A1 = 0 THEN X2 = X1:Y2 = Y1: GOTO 350
300  HPLOT X1,Y1 TO X2,Y2
310 X2 = X1:Y2 = Y1
350 A1 = A1 + (PI / 72)
360  IF A1 < 6.3 GOTO 220
1000  STOP
```

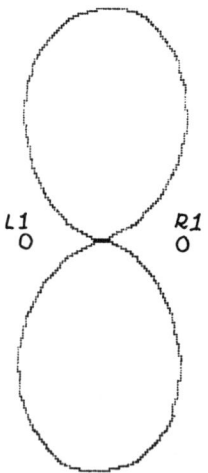

Program 4.1

```
125 X0 = 140:Y0 = 96
130 PI = 3.14159
140 DL = 2 * PI / 4
150  REM   DL=2*PI*D/WAVELENGTH
160  HGR2
170  HCOLOR= 3
180  POKE  - 12524,0
190  POKE  - 12525,64
200  POKE  - 12529,255
205  REM   %%%%%%%%%%%%%%%%%%%%%%%%%%%%%%%%%%%%%%
210  REM   THAT COMPLETES THE GRAPHICS SETUP
215  REM   %%%%%%%%%%%%%%%%%%%%%%%%%%%%%%%%%%%%%%
220 PH = DL *  SIN (A1)
230 VP =  ABS (32 * ( COS (PH) +  COS (3 * PH) +  COS (5 * PH)))
240  REM   ##########################################
250  REM   THAT DOES THE CALCULATION
255  REM   FOR VP
260  REM   ##########################################
270 X1 = X0 + (VP *  SIN (A1))
280 Y1 = Y0 + (VP *  COS (A1))
290  IF A1 = 0 THEN X2 = X1:Y2 = Y1: GOTO 350
300  HPLOT X1,Y1 TO X2,Y2
310 X2 = X1:Y2 = Y1
350 A1 = A1 + (PI / 72)
360  IF A1 < 6.3 GOTO 220
1000  STOP
```

Program 4.2

of Fig. 4.2(a) was 6D and of Fig. 4.2(b), 10D. Each element is an odd number of quarter wavelengths from the center.

It seems logical to ask what would happen if the radiators were all placed together? The answer is to be found at the bottom of Program 4.2. Since the only place in which they truly add is in the direction of the normal to the line of the array, we see a large peak in this direction and a considerable suppression of radiation in all other directions. Following Huygens' principle, the straight-fronted wave of the array is reproduced by a nearly straight-fronted wave at a distance. The curvature is a function of the finite size of the array. The summation is given in line 230 of Program 4.2 where we see that the term for the original pair is followed by terms for the middle and outer pairs. Looking once again at the patterns, it can be seen that the pattern for the entire ensemble is dominated by the pattern for the outermost pair. The beam width is largely a function of the span, in wavelengths, across the outer elements. Figure 4.3 shows the effect of adding yet another outer array pair. It may be seen that the beamwidth is noticeably narrower; as a matter of fact, the pattern has almost reached the point at which the screen resolution will no longer support the polar graphics.

A point worthy of note is the use of the *ABS* function in line 230. It is instructive to remove this term and watch how the image takes shape on the screen, particularly one of the multipetaled patterns of Fig. 4.2. With or without the *ABS* function, the finished figure will be identical. When the *ABS* function is in place, however, the pattern is drawn in sequence, going counterclockwise from the bottom center. When the *ABS* function is missing, the pattern is drawn with alternate lobes on opposite sides of

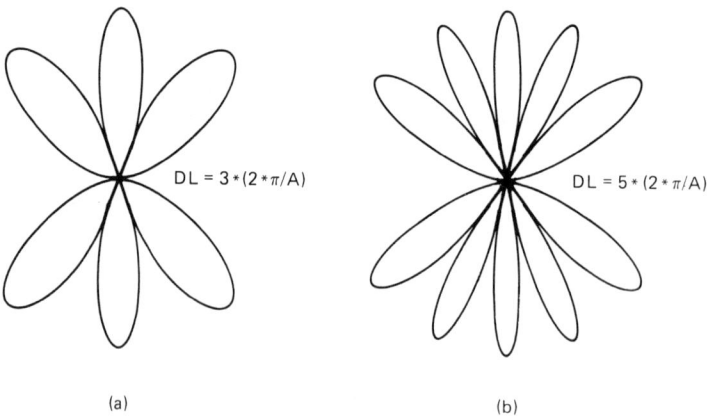

Fig. 4.2 Patterns for two sources with wider spacing.

the pattern. The reason for the latter is that the summation of line 230 is providing alternate negative and positive answers because of the algebraic sign of the *COS* function. The program responds by plotting the negative radius on the far side of the pattern.

For many people, the fact that angles are used to represent distance and electrical phase in equations that also involve real physical angles is one of the most confusing parts of the art of antennas. It might be instructive to spend a bit of time reviewing this matter before we go on.

Harking back to Fig. 4.1, let us review the significance of the terms involved. At the top of the figure, we observe that, as a function of a *real angle* in space, $A1$, the radiation from $R1$ would reach point P earlier than hypothetical radiation from the nonexistent radiator at point 0 and that the radiation from $L1$ would arrive later and by precisely the same period of time. At this level, the problem does not involve frequency dependence. The earliness of $R1$ and the lateness of $L1$ could be described precisely by

$$DT = (D*SIN(A1)/3E8 \text{ SECONDS}$$

where DT = the difference in arrival times. Conventional *ac* algebra, however, adds signals by ignoring the actual phase and accounting only for the differential phase when only a single frequency is considered (the period of one cycle is the reciprocal of the frequency in Hz). The phase difference, PH, can be calculated by

$$PH = 2*PI*DT*F \text{ RADIANS}$$

Thus we see that a difference in time of arrival is typically converted into a phase angle for the convenience of the manipulations required to make the vector addition. The phase angle is obviously not a real angle in space, but it can be seen that it is a *function* of a real angle in space. As $A1$ changes, so does PH. As D becomes larger and larger in terms of wavelengths, moreover, PH can go through many full revolutions, as evidenced by the multipetaled patterns.

A few words about the rotating vectors are also in order. Whenever we have two vectors of equal magnitude rotating in opposite directions, the resultant is a straight line. (Note that a rotation at different speeds would represent two different frequencies.) In the situation shown at the bottom of Fig. 4.1, it can be seen that the broken-line components on the Y axis cancel one another and that only the component along the X axis—the resultant VP—remains. (VP = the voltage at distant point P.) When PH reaches PI/2, the resultant itself is cancelled. Thereafter, VP will be

44 EXPLORING ANTENNAS AND TRANSMISSION LINES

negative until *PH* reaches 3∗PI/2, whereupon it becomes positive again.

Programs 4.3 and 4.4 were included to provide a little more visual exposition. In the former, a "bite" was taken out of the middle of one of the vectors to permit identification. The plot on the screen goes from upper right to lower left, erases, and then starts over. The motion itself is considerably more impressive than the static.

On a larger scale and with no attempt at a three-dimensional illusion, Program 4.4 continuously rotates the vectors. Note that a reversal of the algebraic sign of either one of the vectors will shift the resultant to the *Y* axis; a reversal of both, however, simply flips the phase by PI radians.

You can easily see that the immediately preceding programs run much faster than Program 3.2. One of the reasons is that the actual distance is not calculated. Another is that the cancelling vectors make it possible to drop the imaginaries and calculate only the reals. This simplifications cuts the work in half. A second reduction by two is made possible by the symmetricality of the vectors, which permits us simply to calculate for one and then multiply by two to obtain the resultant. Of course, if the vectors are not equal in absolute magnitude (as they will not be in the near field because of the different ranges involved), the two simplifications will not apply. At the very least, the real and the imaginary must be separately summed and then properly combined.

Program 4.5 illustrates on the screen what happens when one vector is twice the length of the other. The complete rectangle is shown, and the tip of vector *VP* is outlined in a small box for ease of following. It may be seen that *VP* has both real and imaginary parts and that it goes from

```
230 VP = ABS (24 * ( COS (PH) + COS (3 * PH) + COS (5 * PH) + COS (7 * PH)))
```

Fig. 4.3 The eight-element array.

```
150   HGR2
160   HCOLOR= 3
170   POKE  - 12524,0: REM  PRINT BLACK ON WHITE
180   POKE  - 12525,64: REM  PRINT HI-RES PAGE 2
190   POKE  - 12529,255: REM  PRINT UNI-DIRECTIONAL
200 PI = 3.14159
210 Y0 = 5
220 X0 = 259
230 XV = 20 *  COS (A)
240 YV = 20 *  SIN (A)
250 X1 = X0 + XV
252 X2 = X0 + (XV / 3)
254 X3 = X0 + (2 * XV / 3)
260 Y1 = Y0 + YV
265 Y2 = Y0 - YV
267 Y3 = Y0 + (YV / 3)
269 Y4 = Y0 + (2 * YV / 3)
270   HPLOT X1,Y1 TO X3,Y4
272   HPLOT X0,Y0 TO X2,Y3
275   HPLOT X1,Y2 TO X0,Y0
280 A = A + (PI / 18)
290 Y0 = Y0 + 5
300 X0 = X0 - 6.5
310   IF A <  = (2 * PI) GOTO 230
315 A = 0
318   HOME
320   GOTO 150
1000   STOP
```

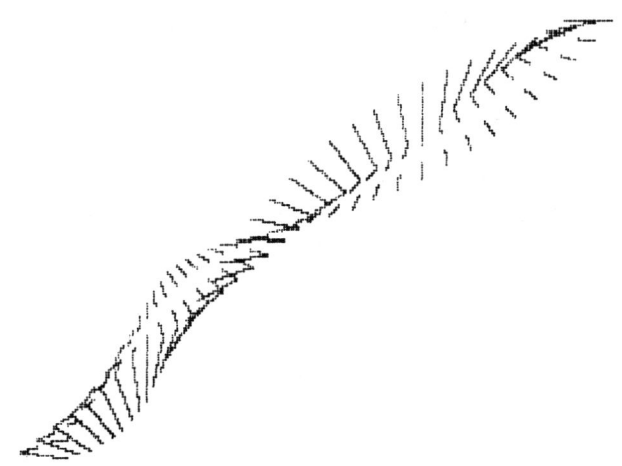

Program 4.3

46 EXPLORING ANTENNAS AND TRANSMISSION LINES

```
150  HGR2
200  PI = 3.14159
210  Y0 = 96
220  X0 = 140
230  XV = 90 * COS (A)
240  YV = 90 * SIN (A)
250  X1 = X0 + XV
260  Y1 = Y0 + YV
270  Y2 = Y0 - YV
290  IF A = 0 THEN A = A + (PI / 72): GOTO 230
300  A = A + (PI / 144)
310  HCOLOR= 0
320  HPLOT X2,Y3 TO X0,Y0 TO X2,Y4
330  X2 = X1
340  Y3 = Y1:Y4 = Y2
350  HCOLOR= 3
360  HPLOT X1,Y1 TO X0,Y0 TO X1,Y2
370  GOTO 230
1000 STOP
```

Program 4.4

positive to negative in both parts without ever passing through zero magnitude.

PRINTING WITH THE SILENTYPE

At this point, it might be fruitful to make room for a brief digression about printing with the silentype. If you are running on a machine other than an Apple or are using a plotter other than the silentype, you may rejoin us in the next section.

One of the peculiarities of the Apple II+ that this writer has noticed is the fact that the silentype cannot be coerced to plot unless it has printed something first. Not only won't it plot, it evidently sets some switchs in such a way that it cannot be made to plot thereafter. Although, there may be a better solution, I usually to store the program and shut down.

Lines 170 through 190 of Program 4.3 inform the printer just what it is that you want. If you attempt to plot with a CTRL-Q before the Apple has printed something, you will find that it does not plot what is on the screen, and a PRINT PEEK (−12525) will reveal that the contents are not 64.

Always have the Apple print something before trying to plot. After power-up and loading a plotting program, have the machine LIST it once. On any power-up after listing, it will plot just fine! This peculiarity does not seem to be confined to any one Apple, but if there is another remedy, I have not found it.

```
150  HGR2
160  HCOLOR= 3
200  PI = 3.14159
210  Y0 = 96
220  X0 = 140
230  XV = 22.5 * COS ( - A)
240  YV = 22.5 * SIN ( - A)
250  XW = 45 * COS (A)
260  YW = 45 * SIN (A)
270  X1 = X0 + XV
280  X2 = X0 + XW
290  X3 = X0 + XV + XW
300  Y1 = Y0 + YV
310  Y2 = Y0 + YW
320  Y3 = Y0 + YV + YW
330  IF A = 0 THEN A = A + (PI / 144): GOTO 230
340  A = A + PI / 72
350  HCOLOR= 0
355  HPLOT M,Q TO N,Q TO N,R TO M,R TO M,Q
360  HPLOT X0,Y0 TO X4,Y4 TO X6,Y6 TO X5,Y5 TO X0,Y0
370  X4 = X1:X5 = X2:X6 = X3
380  Y4 = Y1:Y5 = Y2:Y6 = Y3
390  HCOLOR= 3
400  HPLOT X0,Y0 TO X1,Y1 TO X3,Y3 TO X2,Y2 TO X0,Y0
420  M = X3 + 2:N = X3 - 2
430  Q = Y3 + 2:R = Y3 - 2
440  HPLOT M,Q TO N,Q TO N,R TO M,R TO M,Q
450  GOTO 230
1000 STOP
```

Program 4.5

ODD-ORDER ARRAYS

The odd-order array is characterized by a central element that would be located at point 0 in Fig. 4.1. Since the spacing between elements in such an array is *D*, the outlying elements run in the order *PH*, 2∗*PH*, 3∗*PH*..., with both even and odd multipliers present. To obtain the same element spacing as one has in an even-order array. *D*, of course, would have to be twice as great.

Program 4.6 produces the odd-order array. The principal difference between it and Program 4.2 appears in line 230, where the constant 0.5 has been added into the summation. If you examine the pattern below it, you will notice that the center element, being the reference, is not phase-shifted in any direction.

In previous programs we have simply altered the constant to keep the plot from going off scale. In this program, line 225 adds a divisor to do the scaling automatically. The radius of 96 has been doubled to make up for the fact that all terms of the summation have been factored by two. Doing so saves a certain amount of computation time by eliminating three multiplications.

One difference exists between this pattern and the pattern of the even-ordered array. With an even number of elements spaced a half wave apart and equal in excitation, there will be no end-fire lobe along the line of the array. With an odd number of elements, however, there will be an end-fire lobe because there will always be one uncancelled element. It may be seen in the proof pattern below Program 4.6.

YOUNG'S WAVE INTERFERENCE DEMONSTRATION

The corpuscular theory of Newton had a tough time trying to explain the phenomena of Newton's Rings and the colors that appear in soap bubbles and oil films. Newton's Rings are the pattern of concentric circular rainbows that may be observed when a very shallowly curved lens is placed against an optical flat. In 1800, Thomas Young, a professor at the Royal Institute in London, produced a theory that purported to explain both phenomena. Young proposed that both were a demonstration of the wave nature of light.

In the case of Newton's Rings, Young noted that the spacing between the curved face of the lens and the optical flat gradually increased from zero and that if light were reflected from both interfaces, the two reflections could either add or cancel. He reasoned that the color green was the shorter wavelength, since it added closer to the center, and the color red the longer, since it added further out.

In the case of bubbles and oil films, he reasoned that the peculiar purple often seen in these films was due to the fact that the red and green were reinforced in reflection whereas the yellow was cancelled. Young's arguments were perceptive and accurate, but they made little progress against the awesome reputation of Newton and the general distaste that philosophers of the day had for all of the colorless, weightless fluids like the Aether and Caloric. Nevertheless, Young persisted, and in 1808 he demonstrated wave interference that could not be explained by the corpuscle theory. The apparatus he used is shown in Fig. 4.4.

Nowadays, this experiment can be done very easily with a laser, but in Young's day the only way one could get more-or-less monochromatic light was by burning salts of metals on the wick of a spirit lamp. Copper yielded green; lithium, red; and sodium, yellow. There were two problems: The light was not terribly bright to begin with, and, for wave interference to take place, the light had to be phase coherent. Until the invention of the laser in 1958 by Townes, the only way one could obtain coherent light was through a pinhole. A pinhole sufficiently removed from the source would allow a small amount of coherent light to pass through. The light was coherent simply because far-field light loses coherence slowly and a pinhole is small.

ARRAYS OF SOURCES

```
125 X0 = 140:Y0 = 96
130 PI = 3.14159
140 DL = PI
150   REM   DL=2*PI*D/WAVELENGTH
160   HGR2
170   HCOLOR= 3
180   POKE  - 12524,0
190   POKE  - 12525,64
200   POKE  - 12529,255
205   REM   %%%%%%%%%%%%%%%%%%%%%%%%%%%%%%%%%%%
210   REM   THAT COMPLETES THE GRAPHICS SETUP
215   REM   %%%%%%%%%%%%%%%%%%%%%%%%%%%%%%%%%%%
220 PH = DL *  SIN (A1)
225 N = 7: REM  TOTAL NUMBER OF ARRAY ELEMENTS
230 VP =  ABS ((192 / N) * (.5 +  COS (PH) +  COS (2 * PH) +
     COS (3 * PH)
     ))
240   REM   ########################################
250   REM   THAT DOES THE CALCULATION
255   REM   FOR VP
260   REM   ########################################
270 X1 = X0 + (VP *  SIN (A1))
280 Y1 = Y0 + (VP *  COS (A1))
290   IF A1 = 0 THEN X2 = X1:Y2 = Y1: GOTO 350
300   HPLOT X1,Y1 TO X2,Y2
310 X2 = X1:Y2 = Y1
350 A1 = A1 + (PI / 72)
360   IF A1 < 6.3 GOTO 220
1000  STOP
```

Program 4.6

50 EXPLORING ANTENNAS AND TRANSMISSION LINES

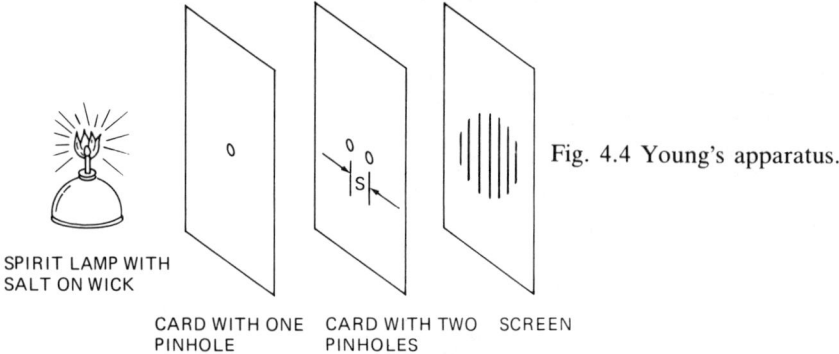

Fig. 4.4 Young's apparatus.

From the pinhole source, a second card having two pinholes was illuminated—a situation similar to that of the two-element array of Fig. 4.1. When a viewing screen was placed about a meter behind the two-pinhole card, bands of light could be seen very faintly upon it.

Program 4.7 will show the result obtained from Young's experiment for two pinholes spaced 1000-wavelengths apart. The fringes cannot be explained in terms of corpuscles since a corpuscle cannot interfere with another corpuscle. Only waves will interfere.

With this experiment, Young was able, for the first time, to compute the wavelength of light. Between two nulls, PH changes by a wavelength. Thus, if S is known, wavelength may be calculated. (See Figs. 4.4 and 4.5.)

The Young experiment could be said to represent the first array antenna.

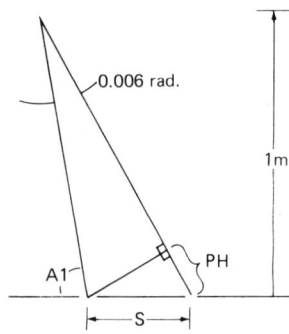

Fig. 4.5 Geometry for Program 4.7.

ARRAYS OF SOURCES 51

```
125 Y = 190
130 PI = 3.14159
140 S = 2E3 * PI
145 A1 = PI / 2 + .0006
150 REM  S=SPACING IN WAVELENGTHS
160 HGR2
170 HCOLOR= 3
180 POKE  - 12524,0
190 POKE  - 12525,64
200 POKE  - 12529,255
205 REM  %%%%%%%%%%%%%%%%%%%%%%%%%%%%%%%%%%%
210 REM  THAT COMPLETES THE GRAPHICS SETUP
215 REM  %%%%%%%%%%%%%%%%%%%%%%%%%%%%%%%%%%%
220 PH = S * ((PI / 2) - (A1 - .0006))
222 V = 1 + COS (PH)
224 V1 = SIN (PH)
226 VP = SQR ((V * V) + (V1 * V1))
228 VP = 95 * VP
240 REM  ########################################
250 REM  THAT DOES THE CALCULATION
255 REM  FOR VP
260 REM  ########################################
270 IF N = 0 THEN N = N + 1:Y1 = Y - VP: GOTO 220
280 A1 = A1 + .00002
290 Y0 = Y - VP
300 X = X1 + 1
305 IF X = 279 GOTO 340
310 HPLOT X1,Y1 TO X,Y0
320 X1 = X:Y1 = Y0
330 GOTO 220
340 HPLOT 0,190 TO 279,190
350 HPLOT M,185 TO M,190
360 M = M + 50
370 IF M < 279 GOTO 340
1000 STOP
```

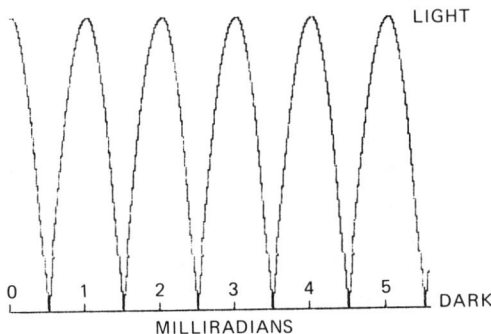

Program 4.7

5
Aperture Distributions

In the preceding two chapters we have assumed that our Huygens wavelets are all precisely equal and all precisely in phase. In this chapter we shall investigate some broader instances where these restrictions do not apply. Such instances are usually more representative of practical antenna systems.

In Program 4.2 and Fig. 4.3, the beam width of an array was shown to be related to the span of the outer elements in wavelengths. We shall shortly see that the side-lobe or secondary lobe responses are related to the discontinuity at the outer element.

To begin with, it is instructive to get a feel for the effect of the element spacing. In Fig. 4.2, the spacings were all multiples of PI/2—a more or less "golden" spacing since it does not "end-fire". Along the axis of the array, $A1$ = PI/4 or $A1$ = 3*PI/4, in Fig. 4.1, we see that the radiation will cancel if $2*D$ = PI/2 since the contributions from the two elements are in phase opposition. This is not true of all spacings, however, in fact, it is a very special case.

The illustrations of Fig. 5.1 show the progression of the pattern for a two-element array with various spacings. For spacings smaller than a half wavelength, the array tends to act more like a single radiator. Very shortly beyond a half-wave spacing the end-fire lobes begin to develop. These lobes are fully developed when DL = PI (a full wave spacing). At this stage, the in-phase condition obtains not only for the broadside, but also for the end-fire, direction.

It is noteworthy that the end-fire lobe is much broader than the broadside lobe. This is characteristic of endfire antennas. If one considers the arrangement of Fig. 4.1, it is fairly obvious that the phasing does not change rapidly in the vicinity of $A1$ = PI/2.

We will see shortly that the beam width is proportional to the aperture width for broadside arrays and proportional to the square root of the aperture for end-fire antennas.

APERTURE TAPER

In most practical antennas, the distribution of energy throughout the aperture will not be constant as it has been in our previous examples. The variation may be a function of the geometry of the device under consideration or a result of deliberately tapering to obtain the overall per-

APERTURE DISTRIBUTIONS 53

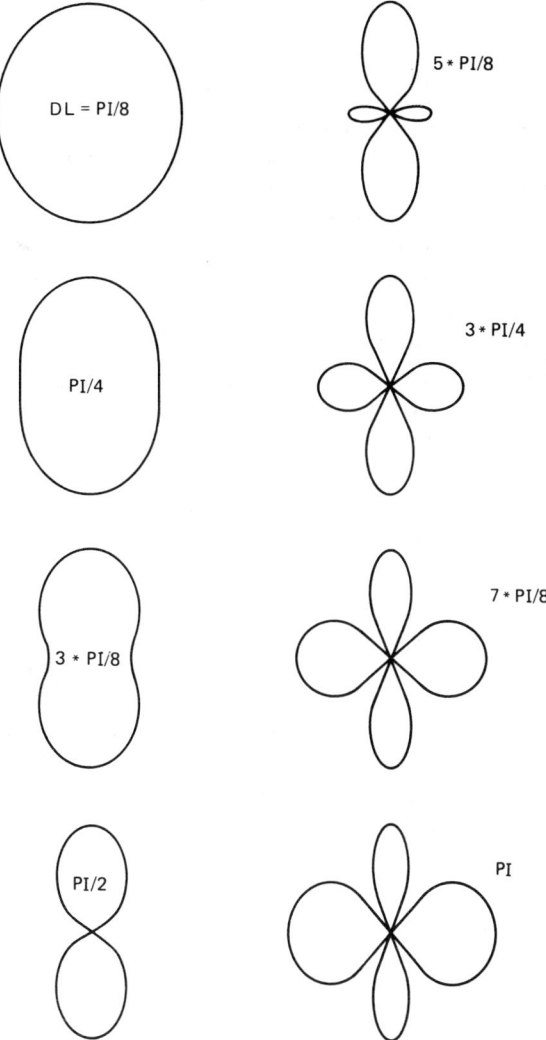

Fig. 5.1 Two-element array spacing effects.

formance desired. An example of geometric tapering is to be found in the excitation of a parabolic antenna. When illuminated from its focus, a parabola is more distant from the feed as it departs from the vertex. As a result, the energy undergoes a "space attenuation" or Inverse Square Law spreading in transit between the feed horn and the reflector. Stated differently, a square inch of reflector surface near the vertex will receive more energy than a square inch near the edge simply because it is closer to the feed.

Another source of geometric tapering is due to integration of the distribution over a three-dimensional reflector. The problems we have encountered thus far have been essentially two-dimensional problems. For pattern calculation and measurement, it is quite common to simplify the complexities of a three-dimensional antenna into a two-dimensional situation by integrating the dimension normal to the plane of the paper into the plane of the paper. If the aperture is circular in three dimensions, the distribution will obviously be lower near the periphery of the circle if it is integrated into a line.

A deliberate taper can be built into the radiation pattern of the feed used to illuminate the antenna. As a matter of fact, it is virtually impossible to avoid a taper when a paraboloid is illuminated with a practical feed. Since it is usually necessary to try to keep as much of the feed energy as possible on the paraboloid, the feed must be directive. The energy cannot fall arbitrarily fast if the size of the feed is restricted. Accordingly, the illumination must taper off before it gets to the edge of the aperture.

Program 5.1 was written to demonstrate calculations for an odd-element array with a large number of elements ($N \leq 64$). The summation is found in lines 330 to 350. Line 310 indicates that the physical angle $A0$ is divided into angular steps of PI/3600, or 0.05-degrees. Each of these is accorded two increments on the X axis by line 475. In line 540, since each angle marker on the X axis is accorded 20 increments, or 10 angular steps of 0.05-degrees, each marker represents 0.5-degrees.

Line 200 tells us that the array has 2*64, or 128, elements plus a center element, for a total of 129-elements. The width of the array, from line 220, is therefore 128*0.313, or 40-wavelengths.

Figures 5.2 and 5.3 show four different distributions of the aperture excitation. In Fig. 5.2(a), we see the result of a uniform distribution, that is, a distribution where every element is excited equally. The value of $X(I,0)$ comes from line 250 in the program. This illumination obviously has an abrupt termination. In Fig. 5.2(b), the distribution follows a cosine law. It goes from zero at one aperture edge to a maximum at the center and back to zero at the other aperture edge. Although the excitation at the aperture edge for this distribution is zero, the slope is considerable.

It may be seen that, compared to the uniform distribution, the cosine distribution reduces the first side lobe from about -13 to -23-dB. This is accomplished at the cost of the decreased gain that results from the increase in beam width, both at the half-power (-3-dB) point and at the first null. Side-lobe levels are usually cited in terms of the highest side

```
200 N = 64: REM   NUMBER OF ARRAY ELEMENTS
210 PI = 3.14159
220 DL = .313: REM   ELEMENT SPACING IN WAVELENGTHS
230  DIM X(N,1)
235 AH = PI / (2 * N)
240  FOR I = 1 TO N
250 X(I,0) = 1 - ((1 / N) * I)
260  NEXT I
295 E0 = 1
296  REM ))))))))))))))))))))))))))))))))))))))))))
297  REM     THIS COMPLETES APERTURE DISTRIBUTION CALCULATION
298  REM &&&&&&&&&&&&&&&&&&&&&&&&&&&&&&&&&&&&&&&&&
300  FOR J = 0 TO 138
310 A0 = (PI / 3600) * J
320 AI = 2 * PI * DL *  SIN (A0)
330  FOR I = 1 TO N
335 AJ = AI * I
340 E0 = E0 + X(I,0) * 2 *  COS (AJ)
350  NEXT I
352  REM #############################################
354  REM     THIS COMPLETES THE SUMMATION
356  REM      FOR ONE ANGLE
358  REM #############################################
360  IF J = 0 THEN EN = E0: GOTO 380
370  GOTO 430
380  HGR2
390  HCOLOR= 3
400  POKE  - 12524,0
410  POKE  - 12525,64
420  POKE  - 12529,255
422  REM #############################################
424  REM     THE GRAPHICS AND PRINTER SETUP
426  REM %%%%%%%%%%%%%%%%%%%%%%%%%%%%%%%%%%%%%%%%%%
430 YA =  ABS (EN / E0)
432 YB = 20 * ( LOG (YA)) / 2.30259
434 Y = 4.75 * YB
436  IF Y > 191 THEN Y = 191
450  IF J = 0 THEN X0 = X:Y0 = Y
460  HPLOT X0,Y0 TO X,Y
470 X0 = X:Y0 = Y
475 X = X + 2
480 E0 = 1
490  NEXT J
492  REM $$$$$$$$$$$$$$$$$$$$$$$$$$$$$$$$$$$$$$$$$$$$
494  REM     ADD THE SCALES
496  REM $$$$$$$$$$$$$$$$$$$$$$$$$$$$$$$$$$$$$$$$$$$$
500  HPLOT 0,1 TO 0,191
510  HPLOT 0,191 TO 279,191
530  HPLOT M,186 TO M,191
540 M = M + 20
560  IF M < 280 THEN  GOTO 530
570  HPLOT 0,14 TO 3,14
580  HPLOT 0,48 TO 5,48
590  HPLOT 0,95 TO 5,95
600  HPLOT 0,143 TO 5,143
610  STOP                        Program 5.1
```

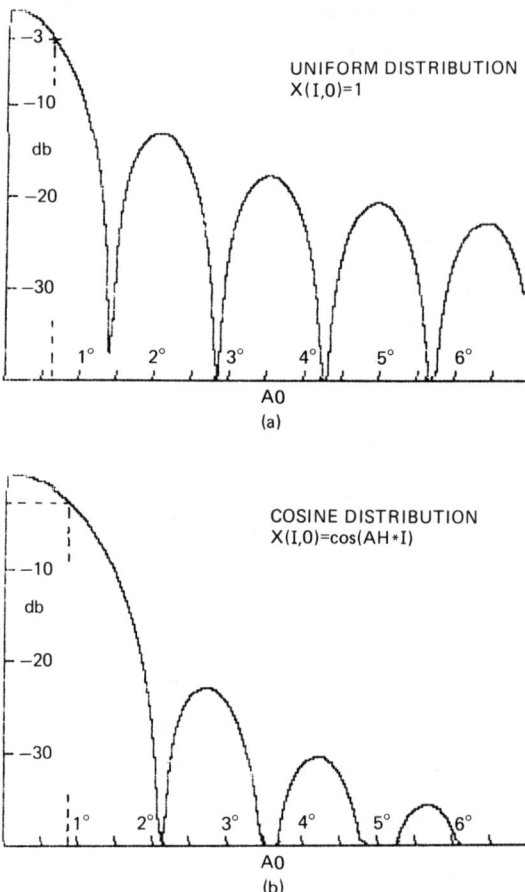

Fig. 5.2 $N = 64$ array patterns: (a) Uniform distribution, and (b) cosine distribution.

lobe, which is usually, but not always, the one closest to the main beam.

Figure 5.3(a) shows a cosine-squared illumination, which has not only a zero value at the aperture edge but also a zero slope. For this distribution, it can be seen that the first side lobe is reduced to about -32-dB. The beam width is also broader by a considerable amount.

Figure 5.3(b) shows the linear or "roof" distribution, in which the excitation of the elements proceeds linearly downward from the center to the outside edge. This distribution bears a special relationship to the uniform distribution, as shown by Fig. 5.3(c). This figure represents a square aperture with a uniform array of radiators that are uniformly excited.

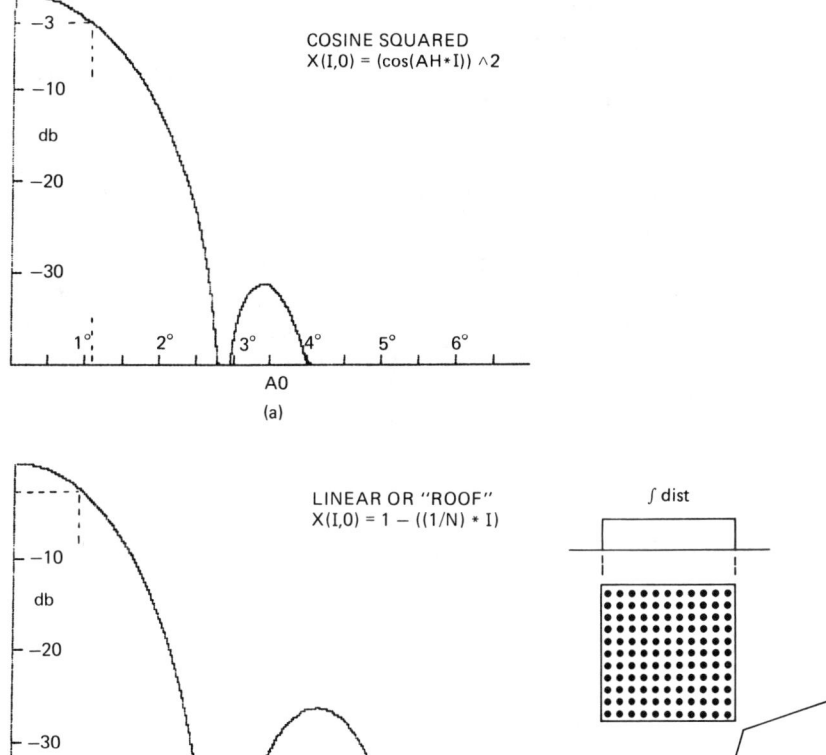

Fig. 5.2 (*continued*)

The integral along the diagonal is the "roof" distribution. The actual beam width is the same in both planes since the aperture along the diagonal is longer than along the face by the square root of 2 but the side lobes of the diagonal are much lower. In fact, the side lobes are approximately 13-dB higher for the uniform distribution than for the "roof" distribution.

This phenomenon can be visually verified by looking through a flyscreen at a distant light on a dark night. You will perceive a dot for the light and a pair of rainbow-string side lobes normal to the faces of the flyscreen opening. Along the diagonal, the side lobes disappear because the "roof" distribution makes them much fainter than the side lobes in the principle planes.

58 EXPLORING ANTENNAS AND TRANSMISSION LINES

The reason for the rainbow in the side lobes is the fact that different wavelengths of light find the apperture to be of different dimensions. The main beam is white because all of the light focuses along the axis. Since the beam widths and first side-lobe positions are narrower for the green and shorter wavelengths than they are for the red and longer wavelengths, the light at the near side lobes is broken down into a spectrum, with the shorter wavelengths on the inside. At the wider side lobes, the color tends to return to white because the lobes overlap.

Table 5.1 shows the responses obtained from several distributions in relatively common configurations and provides an easy means to estimate the performance of a given configuration. The linear or roof distributions can typically be obtained for an array antenna wherever control of individual element excitation is possible. Geometric collimators such as paraboloids are typically fed with distributions on the order of the cosine series. It is difficult to feed such antennas without using a considerable taper.

Table 5.1 Pattern Characteristics for Various Distributions

$X(I,0)$	GAIN[a]	HPBW[b]	FULL WIDTH[c] (RAD)	SIDE-LOBE LEVEL[d] (dB)
1	1	0.88	1	13.2
COS(AH*I)	0.81	1.2	1.5	23
(COS(AH*I))^2	0.667	1.45	2	32
(COS(AH*I))^3	0.575	1.66	2.5	40
1 − ((1/N)*I)	0.75	1.28	2	26.4

[a]The gain figures are relative to a uniformly illuminated aperture of the same dimension. Gain for any of the patterns can be computed as described in Chap. 3.
[b]The half-power beam width (HPBW) is the angle, in radians, between the −3-dB points of the pattern. This double-sided angle is measured clear across the pattern and is relative, like all of the beam widths, to (1/N*DL)), or the width of the aperture, in wavelengths. For example, since the aperture width for the uniform illumination in Fig. 5.2a is 40-wavelengths, its HP beam width is 0.88/40, or 0.022-radians (1.26-degrees).
[c]The full width is the angle, in radians, between the centerline and the first null of the pattern. This is a single-sided angle. The width between nulls is twice the value quoted.
[d]The side-lobe level is the height, in dB, of the first side lobe, relative to the main lobe.

PHASING AND THE END-FIRE CONDITION

If we imagine that the incoming wave of Fig. 3.6 is not exactly normal but has some inclination, then it is not difficult to visualize that every successive point on the aperture will have a phase shift relative to its position. Likewise the wave in Fig. 4.1 could be considered as not going to distant point P but rather as coming from it. Under such conditions, it may be seen that a uniform linear phase shift across the aperture could be included in Program 5.1 by adding the line

APERTURE DISTRIBUTIONS 59

325   AI = AI − 0.089

Figure 5.4 shows that the principle effect of this this phase shift is to steer the beam. In Fig. 5.4a, the steering increment is 0.089-radians. Since *DL* is equal to 0.313-wavelengths, or 1.9666-radians, this corresponds to a steering of .089/1.9666, or 0.04525-radians = (2.59-degrees). Phase steering is a common feature of many modern radars. The required phase shift can be obtained with ferrite phase shifters, with pin diode phase shifters, or with a change of frequency on a serpentine line designed to maximize phase shift with change in frequency.

The unlabeled arrows in Program 5.2 flag out two other housekeeping changes in the program. In the original Program 5.1, line 360 established

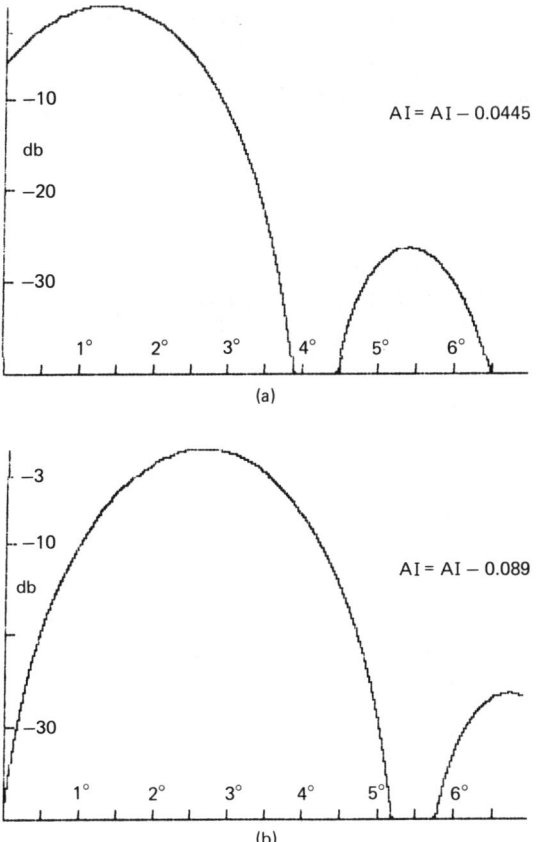

Fig. 5.4 Beam steering.

```
200 N = 64: REM  NUMBER OF ARRAY ELEMENTS
210 PI = 3.14159
220 DL = .313: REM  ELEMENT SPACING IN WAVELENGTHS
230  DIM X(N,1)
235 AH = PI / (2 * N)
240  FOR I = 1 TO N
250 X(I,0) = 1 - ((1 / N) * I)
260  NEXT I
295 E0 = 1
296  REM )))))))))))))))))))))))))))))))))))))))))
297  REM     THIS COMPLETES APERTURE DISTRIBUTION CALCULATION
298  REM &&&&&&&&&&&&&&&&&&&&&&&&&&&&&&&&&&&&&&&&
300  FOR J = 0 TO 138
310 A0 = (PI / 3600) * J
320 AI = 2 * PI * DL *  SIN (A0)
325 AI = AI - .089                              ←——— REM PHASE
330  FOR I = 1 TO N                                   STEERING
335 AJ = AI * I
340 E0 = E0 + X(I,0) * 2 *  COS (AJ)
350  NEXT I
352  REM ##########################################
354  REM     THIS COMPLETES THE SUMMATION
356  REM       FOR ONE ANGLE
358  REM ##########################################
360  IF J = 0 GOTO 380                          ←———
370  GOTO 430
380  HGR2
390  HCOLOR= 3
400  POKE  - 12524,0
410  POKE  - 12525,64
420  POKE  - 12529,255
422  REM ##########################################
424  REM     THE GRAPHICS AND PRINTER SETUP
426  REM %%%%%%%%%%%%%%%%%%%%%%%%%%%%%%%%%%%%%%%%
430 EN = 1: FOR I = 1 TO N:EN = EN + 2 * X(I,0): NEXT I  ←———
431 YA =  ABS (EN / E0)
432 YB = 20 * ( LOG (YA)) / 2.30259
434 Y = 4.75 * YB
436  IF Y > 191 THEN Y = 191
450  IF J = 0 THEN X0 = X:Y0 = Y
460  HPLOT X0,Y0 TO X,Y
470 X0 = X:Y0 = Y
475 X = X + 2
480 E0 = 1
490  NEXT J
492  REM $$$$$$$$$$$$$$$$$$$$$$$$$$$$$$$$$$$$$$$$$
494  REM     ADD THE SCALES
496  REM $$$$$$$$$$$$$$$$$$$$$$$$$$$$$$$$$$$$$$$$$
500  HPLOT 0,1 TO 0,191
510  HPLOT 0,191 TO 279,191
530  HPLOT M,186 TO M,191
540 M = M + 20
560  IF M < 280 THEN  GOTO 530
570  HPLOT 0,14 TO 3,14
580  HPLOT 0,48 TO 5,48
590  HPLOT 0,95 TO 5,95
600  HPLOT 0,143 TO 5,143
610  STOP
```

Program 5.2

APERTURE DISTRIBUTIONS

the normalizing voltage *EN* as the value of *E0* when *J* = 0. For the steered pattern, this equivalency no longer applies, and thus the routine of line 430 has been substituted to calculate *EN*.

At the present writing, pin diode phase shifting is probably the most popular. Ferrite phase shifters are heavy and consume power in the magnetic field coils that do the steering. Frequency steering is not power-consuming, but the serpentine line tends to make the array heavy and bulky. In addition, the frequency scanner is constrained in operation by frequency and is thus deprived of one-degree of freedom from jamming. The pin diode phase shifter constitutes the smallest and lightest method and consumes only modest power.

A whole family of radars has appeared in recent years and promises to dominate certain applications eventually. This group is sometimes described as "solid state" radar. Microwave transistors have not developed to the point that megawatt pulse powers are directly available from single devices or simple ensembles. One approach to obtaining enough power for radar applications has been to supply each radiator or each column or set with a separate power amplifier. The power is combined only in the far field of the antenna pattern, thus eliminating the power-combining network and permitting an easy implementation of the phase-shift scanning operation since the phase shifting can be accomplished at low power before the amplification.

END-FIRE ARRAYS

If the phase shift per radiator is increased sufficiently, we finally arrive at the point at which the beam is pointed in the end-fire direction. This type of antenna is probably the most common type of directive antenna in terms of sheer numbers since most home TV and FM antennas fall in this category. In terms of dollars per decibel, the end-fire antenna is the least expensive type to build for low to moderate gains.

Program 5.3 gives the antenna pattern calculation for an end-fire array. The antenna patterns could be solved with the geometry used in Program 5.2, but the operation is easier to visualize with a different geometry.

Figure 5.5 shows the geometry used for Program 5.3. The equations involved are as follows:

$$T_{(n)} = N*2*PI*DL \tag{5.1}$$

$$PH_{(n)} = N*2*PI*DL*COS(A\theta) \tag{5.2}$$

$$E_p = \sum_{1}^{N} 1 + X_n \varepsilon^{j(T_n - PH_n)} \tag{5.3}$$

$$E_p = \sum_{1}^{N} 1 + COS(N*(T - PH)) + jSIN(N*(T*PH)) \tag{5.4}$$

$$(T - PH) = (2*PI*DL)*(1 - COS(A\theta)) \tag{5.5}$$

```
200 N = 128: REM  NUMBER OF ARRAY ELEMENTS
210 PI = 3.14159
220 DL = .313: REM  ELEMENT SPACING IN WAVELENGTHS
230  DIM X(N,1)
235 AH = PI / (2 * N)
240  FOR I = 1 TO N
250 X(I,0) = 1
260  NEXT I
295 E0 = 1
296  REM ))))))))))))))))))))))))))))))))))))))))))))
297  REM     THIS COMPLETES APERTURE DISTRIBUTION CALCULATION
298  REM &&&&&&&&&&&&&&&&&&&&&&&&&&&&&&&&&&&&&&&&&
300  FOR J = 0 TO 138
310 A0 = (PI / 720) * J
320 AI = (2 * PI * DL) * (1 -  COS (A0))
330  FOR I = 1 TO N
335 AJ = AI * I
340 E0 = E0 + X(I,0) * 2 *  COS (AJ)
345 EI = EI + X(I,0) * 2 *  SIN (AJ)
350  NEXT I
351 E0 =  SQR ((E0 * E0) + (EI * EI))
352  REM ############################################
354  REM     THIS COMPLETES THE SUMMATION
356  REM      FOR ONE ANGLE
358  REM ############################################
360  IF J = 0 THEN EN = E0: GOTO 380
370  GOTO 430
380  HGR2
390  HCOLOR= 3
400  POKE  - 12524,0
410  POKE  - 12525,64
420  POKE  - 12529,255
422  REM ############################################
424  REM     THE GRAPHICS AND PRINTER SETUP
426  REM %%%%%%%%%%%%%%%%%%%%%%%%%%%%%%%%%%%%%%%%%%
430 YA =  ABS (EN / E0)
432 YB = 20 * ( LOG (YA)) / 2.30259
434 Y = 4.75 * YB
435  IF Y < 0 THEN Y = 0
436  IF Y > 191 THEN Y = 191
450  IF J = 0 THEN X0 = X:Y0 = Y
460  HPLOT X0,Y0 TO X,Y
470 X0 = X:Y0 = Y
475 X = X + 2
480 E0 = 1
485 EI = 0
490  NEXT J
492  REM $$$$$$$$$$$$$$$$$$$$$$$$$$$$$$$$$$$$$$$$$$$$
494  REM     ADD THE SCALES
496  REM $$$$$$$$$$$$$$$$$$$$$$$$$$$$$$$$$$$$$$$$$$$$
500  HPLOT 0,1 TO 0,191
510  HPLOT 0,191 TO 279,191
530  HPLOT M,186 TO M,191
540 M = M + 20
560  IF M < 280 THEN  GOTO 530
570  HPLOT 0,14 TO 3,14
580  HPLOT 0,48 TO 5,48
590  HPLOT 0,95 TO 5,95
600  HPLOT 0,143 TO 5,143
610  STOP
```

Program 5.3

Since the radiators are no longer in symmetrical pairs, no term of Euler's equation can be omitted from Eq. 5.4. As a matter of fact, with no terms eliminated, it may be seen that this expression is a Fourier Series. It follows that the radiation pattern is the inverse Fourier Transform of the distribution, and the distribution is the Fourier Transform of the pattern.

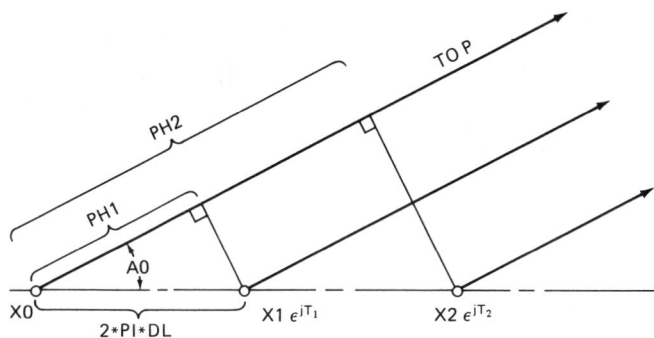

Fig. 5.5 Geometry of the end-fire array.

Figure 5.6 shows some end-fire radiation patterns. In Fig. 5.6(a) is the pattern for a 40-wavelength array. Note that the beamwidth is much broader than that of a broadside array of the same size. End-fire arrays may be either "all driven" or "parasitic". In the latter case, only one element is driven and the remaining elements are excited by mutual coupling. With mutually coupled arrays, the excitation tapers from the exciting end to the open end. This is shown in Fig. 5.6(b).

The 40-wavelength array is much longer than a typical end-fire antenna. The 10-wavelength array in Fig. 5.6(c) comes close to the limit usually employed for an antenna of this type. It may be seen that the beam width of an end-fire array is proportional to the square root of the length. Such arrays are usually limited in length, therefore, because additional radiators will not add to the directivity economically. Note the differences in scale between these graphs and earlier ones.

Lower gain types of end-fire array have an advantage in that they have a relatively small projected area and can usually by constructed from a few pieces of tubing. The small projected area means minimum wind resistance and economys of material and support. This array has additional the advantage of shaping the pattern in both planes as compared to a line array, which shapes in only one plane.

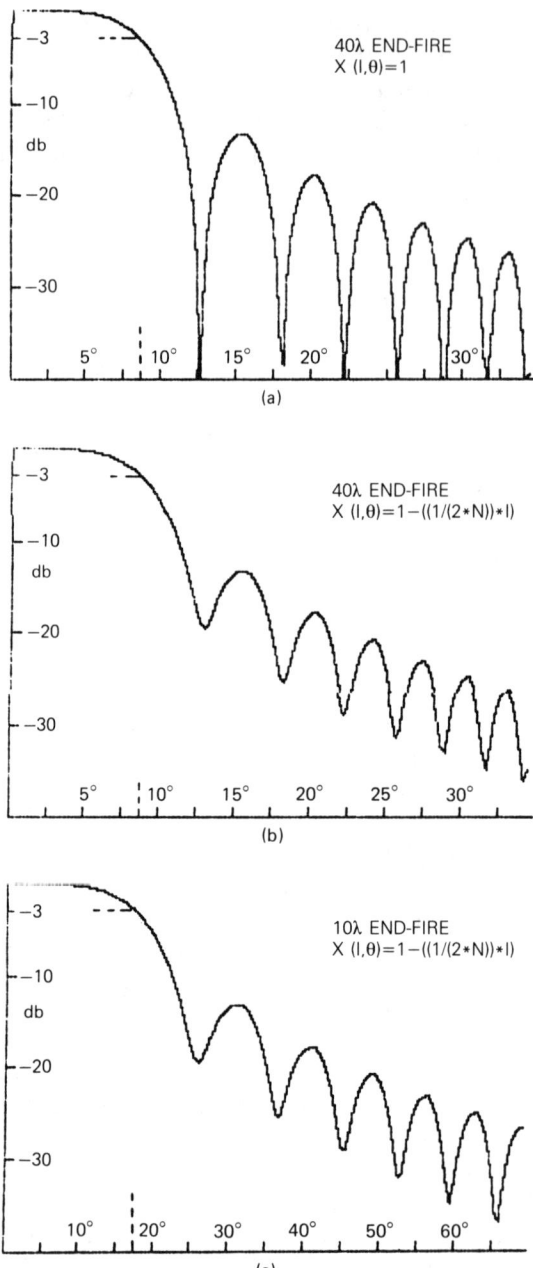

Fig. 5.6 End-fire radiation patterns.

APERTURE DISTRIBUTIONS 65

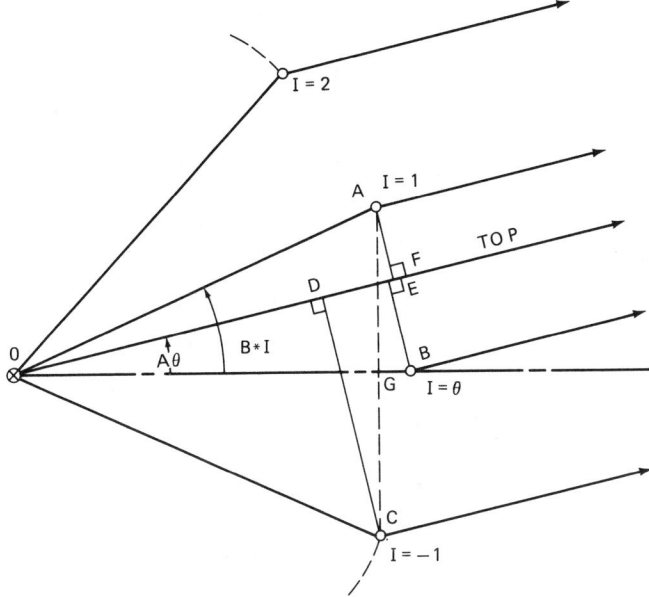

Fig. 5.7 Geometry of the Wullenweber array.

RING ARRAYS

The ring array comes in two varieties, both of which take advantage of end-fire characteristics, although in different ways. In the omnidirectional ring array, all the radiators are excited in phase, thus radiating a pattern with little or no variation in azimuth angle. In this array, the end-fire action is active in the elevation plane only. The second variety is the Wullenweber array, which is deliberately made directive in azimuth by appropriate phasing. Since the radiators are spaced about the periphery of a circle and excited to add in an end-fire timing, this antenna has advantages of both end-fire and broadside characteristics. The biggest advantage of the Wullenweber, in fact, is that it can be scanned through an entire circle without moving the antenna and without beam degradation. Furthermore, it can be arranged to provide multiple simultaneous beams, either at different or the same frequencies.

Figure 5.7 shows the basic geometry of the system, which can be summarized as follows:

$$R = OA \text{ RADIANS}$$

Initial Phase:

$$PJ = R\cos(B*I)$$

Propagation Delay:

$$PH = R\cos((B*I) - A\theta)$$

Then,

$$EP = \sum_{-N}^{+N} AI * \varepsilon^{j(PJ - PH)} \quad (5.6)$$

$$= \sum_{-N}^{+N} AI * (\cos(PJ - PH) + j\sin(PJ - PH)) \quad (5.7)$$

Program 5.4 gives the mechanism for the calculation. Angle *PJ* is the end-fire timing, and angle *PH* is the phase shift resulting from the target off-axis position. In this case, the quadrant of the array closest to the target is the only portion excited. From the results shown in Fig. 5.8, we see that a slight benefit is obtained from the end-fire action in terms of beam width, but the side lobes are noticeably worse, also because of the end-fire action.

Figure 5.8(b) shows the effect of a cosine taper, and Fig. 5.8(c) shows the effect of a cosine-squared taper. In each case, the results are somewhat worse in terms of side lobes than those obtained from a plane aperture. In Fig. 5.4(c), however, the side lobes are good enough for many radar applications.

It seems logical to question whether the use of a larger segment of the circle would provide a narrower beam width without destroy- ing the sidelobe properties. The changes shown in Program 5.5 and Fig. 5.9 provide a partial answer. In this case, a third of the aperture is used. Some damage is done to the close-in side lobes, but the wider-angle side lobes tend to suffer even worse. This problem will be investigated later.

A modern usage of the Wullenweber expands the antenna into a sphere, or a major portion of a sphere, to serve as a multibeam shipboard defense radar. In the *HF* range, the University of Illinois has constructed a very large Wullenweber array outside Champaign, Illinois. This unit provided a high-speed scan for *HF* direction finding with a narrow-beam antenna.

In subsequent chapters, we shall see the effects of random-phase errors on antenna patterns. We shall also examine the effects of the phase errors induced by steering of geometric collimators.

APERTURE DISTRIBUTIONS 67

```
200 N = 40: REM   THE # OF THE WIDEST ELEMENTS
210 PI = 3.14159
220 R = (8 * N * PI) / (2 * PI): REM   RADIUS IN RADIANS
225 M = (2 * N) + 1
230   DIM X(M,1)
240   FOR I = - N TO N
245 K = N + I
250 X(K,0) = 1
260   NEXT I
270 B = 2 * PI / (8 * N)
296   REM )))))))))))))))))))))))))))))))))))))
297   REM      THIS COMPLETES APERTURE DISTRIBUTION CALCULATION
298   REM &&&&&&&&&&&&&&&&&&&&&&&&&&&&&&&&&&&&&&&
300   FOR J = 0 TO 138
310 A0 = (PI / 3600) * J
315   FOR I = - N TO N
320 K = N + I
322 PJ = R *   COS (B * I)
324 PH = R *   COS ((B * I) - A0)
330 ER = ER + X(K,0) *   COS (PJ - PH)
335 EI = EI + X(K,0) *   SIN (PJ - PH)
340   NEXT I
350 E0 =    SQR ((ER * ER) + (EI * EI))
352   REM ##############################################
354   REM      THIS COMPLETES THE SUMMATION
356   REM      FOR ONE ANGLE
358   REM ##############################################
360   IF J = 0 THEN EN = E0: GOTO 380
370   GOTO 430
380   HGR2
390   HCOLOR= 3
400   POKE  - 12524,0
410   POKE  - 12525,64
420   POKE  - 12529,255
422   REM ##############################################
424   REM      THE GRAPHICS AND PRINTER SETUP
426   REM %%%%%%%%%%%%%%%%%%%%%%%%%%%%%%%%%%%%%%%
430 YA =   ABS (EN / E0)
432 YB = 20 * ( LOG (YA)) / 2.30259
434 Y = 4.75 * YB
436   IF Y > 191 THEN Y = 191
450   IF J = 0 THEN X0 = X:Y0 = Y
460   HPLOT X0,Y0 TO X,Y
470 X0 = X:Y0 = Y
475 X = X + 2
480 E0 = 0:ER = 0:EI = 0
490   NEXT J
```

Program 5.4

68 EXPLORING ANTENNAS AND TRANSMISSION LINES

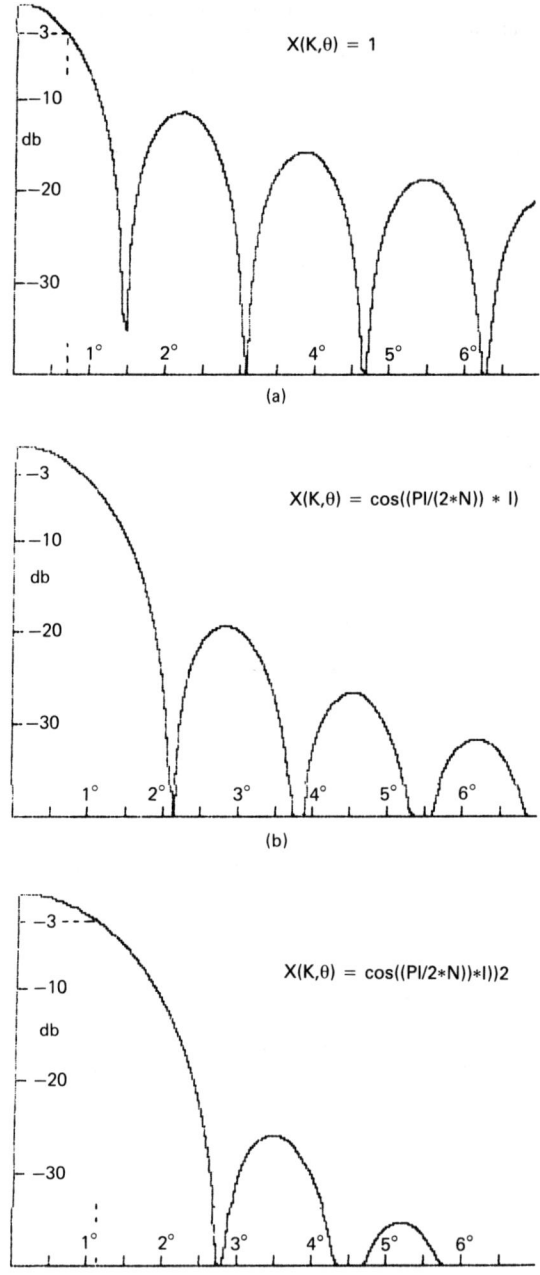

Fig. 5.8 $N = 40$ Wullenweber array ($D = 36\lambda$; $R = 160$ radians): (a) One-fourth of array active; (b) cosine taper; and (c) cosine-squared taper.

APERTURE DISTRIBUTIONS

```
200 N = 54
210 PI = 3.14159
220 R = (6 * N * PI) / (2 * PI): REM   RADIUS IN RADIANS
225 M = (2 * N) + 1
230 DIM X(M,1)
240 FOR I = - N TO N
245 K = N + I
250 X(K,0) = ( COS ((PI / (2 * N)) * I)) ^ 2
260 NEXT I
270 B = 2 * PI / (6 * N)
296 REM )))))))))))))))))))))))))))))))))))))))
297 REM     THIS COMPLETES APERTURE DISTRIBUTION CALCULATION
298 REM &&&&&&&&&&&&&&&&&&&&&&&&&&&&&&&&&&&&&&&&
300 FOR J = 0 TO 138
```

Program 5.5

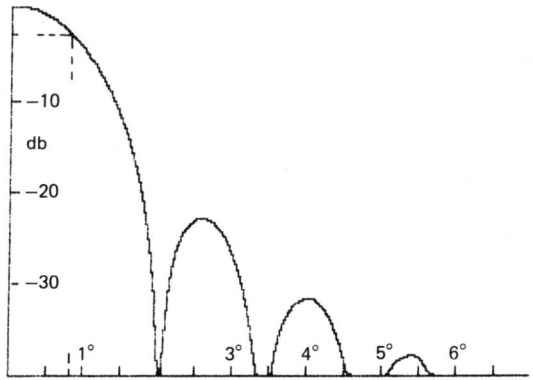

Fig. 5.9 $N = 54$ Wullenweber array with one-third of array active.

6
Geometric Collimators

In the text to this point we have seen that a straight-fronted wave is required if the energy is to be confined in a "beam" or tight bundle. From the field of optics we borrow the verb *to collimate*, meaning "to render parallel." Although this concept really has more to do with "ray optics" than with "diffraction optics," it is sometimes used with respect to antennas, and we shall do so here.

The "ray" concept, in which the behavior of an optical system is examined by tracing straight lines or "rays" through a system, is more closely suited to optical work than to radio work. In optics, the smallest pencil of light is still many wavelengths wide, and thus spreading takes place very slowly. To a first-order approximation, the effects of diffraction may be ignored. For a wavelength of $0.6E - 6$ meters (red light), an aperture that is 1000-wavelengths broad will be only 0.6-mm in diameter.

From Table 5.1 of the previous chapter, we see that a round aperture of 1000-wavelengths with uniform illumination would have a half-power beam width of only $1.2E - 3$-radians, or 0.069-degrees! The ray would have to travel 0.83-km before the beam had spread to a width of 1 meter.

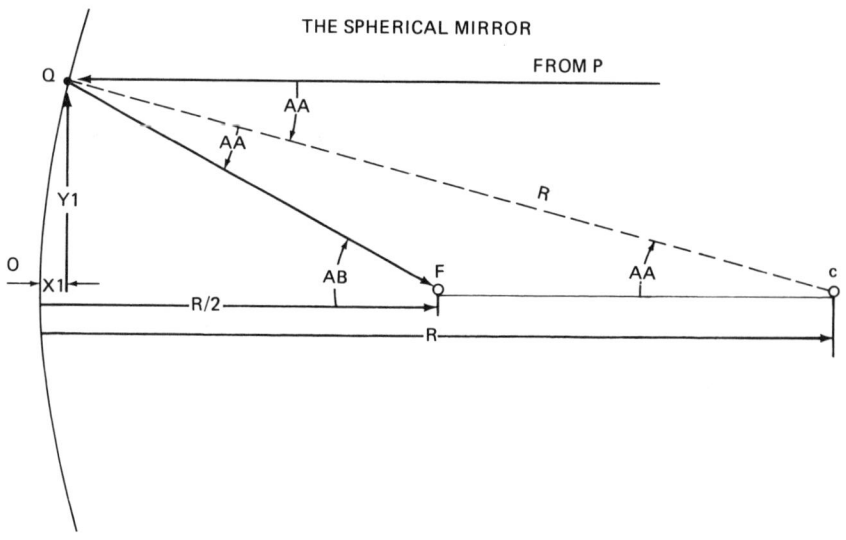

Fig. 6.1 The spherical mirror.

To a first-order approximation, this could be considered to be a "ray" that never spreads, has no side lobes, etc. By comparison, an "S" band radar operating at a wavelength of 10-cm (3-GHz) would require an antenna aperture of 100-meters for the same beamwidth—more than the length of a football gridiron! Except for millimeter wave devices, antennas are seldom built with an aperture diameter in excess of a hundred wavelengths because of the huge sizes involved. This is one of the principal differences between antennas and optical devices. At radio frequencies, even to ultra microwaves, we are able to handle individual dipole radiators. In optics, even the most tightly collimated laser has an effective apperture hundreds or thousands of wavelengths across. There is no optical equivalent of a dipole antenna or parasitic end-fire array since light waves are too small to be handled individually.

This situation is changing gradually as a result of fiber optics and integrated optical devices, but it will be some time before we will deal with single optical dipoles.

THE SPHERICAL MIRROR

To begin with, we shall investigate the operation of a sherical mirror using a ray optics approach. As we progress to other collimators, we shall diverge to a more "diffraction optics" approach.

Consider the geometry of Fig. 6.1. At the left, we have a mirror whose surface is a portion of a sphere with a center at C. One of many parallel rays from a distant source P is incident upon the reflector at point Q. Path PCO is the ray from point P through the center of curvature. P is sufficiently distant so that PQ is parallel to CO (P might be a star). Angle AA formed by PQC is identical to angle QCO because complementary angles formed by a line cutting parallel lines are equal. Since the angle of incidence is equal to the angle of reflection (the radius is normal to the circle), we see that angle AB is equal to twice angle AA. Now, if $X1$ is insignificantly small because $Y1$ is small, we see, using the small-angle rule (the sin−tan angle, in radians) that OF is one-half OC, or half the radius of curvature. In the figure, angle AB is purposely *not* a small angle so that the basic geometry may be observed.

As $Y1$ increases, the crossing of the reflected ray tracks inside point F, despite the increase in $X1$, as a result of the breakdown of the small-angle assumption. Figure 6.2 illustrates this effect. The shift of focus is termed *spherical aberration*, and the locus traced by the arrays is called the *caustic surface*. This surface can be seen displayed on the surface of a glass of milk which is strongly illuminated from the side.

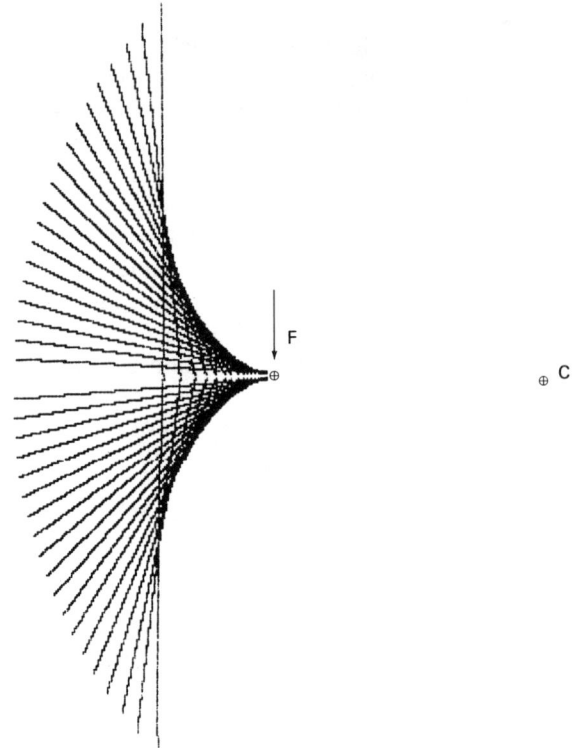

Fig. 6.2 Spherical aberration.

The amount of spherical aberration can be kept to manageable levels by restricting the maximum value of $Y1$ that can be illuminated. If $Y1$ is restricted to one-thirtieth the length of OF, the shortest focal zone will be 0.9998 R/2. Since the diameter of the dish would be one-fifteenth the focal length, this would be described as an "F15" dish. Such "slow" (in photographic terms) reflectors are used in a certain class of astronomical telescopes.

For radio telescopes, the spherical reflector is occasionally used with a feed that illuminates only a portion of the reflecting surface. It is possible to design a type of end-fire feed that makes corrections for a certain amount of spherical aberration. The advantage of such an arrangement is that a telescope can be constructed that can be steered over a limited range by moving only the feed. This is the scheme used on the giant telescope at Arecibo in Puerto Rico. The dish itself is 1000-ft wide and located in a natural sinkhole. Since it cannot be moved, all scanning must come from feed motion.

THE NEWTONIAN SYSTEM (THE PARABOLA)

From a series of measurements, Newton concluded, incorrectly, that all glasses had the same dispersion and that it would therefore be impossible to construct an object lens for a telescope that did not suffer from chromatic aberration, that is, that did not have a different focal length for each color. He reasoned that a reflecting telescope could be constructed in which at least the objective (the main collimating lens) would be completely free of chromatic aberration. He also showed that such a telescope would exhibit no spherical aberration if the objective mirror were in the form of a paraboloid.

If one were to take a line that crosses another line at an oblique angle and rotate the assembly about the first line, the second line would sweep out a volume in space consisting of a pair of identical cones arranged point to point. If one were then to saw one of the cones along a plane parallel to the inclined generating line, the exposed face would have the outline of a *parabola* where it intersected the surface of the cone.

Another definition of the parabola that is more to the point for antenna work is the one that defines it as the locus of points equidistant from a line AB, called the *directrix*, and a point F, called the *focus*. Figure 6.3 digrams the relationship. By this definition, the parabola is also the locus of points such that $SQ = QF$ (understand that S can move up and down AB). Now, if CD is parallel to AB, then the length of line RQS is a constant, and if RQS is a constant, then length RQF must be constant as well. Thus, a spherical wave launched from F toward the parabola will wind up as a straight line at CD. All points on CD will be in-phase since all paths to it are of equal length. In other words, the parabola has accomplished the task of collimating the wave. It can also be shown that the angle-of-incidence and angle-of-reflection condition has been satisfied.

As indicated by Fig. 6.3, the parabola is determined by the equation, $y^2 = 4FX$.

Usually, the most convenient way to design a parabolic antenna is to reference most of the dimensions to the focus and place some kind of feed there. The maximum subtended angle at the edge of the aperture then becomes a matter of interest. This can be calculated with the formula,

$$\angle AB = ATN(Y/(F - X)) = ATN(4FY/(4F^2 - Y^2)) \qquad (6.1)$$

We had previously noted that the distance from the feed to a point on the reflector increases with distance from the vertex (V). When calculating the energy at any point on an aperture, we must account for the Inverse

74 EXPLORING ANTENNAS AND TRANSMISSION LINES

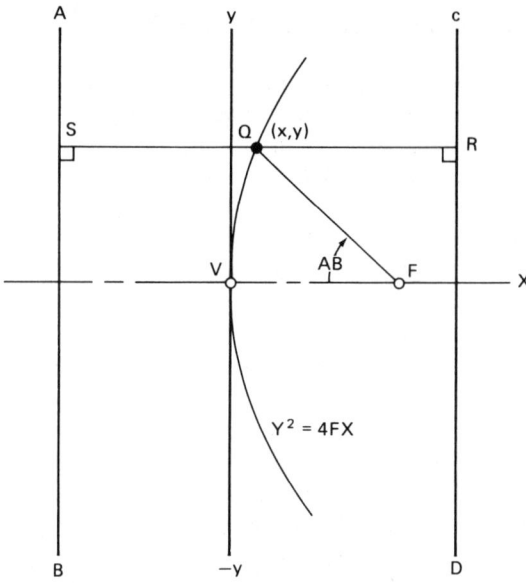

Fig. 6.3 The parabola.

Square Law spreading. The field striking the aperture is derated by the factor (F/FQ), where

$$FQ = 2*F/(1 + COS(AB)) \tag{6.2}$$

The pattern of the feed is generally known as a pattern function of angle AB. The following equation can be used to translate this pattern function into a distribution on the aperture, which should then be derated for the reason stated above:

$$Y = (2*F/(1 + COS(AB)))*SIN(AB) \tag{6.3}$$

The parabola can take either the form of a parabolic cylinder with a line feed or the form of a paraboloid of revolution. The latter is obtained by rotating the parabola about axis FV. The paraboloid of revolution can be fed with a point source such as a waveguide horn or a dipole and reflector.

In a parabolic cylinder antenna, the shaping in the azimuth plane is accomplished separately from the beam shaping in the elevation plane in order to provide the designer with a certain degree of freedom. For example, scanning is always easier to accomplish in one plane (two

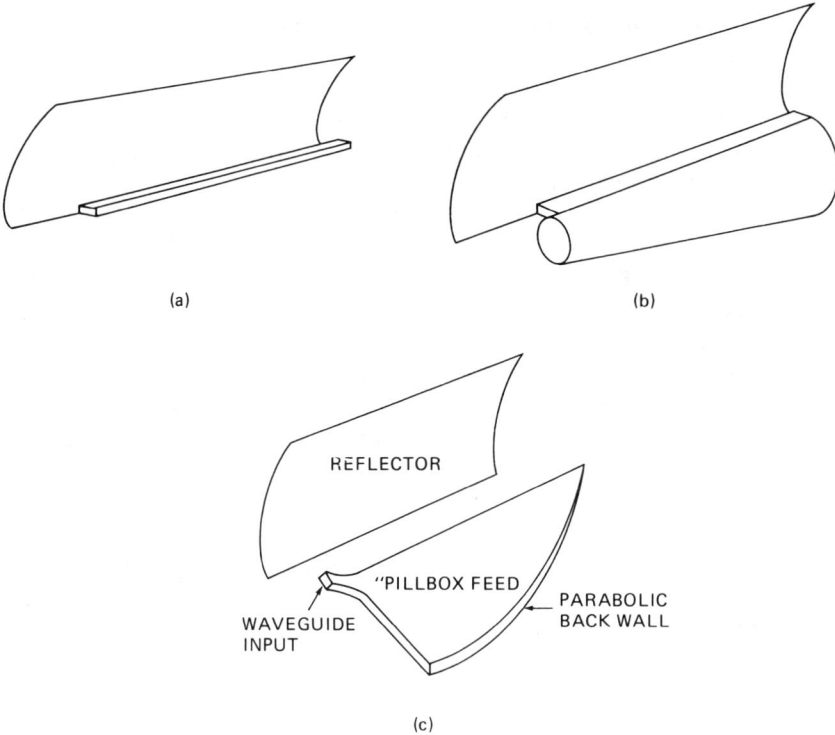

Fig. 6.4 Parabolic cylinder antennas: (a) Waveguide line feed, (b) electromechanical scanner feed, and (c) Foster scanner with pillbox feed and collimation in vertical and horizontal planes performed separately.

dimensions) than in two. An antenna capable of such scanning can be formed by using a scanning line feed to illuminate a parabolic cylinder, which, in turn, takes care of the beam shaping in the other plane. Moreover, parabolic cylinders are mechanically easier to produce than paraboloids of revolution. The AN/CPS−1, or Microwave Early Warning (MEW), radar of World War II used a parabolic cylinder fed by a waveguide line feed. The reflector was made of cast iron and was 30-feet across, or 82-wavelengths! The waveguide feed was excited in a "roof" distribution to provide the unit with modest side lobes.

Figure 6.4(a) shows an example of an early parabolic clinder antenna such as the MEW, and Fig. 6.4(b) depicts a scanning antenna using a parabolic cylinder array and an electromechanical scanner. The Foster scanner depicted in Fig. 6.4(b) consists of a parabolic "pillbox" (two-dimensional) antenna wrapped around the surface of a cone as in 6-4(c) and mechanically rotated at high speed. During rotation, the distance from the equiphase mouth of the "pillbox" to the mouth of the wave guide horn obviously increases more rapidly at the large diameter of the cone than at the smaller diameter and thus causes the beam to swing in azimuth at a rate determined by the rotation speed. This principle was used in a number of mortar and artillery locating radars such as the AN/TPQ − 10 and certain blind-landing and gunfire-directing radars that required a rapid scan over a limited azimuth angle.

A function frequently required of a radar is to determine the height of an aircraft target. This is sometyimes accomplished with an antenna that flaps the beam up and down in a regular pattern. A number of 3-D radars have been constructed in which the line feed is excited by a frequency-scanning feed so that the elevation of the beam can be electronically steered just by changing frequency. In such radars, azimuth beam shaping is provided by a parabolic cylinder with a vertical axis, and azimuth searching is achieved by a continuous mechanical rotation of the antenna about this axis.

The antennas depicted in Fig. 6.4 make use of what is known as a "cut parabola." It can be seen that focusing or collimating properties apply to any given fragment of a parabola or paraboloid. We shall also shortly see that there are certain advantages to "cutting" a paraboloid in such a way that its vertex is not included on the surface used. We shall first turn our attention, however, to some other forms of collimating systems, those using "all-reflecting" components.

THE ELLIPSE AND THE HYPERBOLA

Let us go back to our two point-to-point cones that were generated by rotating a pair of obliquely crossing lincs about onc of them. A cut through either of the cones at an angle steeper to this axis line than to the other line will cut completely through it, and the intersection of the plane of the cut and the cone will produce an *ellipse*. The ellipse and some of its relationships are shown in Fig. 6.5(a). One of the properties of the ellipse is that $(F'P + PF)$ is a constant. Others are as follows:

$$\frac{x^2}{a^2} + \frac{y^2}{b^2} = 1 \tag{6.4}$$

$$b^2 = a^2 - c^2 \tag{6.5}$$

If we cut the cones *parallel* to the axis of *rotation*, on the other hand, we will obtain a *hyperbola*, which is illustrated in Fig. 6.5(b). One of the

GEOMETRIC COLLIMATORS 77

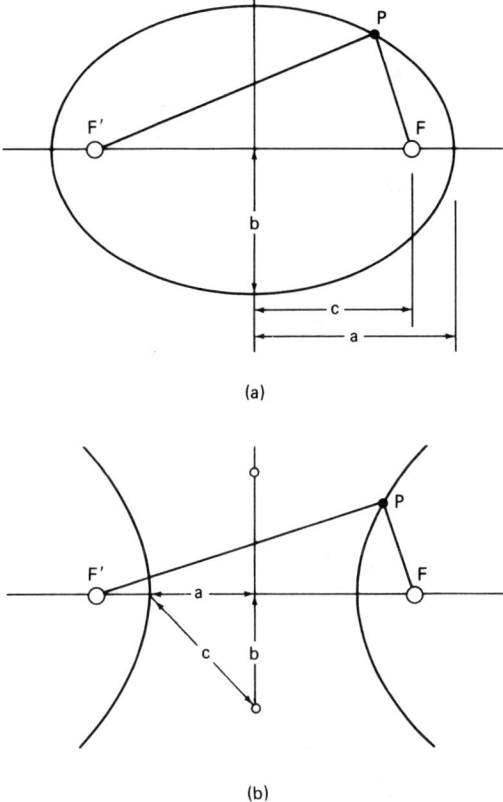

Fig. 6.5 The ellipse and the hyperbola.

properties of interest of the hyperbola is that the difference between $F'P$ and PF is a constant. Since the cut is parallel to the axis of rotation, it will cut *both* cones, and the curves formed on opposite sides of the vertex are identical. Other properties are as follows:

$$\frac{x^2}{a^2} - \frac{y^2}{b^2} = 1 \tag{6.6}$$

$$F'P - PF = 2a \tag{6.7}$$

$$b^2 = c^2 - a^2 \tag{6.8}$$

78 EXPLORING ANTENNAS AND TRANSMISSION LINES

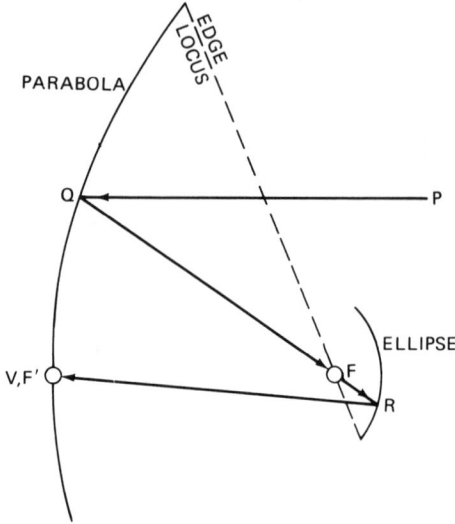

Fig. 6.6 The Gregorian telescope.

Before considering how these curves may be put to use, let us first consider the Gregorian telescope shown in Fig. 6.6. Here we see a parabolic antenna that has been fitted with a small segment of an ellipse. The ellipse is selected so that one focus (F) coincides with the focus of the parabola and the second focus (F') coincides with the vertex (V). From previous considerations, we know that path PQF is a constant, and, from the definition of the ellipse, we know that FRF' is also a constant. Therefore, the path from P to F' is a constant as well, thus satisfying the necessary condition for collimation.

It may be seen that the Gregorian telescope has the advantage of transferring the focus from in front of the main mirror to a point on its vertex. By using a different second focus, of course, this focus could be placed either behind or in front of the vertex. Such adaptability is one of the principle reasons for employing a system of this type. In any optical system, the lengthening of the apparent focal length of the primary element has the advantage of providing higher magnifications without having to resort to oculars of very short focal length. In radio work, such lengthening can be advantageous for monopulse feeding of the dish, etc.

The Cassagrainian telescope shown in Fig. 6.7 employs a hyperbola to increase the effective focal length. The actual path for this unit is $PQRF'$ but the path can also be written as $(PQ + (QF - RF) + (RF')$, or

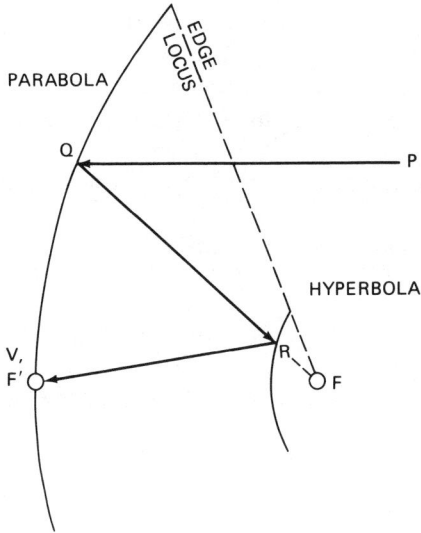

Fig. 6.7 The Cassagranian telescope.

$(PQ + QF) + (RF' - RF)$. From the definition of a hyperbola in Eq. 6.7, we know that $(RF' - RF)$ is a constant, and, since we know from the properties of the parabola $(PQ + QF)$ is also a constant, we see that PQRF' is a constant as well—once again the necessary condition for collimation.

The curves referred to as parabolas, ellipses and hyperbolas can also take the form of three-dimensional figures: paraboloids, ellipsoids, and hyperboloids.

In both Figs. 6.6 and 6.7, we have shown by a broken line the locus of the extreme "ray" in the system that is sometimes used to define the limit of the ellipsoidal or hyperboloidal subreflector.

The Cassagrainian system has the virtue of offering a long focal length without imposing the penalties of long physical length. In modern practical usage, the Gregorian system is seldom preferred to the Cassagranian. The latter provides not only a shorter physical length for the same effective focal length but certain other performance advantages as well. We shall have more to say about some of these characteristics shortly.

APERTURE INTEGRATION

We have seen that the line-fed parabolic-cylinder antenna makes it possible to separate the functions of collimation into two separate orthogonal planes on a true physical basis. It is also possible to simplify the pattern calculations by separating them into two slightly fictitious planes. To do so, we collapse the real two-dimensional distribution of illumination into an equivalent line-array distribution. In the resulting line array, the radiation pattern in the plane of the line will be the same as the two-dimensional distribution of the original paraboloid.

```
140  REM     SPACE ATTENUATION=E0/E0=(1+COS(0))/2
150  REM  ANGLE P0=ATN((4FY)/(4(F*F)-(Y*Y)))
155  REM     FOR THE PARABOLA
156  REM     (Y*Y)=4FX
160  N = 64: REM  THE NUMBER OF APERTURE ELEMENTS TO EACH SIDE OF THE CENTE
     R ROW
170  PI = 3.14159
180  DL = .313: REM  THE ELEMENT SPACING IN WAVELENGTHS
185  HOME
190  DIM X(N,1)
200  PRINT "INPUT F/D"
210  INPUT FD
220  YM = 1 / (2 * FD)
230  TH = (4 * YM) / (4 - (YM * YM))
235  TE =  ATN (TH)
237  TH = TE * 180 / PI
240  PR# 1
242  PRINT ""
250  PRINT "FOR FOCAL RATIO =",FD
260  PRINT "HALF SUBTENDED ANGLE IN DEGREES=",TH
270  PR# 0
280  PRINT "ENTER DESIRED EDGE ILLUMINATION"
290  PRINT " IN DB BELOW FEED PEAK"
300  INPUT DE
301  PR# 1
302  PRINT "FOR EDGE ILLUMINATION-",DE,"DB"
303  PR# 0
310  DF = DE / 20
320  DG = 1 / (10 ^ DF)
330  DH = DG * 2 / (1 +  COS (TE))
340  TI =  -  ATN (DH /  SQR ( - DH * DH + 1)) + 1.5708
345  REM  CALCULATE THE ARCCOS(DH)
350  PF = TE / TI
360  TJ = 44.93 * PF
362  GG = 41253 / (TJ * TJ)
364  GH = 10 *  LOG (GG) / 2.3026
370  PR# 1
380  PRINT "AFTER ALLOWANCE FOR SPACE ATTENUATION"
390  PRINT "THE PRIMARY PATTERN FACTOR IS",PF
400  PRINT "WHICH CALLS FOR A PRIMARY HALF-POWER"
410  PRINT "BEAMWIDTH OF",TJ,"DEGREES"
420  PRINT "WITH A HORN GAIN OF",GH,"DB"
430  PR# 0
440  REM  ******************************************
450  REM  NOW TO CALCULATE THE CENTER STRIP INTEGRAL
460  REM  ******************************************
470  TK = TE / N: REM   THE STEP ANGLE PER ELEMENT
480  EN = 1
490  FOR I = 1 TO N
500  TL = I * TK / PF
```

Program 6.1

GEOMETRIC COLLIMATORS 81

```
510 TM = I * TK
520 EM = ( COS (TL) * (1 + COS (TM)) / 2)
525 EN = EN + 2 * EM
530 NEXT I
540 E0 = EN
542 PR# 1
543 PRINT "CENTER STRIP TOTAL=",E0
544 PR# 0
550 REM $$$$$$$$$$$$$$$$$$$$$$$$$$$$$$$$$$$$$$$$$$
555 REM      NOW CALCULATE THE STRIPS
560 REM >>>>>>>>>>>>>>>>>>>>>>>>>>>>>>>>>>>>>>>>>>
561 PR# 1
562 PRINT "STRIP";"  ";"EXCITATION"
563 PRINT "    0     ";"1.0"
564 PR# 0
570 FOR J = 1 TO 64
580 K% = SQR (4096 - (J * J))
585 EN = 0
590 FOR I = 0 TO K%
600 TM = TK * ( SQR ((I * I) + (J * J)))
610 TL = TM / PF
620 EM = ( COS (TL) * (1 + COS (TM)) / 2)
630 IF I = 0 THEN EN = EN + EM: GOTO 650
640 EN = EN + 2 * EM
650 NEXT I
660 PR# 1
670 PRINT J;"   ";(EN / E0)
680 NEXT J
5000 STOP
```

```
FOR FOCAL RATIO =              .33
HALF SUBTENDED ANGLE IN DEGREES=          74.2934361

FOR EDGE ILLUMINATION         -10        DB
AFTER ALLOWANCE FOR SPACE ATTENUATION
THE PRIMARY PATTERN FACTOR IS   1.23511241
WHICH CALLS FOR A PRIMARY HALF-POWER
BEAMWIDTH OF     55.4936004     DEGREES
WITH A HORN GAIN OF             11.2696244    DB
CENTER STRIP TOTAL=             94.600883
STRIP   EXCITATION
    0      1.0
 1   .993059286    17   .908531451    33   .690578067    49   .40089226
 2   .992293839    18   .900240664    34   .677948396    50   .375906221
 3   .991019018    19   .89153392     35   .658525825    51   .35791778
 4   .989236212    20   .875792279    36   .638984122    52   .340231586
 5   .986947358    21   .866375163    37   .626091203    53   .316204184
 6   .984154949    22   .856574952    38   .606464552    54   .299267808
 7   .980862018    23   .839792055    39   .586798201    55   .276059043
 8   .977072145    24   .829363841    40   .56712118     56   .25330599
 9   .972789444    25   .811925742    41   .554179937    57   .237769742
10   .968018563    26   .80093037     42   .534599959    58   .216104965
11   .962764679    27   .783613938    43   .515085613    59   .188390225
12   .950386951    28   .771421505    44   .495664572    60   .168058714
13   .944241657    29   .759627403    45   .476364053    61   .141842318
14   .937636184    30   .74098702     46   .457210759    62   .109876378
15   .930577659    31   .722096291    47   .438230815    63   .0790972318
16   .916397394    32   .709778508    48   .419449714    64   3.34273562E-03
```

Program 6.1 (*continued*)

Program 6.1 provides a routine that performs this operation after doing a few other things useful to the design of paraboloidal antennas. Line 160, for example, sets up the number of strips to be evaluated in the aperture. In this case, $N = 64$, but it could be any appropriate number provided line 570 is suitably altered.

Because of tooling costs, a paraboloidal antenna is frequently designed with a "stock" reflector. A number of manufacturers offer paraboloidal "dishes" of a specified diameter and focal length. The dish is also frequently specified by diameter and F/D ratio instead. Lines 200 to 237 are used to calculate the half-subtended angle of the reflector from the F/D.

It has been shown earlier that the side-lobe levels attainable with an antenna are a function of the smoothness with which the aperture distribution tapers off at the edges. The main portion of most antenna feeds has a pattern that can usually be simulated by COS($n*$TH), where n is some constant. After entering the desired illumination at the edge of the dish in line 300, the program calculates the required multiple of TH that will give the desired illumination at the edge of the reflector. Line 330 makes allowance for the space attenution. Line 340 is the ARCCOS routine given on page 103 of the applesoft Basic Programming Manual. The program calculates the angle at which the cosine function has decayed to the specified edge level. Dividing this angle by the angle to the reflector edge calculated in 235 gives the pattern multiplier for the primary pattern. When the multiplier is 1, the primary or feed pattern is a cosine function.

Lines 362 and 364 calculate the gain of the feed horn, using the solid angle formula from Chap. 2. This value is useful for estimating the back lobes and "spill-over" energy of the complete antenna. Let us suppose that the gain of the complete feed and reflector assembly is 30-dB and the edge illumination is 10-dB. As a first estimate, the spill-over lobes that are radiated more or less rearward by the feed horn will be 10-dB less than the feed-horn gain. Suppose that the feed-horn gain is 11-dB, as in the example given with Program 6.1, then the spill-over will be 1-dB above isotropic and 29-dB below the peak of the antenna gain.

Lines 470 through 530 calculate the illumination for the first (or 0th) column. Line 520 is the computation for correcting for spreading. The numeric value of the first column is printed out so that it can be used in aperture blocking calculations.

Lines 570 through 680 compute the illumination for the other strips. The computation treats the surface of the reflector as if it were covered by a checkerboard with J columns on each side of the center column and I rows. The number of rows in each column is computed in line 580 for a

GEOMETRIC COLLIMATORS 83

```
200 N = 64: REM  NUMBER OF ARRAY ELEMENTS
210 PI = 3.14159
220 DL = .313: REM  ELEMENT SPACING IN WAVELENGTHS
230 AH = PI / (2 * N)
235  REM  THE DATA STATEMENTS CONTAIN A LIST OF N+1 EXCITATION VALUES
240  DATA  1,.993,.992,.991,.989,.987,.984,.981,.977
245  DATA  .973,.968,.963,.950,.944,.938,.931,.916
250  DATA  .909,.900,.892,.876,.866,.857,.840,.829
255  DATA  .812,.801,.790,.771,.760,.741,.722,.710
260  DATA  .691,.678,.659,.639,.626,.606,.587,.567
265  DATA  .554,.535,.515,.496,.476,.457,.438,.419
270  DATA  .401,.376,.358,.340,.316,.299,.276,.253
275  DATA  .238,.216,.188,.168,.142,.110,.079,.003
295 E0 = 1
296  REM  )))))))))))))))))))))))))))))))))))))))))
297  REM     THIS COMPLETES APERTURE DISTRIBUTION CALCULATION
298  REM  &&&&&&&&&&&&&&&&&&&&&&&&&&&&&&&&&&&&&&&&&
300  FOR J = 0 TO 138
310 A0 = (PI / 3600) * J
320 AI = 2 * PI * DL *  SIN (A0)
330  FOR I = 1 TO N
335 AJ = AI * I
337  READ E
340 E0 = E0 + E * 2 *  COS (AJ)
350  NEXT I
352  REM  ##########################################
354  REM     THIS COMPLETES THE SUMMATION
356  REM        FOR ONE ANGLE
358  REM  ##########################################
360  IF J = 0 THEN EN = E0: GOTO 380
370  GOTO 430
380  HGR2
390  HCOLOR= 3
400  POKE  - 12524,0
410  POKE  - 12525,64
420  POKE  - 12529,255
422  REM  ##########################################
424  REM     THE GRAPHICS AND PRINTER SETUP
426  REM  %%%%%%%%%%%%%%%%%%%%%%%%%%%%%%%%%%%%%%%%%%
430 YA =  ABS (EN / E0)
432 YB = 20 * ( LOG (YA)) / 2.30259
434 Y = 4.75 * YB
436  IF Y > 191 THEN Y = 191
450  IF J = 0 THEN X0 = X:Y0 = Y
460  HPLOT X0,Y0 TO X,Y
470 X0 = X:Y0 = Y
475 X = X + 2
480 E0 = 1
485  RESTORE
490  NEXT J
```

Program 6.2

```
FOR FOCAL RATIO =                    .250001
HALF SUBTENDED ANGLE IN DEGREES=                    89.9999469
FOR EDGE ILLUMINATION           10                  DB
AFTER ALLOWANCE FOR SPACE ATTENUATION
THE PRIMARY PATTERN FACTOR IS    1.77273511
WHICH CALLS FOR A PRIMARY HALF-POWER
BEAMWIDTH OF      79.6489883          DEGREES
WITH A HORN GAIN OF                   8.13089746    DB


FOR FOCAL RATIO =                    .3
HALF SUBTENDED ANGLE IN DEGREES=                    79.6112094
FOR EDGE ILLUMINATION          -10                  DB
AFTER ALLOWANCE FOR SPACE ATTENUATION
THE PRIMARY PATTERN FACTOR IS    1.38213867
WHICH CALLS FOR A PRIMARY HALF-POWER
BEAMWIDTH OF      62.0994906          DEGREES
WITH A HORN GAIN OF                  10.292728      DB


FOR FOCAL RATIO =                    .4005
HALF SUBTENDED ANGLE IN DEGREES=                    63.946501
FOR EDGE ILLUMINATION          -10                  DB
AFTER ALLOWANCE FOR SPACE ATTENUATION
THE PRIMARY PATTERN FACTOR IS    1.00023112
WHICH CALLS FOR A PRIMARY HALF-POWER
BEAMWIDTH OF      44.940384           DEGREES
WITH A HORN GAIN OF                  13.101735      DB


FOR FOCAL RATIO =                    .5
HALF SUBTENDED ANGLE IN DEGREES=                    53.1301472
FOR EDGE ILLUMINATION          -10                  DB
AFTER ALLOWANCE FOR SPACE ATTENUATION
THE PRIMARY PATTERN FACTOR IS     .796356512
WHICH CALLS FOR A PRIMARY HALF-POWER
BEAMWIDTH OF      35.7802981          DEGREES
WITH A HORN GAIN OF                  15.0815786     DB

FOR FOCAL RATIO =                    .6
HALF SUBTENDED ANGLE IN DEGREES=                    45.2397681
FOR EDGE ILLUMINATION          -10                  DB
AFTER ALLOWANCE FOR SPACE ATTENUATION
THE PRIMARY PATTERN FACTOR IS     .663193442
WHICH CALLS FOR A PRIMARY HALF-POWER
BEAMWIDTH OF      29.7972814          DEGREES
WITH A HORN GAIN OF                  16.6709146     DB
```

Fig. 6.8 Subtended angles of primary beam width and gain.

round aperture. If a different aperture shape is used, this line will have to be changed to accommodate it.

Before the printout in line 670, the summation is "normalized" by dividing it by the summation for the zeroth column. The resulting quotient is the illumination factor for N columns to each side of the zeroth column.

The print-out table of Fig. 6.8 shows the subtended angle of feed (or primary) beam width and feed (or primary) gain for a variety of focal lengths with 10-dB edge illumination. It may be seen that as F/D grows larger, the primary beam width becomes smaller, thereby implying the need for a larger horn or feed apparatus.

Common usage describes the source that feeds the major collimating agent as the "primary" or "feed," whereas the major collimating element itself is described as the "secondary." Thus, the beam from the paraboloid is the "secondary" or "main" beam, whereas the illumination of the paraboloid is the "primary" or "feed."

APERTURE BLOCKING

Having performed the operture distribution integration, let us investigate some of the properties of the antenna. The routine of Program 6.2 differs from the earlier center-element array calculations in its use of the DATA statement. The RESTORE in line 485 is intended to reset the data read count. The resulting pattern is shown in Fig. 6.9. It may be seen that this is a well-behaved pattern, with modest side lobes. However, there is a certain amount of fiction involved here. With a paraboloid of revolution, there must be something in front of the paraboloid to feed it, and there must be something to support the feed.

Let us suppose that the feed is a disc 2.8-wavelengths in diameter at the focus and that the feed support consists of a tripod with struts 1-wavelength in diameter. The net effect is a certain amount of *aperture blockage*. If we assume the tripod to have one leg vertical, then one-half of strips 0 and ±1 are blocked. To determine the horn or feed blockage, we can draw a circle on a piece of graph paper and write in the excitation for each square. These can be added to determine the total blockage.

The "blockage" itself can simply be subtracted from the excitation of the appropriate strip. The behavior is just as if the blocked areas were radiating a negative energy. This is called Babinet's Principle. The effect upon a wavefront of blocking one of its areas is exactly as if the blocked area were radiating negative energy along with the original unblocked wavefront.

86 EXPLORING ANTENNAS AND TRANSMISSION LINES

```
492  REM $$$$$$$$$$$$$$$$$$$$$$$$$$$$$$$$$$$$$$$$$$$$$$$$$
494  REM    ADD THE SCALES
496  REM $$$$$$$$$$$$$$$$$$$$$$$$$$$$$$$$$$$$$$$$$$$$$$$$$
500  HPLOT 0,1 TO 0,191
510  HPLOT 0,191 TO 279,191
530  HPLOT M,186 TO M,191
540  M = M + 20
560  IF M < 280 THEN GOTO 530
570  HPLOT 0,14 TO 3,14
580  HPLOT 0,48 TO 5,48
590  HPLOT 0,95 TO 5,95
600  HPLOT 0,143 TO 5,143
610  STOP
```

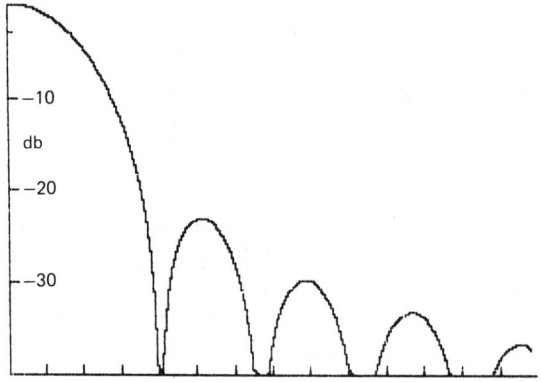

Fig. 6.9 Pattern resulting from aperture distribution of Program 6.1.

The blocked excitation and the resulting waveform are shown in Fig. 6.10. Note in particular that the odd-numbered side lobes—i.e., the first and third—are enhanced, whereas the even-numbered ones—the second and fourth—are supressed. The main beam (which can be considered the zeroth) is also supressed. The allusion to the blockage as "radiating dark" is fairly accurate.

The reason for the choice of a cut paraboloid now becomes a little more evident. If the parabolid is cut so that the used portion of the surface is all to one side of the vertex, then it is possible to eliminate aperture blockage by positioning the feed horn off to one side and out of the way.

```
235 REM  THE DATA STATEMENTS CONTAIN A LIST OF N+1 EXCITATION VALUES
240 DATA .451,.451,.896,.918,.943,.987,.984,.981,.977
245 DATA .973,.968,.963,.950,.944,.938,.931,.916
250 DATA .909,.900,.892,.876,.866,.857,.840,.829
255 DATA .812,.801,.790,.771,.760,.741,.722,.710
260 DATA .691,.678,.659,.639,.626,.606,.587,.567
265 DATA .554,.535,.515,.496,.476,.457,.438,.419
270 DATA .401,.376,.358,.340,.316,.299,.276,.253
275 DATA .238,.216,.188,.168,.142,.110,.079,.003
```

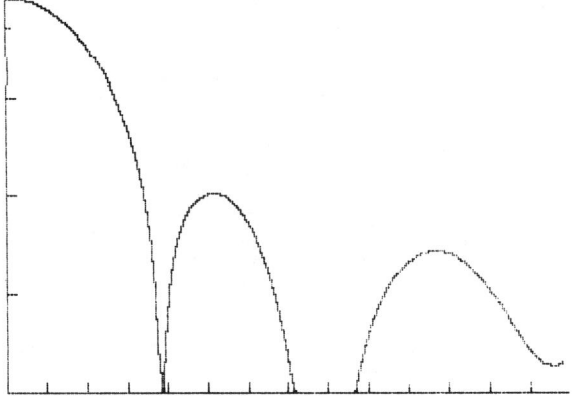

Fig. 6.10 Aperture blocking by a 1-λ strut and a 2.8-λ feed horn.

A more intense blockage is found in both the Cassagrainian and Gregorian antennas. Figure 6.11 shows the waveform when the original distribution is blocked not only by the same 1-wavelength-diameter strut but by a 6.9-wavelength-diameter subreflector. In this case, the blockage effect is large enough to prevent the second and fourth side lobes from ever going positive, and thus there is no axis crossing. The odd order side lobes are increased very significantly.

In answer to the inevitable question of why one would want to use a Cassagranian antenna if its gain and side-lobe levels are deteriorated by aperture blockage, several points should be noted. First of all, an antenna with an aperture of only 40-wavelengths, as in this example, would rarely be designed as a Cassagranian and for just this reason. For a 100-wavelength aperture, however, the blockage might be more acceptable. The second reason stems from the direction of the side lobes. In the Newtonian system used for satellite reception, the spill-over from the feed looks directly at the warm earth, whereas the spill-over and aperture blockage lobes for the Cassagranian system look at the cold sky. The

```
235 REM  THE DATA STATEMENTS CONTAIN A LIST OF N+1 EXCITATION VALUES
240 DATA .383,.383,.767,.769,.767,.776,.783,.801,.822
245 DATA .829,.846,.936,.950,.944,.938,.931,.916
250 DATA .909,.900,.892,.876,.866,.857,.840,.829
255 DATA .812,.801,.790,.771,.760,.741,.722,.710
260 DATA .691,.678,.659,.639,.626,.606,.587,.567
265 DATA .554,.535,.515,.496,.476,.457,.438,.419
270 DATA .401,.376,.358,.340,.316,.299,.276,.253
275 DATA .238,.216,.188,.168,.142,.110,.079,.003
```

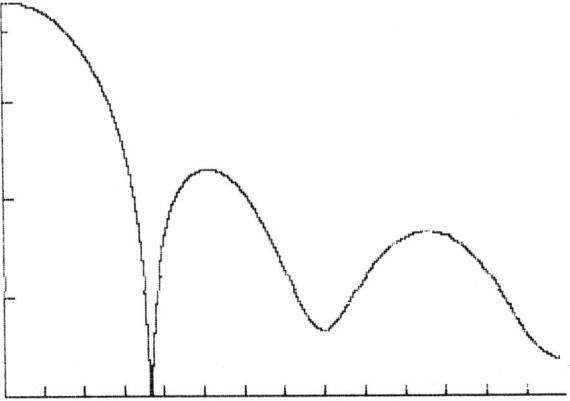

Fig. 6.11 Aperture blocking by a 1-λ strut and a 6.9-λ subreflector.

overall performance of the Cassagranian can therefore exceed the performance of the Newtonian in some applications even though its side lobes may be higher and its gain lower.

In the next chapter, we shall explore some other mechanisms beside aperture blocking that cause the departure of secondary patterns from ideal performance.

7
Off-Axis Feeds and Phase Errors

Newton demonstrated that a paraboloid is free from spherical aberration. When one departs from the geometric focus, however, the paraboloid displays other forms of aberration, and these must be treated in certain applications. In this chapter, we shall examine some of them.

We noted earlier, without proof, that the paraboloid fulfills the Angle-of-Incidence/Angle-of-Reflection Law. Let us begin by adducing the proof for this as a foundation for for some of the material to follow.

In Fig. 7.1 we see a segment of a parabola with some of the usual construction lines. Differentiating the equation for the parabola, $Y^2 = 4FX$, gives us its slope, or $dy/dx = 2F/Y$. The negative reciprocal of the slope, or $-Y/2F$, is the normal to the curve. The normal is marked QR in the figure. In order for the angle of incidence to be equal to the angle of reflection, angle PQR must be equal to angle FQR.

Now, by the Pythagorean Theorem, $FQ^2 = (FV - VS)^2 + SQ^2$, but since $FV = F$, $VS = X$, and $SQ = Y$,

$$FQ^2 = (F - X)^2 + Y^2$$
$$= F^2 - 2FX + X^2 + 4FX$$
$$= (F^2 + 2FX + X^2)$$

Therefore,

$$FQ = F + X \tag{7.1}$$

Working from the known slope of the normal, QR,

$$SR = Y/(-Y/2F)$$
$$SR = 2F \tag{7.2}$$

Since $FR = SR - (FV - VS)$,

$$FR = 2F - (F - X)$$
$$FR = F + X \tag{7.3}$$

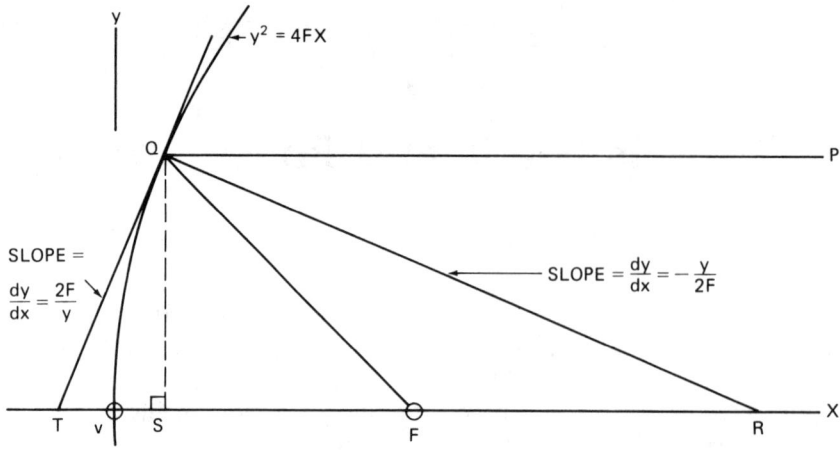

Fig. 7.1 Geometry for proof of Angle of Incidence/Angle of Reflection Law.

Therefore, since $FQ = F + X$ as well,

$$FR = FQ \tag{7.4}$$

This being the case, $\triangle FQR$ is an isoceles triangle; and therefore

$$\angle FQR = \angle FRQ \tag{7.5}$$

Since QP and VR are parallel lines,

$$\angle FRQ = \angle PQR \tag{7.6}$$

Therefore,

$$\angle PQR = \angle FQR \tag{7.7}$$

Thus, the angle of incidence is equal to the angle of reflection. Consequently, both the diffraction condition that the path lengths must be equal and the ray-optics incidence/reflection angle conditions are satisfied by the paraboloid (and parabola), at least from stand point of the focus.

MOVING OFF FOCUS

Figure 7.2 shows the geometry when one moves off-focus by a distance D perpendicular to the focal axis and a distance C parallel to the focal axis

OFF-AXIS FEEDS AND PHASE ERRORS

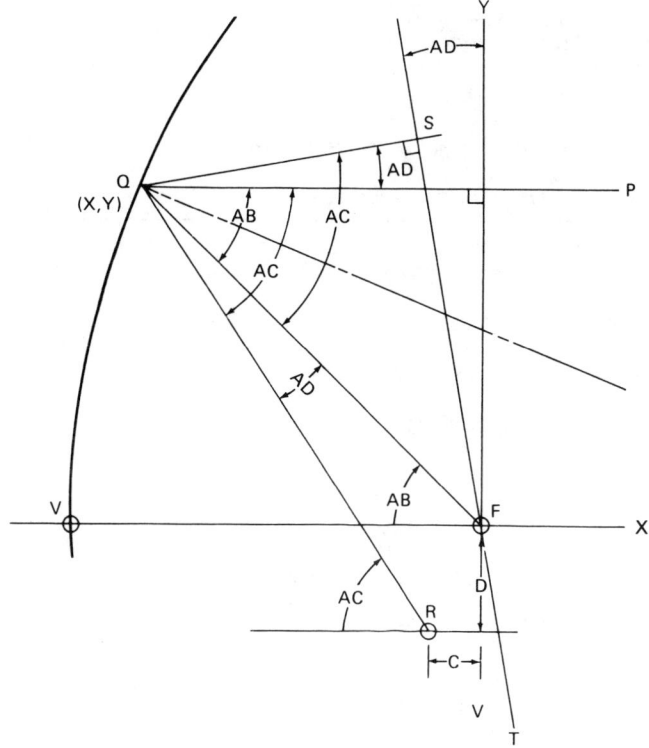

Fig. 7.2 Geometry when moving off focus.

to a new feedpoint R. Although the incidence/reflection relationship is retained in this situation, the "nominal phase front ST is tilted with respect to the "normal phase front" UV. The relevant equations are as follows:

$$\angle AB = ATN(Y/F - X)) = ATN(4FY/(4F^2 - Y^2)) \quad (7.8)$$
$$\angle AC = ATN(4F*(Y - D)/(4F^2 - Y^2 - C)) \quad (7.9)$$
$$\angle AD = \angle AC - \angle AB \quad (7.10)$$
$$QS = (FQ)*COS(AC) \quad (7.11)$$
$$SF = (FQ)*SIN(AC) \quad (7.12)$$
$$FQ = F + X \quad (7.13)$$
$$(RQ)^2 = (Y - D)^2 + (F - X - C)^2 \quad (7.14)$$
$$RQS = RQ + QS \quad (7.15)$$

The path RQS can be evaluated (Eq. 7.15) by using Eqs. 7.11 and 7.14. It is usual to evaluate path lengths normalized to distance $2*FV$ since

```
150 PI = 3.141592654
160  PRINT "INPUT YMAX/F"
170  INPUT YM
180  PRINT "INPUT SQUINT DISPLACEMENT"
190  INPUT D
200  PRINT "INPUT FOCAL DISPLACEMENT"
210  INPUT C
215  IF C = 0 THEN C = .0000001
245  REM  ------------------------------------------------
250  REM    CALCULATE THE ANGLE AC
255  REM  ================================================
260 Y = YM
270 X = (Y * Y) / 4
280 FQ = 1 + X
290 AC = ATN (4 * (Y - D) / (4 - (Y * Y) - C))
300 QS = FQ *  COS (AC)
310 SF = FQ *  SIN (AC)
320 RQ =  SQR ((Y - D) ^ 2 + (1 - X - C) ^ 2)
330 ER = RQ + QS - 2
400   PR# 1
405  IF Y <  > YM GOTO 430
410  PRINT "           FOR YMAX=",YM,"*F"
415  PRINT "AND SQUINT DISPLACEMENT=",D,"*F"
420  PRINT "AND FOCAL DISPLACEMENT=",C"*F"
425  PRINT "POSITION","ERROR"
430  PRINT SF,ER
500 Y = Y - 1 / 20
510  IF Y >  =  - YM THEN  GOTO 270
1000  END
```

Program 7.1

path *RQS* will typically be many, many wavelengths long, and the pattern effects are concerned only with the phase differences across the nominal phase front *ST*.

When one examines Fig. 7.2 closely, it seems likely that the phase front might be steered through some angle like *AD* by the feed movement. Certainly will this be true for the limit where *F/D* becomes very large and the mirror becomes nearly flat.

The use of an optical or radio system in an off-axis fashion is termed "squint," probably relating to the usage of "squint" to describe the condition commonly called "cross-eyed" or "wall-eyed." Any antenna not operating with the beam along the focal axis is referred to as "squinted."

OFF-AXIS FEEDS AND PHASE ERRORS 93

Program 7.1 is arranged to solve the distance (or delay) error. Focal length *FV* of Fig. 7.2 is normalized to unity, and all answers are calculated in terms of focal length. For purposes of illustration, the example shown is for a very "fast" *F* 0.25 reflector. A reflector this fast would seldom be used with an off-axis feed.

The table of Fig. 7.3 shows the result of this computation. It can be seen that the result will be an S-shaped error curve with the distance too short in the upper portion and too long in the low portion. In much of the central portion, the error is tolerably small. The fact that the distance is too short in the upper portion means that the signals (which travel at the speed of light) will arrive early and therefore that the phase will lead. This implies that a section of wavefront will be aimed back toward the axis of the system. A similar but converse effect is exhibited in the lower portion of the curve in which the distance is too long; here, the phase will lag, again creating a phase front aimed closer to the axis of the system. Since this error is governed by a quadratic equation, it is termed either a "quadratic error" or "coma," the latter for reasons we shall see shortly.

Now let us suppose that we are using the same paraboloid dimension that has appeared in some of the earlier computations, namely, an aperture width of 40-wavelengths. This would make *F* equal to 10-wavelengths. For a maximum phase error of a quarter wavelength on the

```
INPUT YMAX/F
?2
INPUT SQUINT DISPLACEMENT
?.1
INPUT FOCAL DISPLACEMENT
?0
               FOR YMAX=         2              *F
AND SQUINT DISPLACEMENT=        .1              *F
AND FOCAL DISPLACEMENT=         1E-07*F
POSITION          ERROR
-1.99999999      -.0999999726
 1.94993064      -.0972991898
 1.89971513      -.0944601046
 1.84934213      -.0914839385
 1.79880015      -.0883729723
 1.74807767      -.0851307293
 1.69716331      -.0817621243
 1.646046        -.0782735846
 1.59471518      -.0746731907
 1.54316108      -.0709707877
 1.49137498      -.0671780976
 1.43934949      -.0633087847
 1.38707896      -.0593785094
 1.33455979      -.0554049346
```

Fig. 7.3 Result of computation in Program 7.1.

94 EXPLORING ANTENNAS AND TRANSMISSION LINES

1.28179084	-.0514076823	-1.03278419	.0355236549
1.22877384	-.0474082273	-1.07934841	.0389965046
1.17551375	-.0434297412	-1.12609173	.0425271485
1.12201921	-.0394968363	-1.17302391	.0460974239
1.0683028	-.0356352515	-1.22015213	.0496898666
1.01438146	-.031871452	-1.26748123	.0532878758
.960276616	-.0282321279	-1.31501392	.0568758305
.906014378	-.0247436478	-1.36275095	.0604391834
.851625545	-.0214314186	-1.41069139	.0639645308
.797145495	-.0183192114	-1.4588328	.0674396418
.74261392	-.0154284481	-1.50717142	.0708534773
.688074391	-.0127774901	-1.5557024	.0741961934
.633573769	-.0103809568	-1.60441998	.0774591118
.579161442	-8.24909238E-03	-1.65331761	.0806346908
.524888419	-6.38725096E-03	-1.70238816	.0837164838
.470806294	-4.79548145E-03	-1.75162404	.0866990844
.416966116	-3.46828578E-03	-1.80101731	.089578066
.363417212	-2.39454489E-03	-1.85055983	.0923499204
.310206004	-1.55764353E-03	-1.90024331	.0950119859
.257374858	-9.3576964E-04	-1.95005946	.0975623923
.204961049	-5.02418261E-04	1.99999999	.100000011
.152995827	-2.27042008E-04		
.101503699	-7.58501701E-05		
.0505018831	-1.27092935E-05	FOR F = 10λ, ERROR ≤ λ/4	
3.07064219E-09	-1.00582838E-07		
-.0499999982	-9.96515155E-08		
-.0995037185	2.46530399E-05		
-.148523951	1.10057183E-04		
-.197080155	2.89733522E-04		
-.245197781	5.94327226E-04		
-.292907473	1.05093047E-03		
-.340244198	1.68266799E-03		
-.38724633	2.50840187E-03		
-.433954721	3.54259461E-03		
-.480411803	4.79528494E-03		
-.526660724	6.27219491E-03		
-.57274456	7.97492824E-03		
-.618705603	9.90124885E-03		
-.664584738	.0120454226		
-.710420923	.0143986139		
-.756250758	.0169492969		
-.802108161	.0196836833		
-.84802412	.0225861622		
-.894026543	.0256397119		
-.940140171	.0288263029		
-.986386573	.0321272705		

Fig. 7.3 (*continued*)

aperture edge, we see that the maximum aperture size would be approximately 2∗0.9∗*F*, calling for an *F* 0.56 dish. The "wings" of a faster dish would accummulate error very rapidly beyond this point.

Another point is worthy of note. The approximate angle of the feed displacement is 0.1-radian (0.1/1), or about three beam widths off-axis.

```
150 PI = 3.141592654
160  PRINT "INPUT YMAX"
165  PRINT "IN WAVELENGTHS"
170  INPUT YM
180  PRINT "INPUT FOCAL LENGTH"
185  PRINT "IN WAVELENGTHS"
190  INPUT F
200  PRINT "INPUT SQUINT DISPLACEMENT"
205  PRINT "IN WAVELENGTHS"
210  INPUT D
215  PRINT "INPUT FOCAL DISPLACEMENT"
220  PRINT "IN WAVELENGTHS"
230  INPUT C
240  IF C = 0 THEN C = .000001
250 Y = YM
260 AD =  ATN (D / F)
265  REM  NOMINAL SQUINT ANGLE
270  PRINT "INPUT STARTING ANGLE"
275  PRINT "IN DEGREES"
280  INPUT A0
285 A1 = A0 * PI / 180
288   GOTO 2020
290   REM  =====================================================
295   REM     CALC. SQUINT PHASE SHIFT
298   REM  =====================================================
300 N = 4 * YM
305  DIM Z(N,5)
310  FOR I = 0 TO N
320 Y = YM - I / 2
330 X = (Y * Y) / (4 * F)
340 FQ = (F + X)
350 AC =  ATN (4 * F * (Y - D) / (4 * (F * F) - (Y * Y) - C))
360 QS = FQ *  COS (AC)
370 SF = FQ *  SIN (AC):Z(I,0) = SF
380 RQ =  SQR ((Y - D) ^ 2 + (F - X - C) ^ 2)
390 PH = (RQ + QS - (2 * F)) * 2 * PI:Z(I,1) = PH
400 AA = (Y / YM) * PI / 2.05
410 Z(I,2) =  COS (AA)
420  NEXT I
430  REM   Z(I,2)IS THE AMPLITUDE
```

Program 7.2

```
450  REM  $$$$$$$$$$$$$$$$$$$$$$$$$$$$$$$$$$$$$$$$$$$$$$$$$$$
455  REM       DO PATTERN SUMMATION
460  REM  $$$$$$$$$$$$$$$$$$$$$$$$$$$$$$$$$$$$$$$$$$$$$$$$$$$
465  FOR J = 0 TO 130
470  AP = A1 + (PI / 1800) * J
472  FOR I = 0 TO N
480  PS = 2 * PI * (Z(I,0)) * SIN (AD + AP)
485  REM  PHASE SHIFT DUE TO SPACE ANGLE
490  ER = ER + Z(I,2) * COS (Z(I,1) + PS)
500  EJ = EJ + Z(I,2) * SIN (Z(I,1) + PS)
510  NEXT I
530  EP = SQR ((ER * ER) + (EJ * EJ))
535  CALL 64477
540  Q = 95 * ( LOG (E0 / EP)) / 2.30259
550  IF Q > 191 THEN Q = 191
555  IF Q < 0 THEN Q = 0
560  IF J = 0 THEN Q0 = Q:M0 = M: GOTO 580
570  HPLOT M0,Q0 TO M,Q
580  M0 = M:Q0 = Q
590  M = M + 2
600  ER = 0:EJ = 0
610  NEXT J
1000 STOP
2000 REM  ***************************************************
2005 REM       CALC. NORMALIZING VOLTAGE
2010 REM  ***************************************************
2020 AA = (Y / YM) * PI / 2.05
2030 A = COS (AA)
2040 E0 = E0 + 2 * A
2050 Y = Y - .5
2060 IF Y > = 0 GOTO 2020
2070 E0 = E0 - 1
2080 Y = YM
2090 HGR2
2100 HCOLOR= 3
2110 POKE - 12524,0
2120 POKE - 12525,64
2130 POKE - 12529,255
2135 HPLOT 0,191 TO 260,191
2140 HPLOT R,186 TO R,191
2150 R = R + 10
2160 IF R < 265 GOTO 2140
2170 GOTO 300
```

Program 7.2 (*continued*)

Since the wavefront is actually tilted with respect to the arbitrary front ST by about 0.02-radian (0.020/1), however, the beam is squinted about 0.08-radian, or only about 80-percent of the feed displacement. This is another form of distortion of the image.

Now let us examine the radiation pattern that results from such an "S-shaped" wavefront. Program 7.2 is arranged to compute the pattern for the squinted case. Provision has been made for the de-focusing error as well as the squint. It may be seen that the distribution has been simplified in lines 410 and 420. This program presents a fairly accurate simulation of the distribution obtained on a circular paraboloid illuminated with a $\cos(N*TH)$ function. Alternatively, a program such as Program 6.1 could be employed at some cost in time, or the table for "Z"(I,2) could be manually loaded. The latter is the alternative frequently used when it is desired to calculate the effects of blockage in the aperture.

The program has two main loops. One of these calculates the phase shift resulting from squint and the amplitude distribution along the tilted wavefront ST. This loop runs from line 290 through line 430. The second loop, which runs from line 465 through line 610, calculates the pattern summation for each azimuth angle over a range of 13-degrees, with the constants shown. The angular marks are at 0.5-degree intervals. The constant determining the step size in the angle appears in line 470, and the angular calibration is established by lines 590 and 2150.

With the program running a 40-wavelength aperture at half-wavelength steps, it will take about 10 seconds to cycle through the loop. Since 0.1-degree steps (from line 470) and a total of 131 steps (from line 465) are involved, it may be seen that the pattern will require about 22 minutes to run. Obviously, this is a "start it and go for coffee" program!

To illustrate the effects of squint, the program was run to obtain the patterns of Fig. 7.4 ($C = O$ for all curves). The feed displacement in Fig. 7.4(a) is 1.5-wavelengths for a paraboloid in which YMAX = 20 wavelengths and F = 20-wavelengths. (The broken line indicates the curve when $D = 0$.) Figures 7.4(b) and (c) show displacements of 1.0 and 0.5-wavelength, respectively.

The reflector was selected for a focal ratio of 0.5 since that would be one of the "fastest" that would be considered for squinted operation. With a squint offset of approximately 0.8-wavelength, the beam would be approximately in the position of crossing over the centered beam at the -3-dB level. This would permit the use of three feed horns to yield a three-beam ("stacked-beam") arrangement, There will be a significant deterioration of the side-lobe levels in the outer beams, however, as can be seen from the figure. For good stacked-beam performance, it would be necessary to use a longer focal length.

98 EXPLORING ANTENNAS AND TRANSMISSION LINES

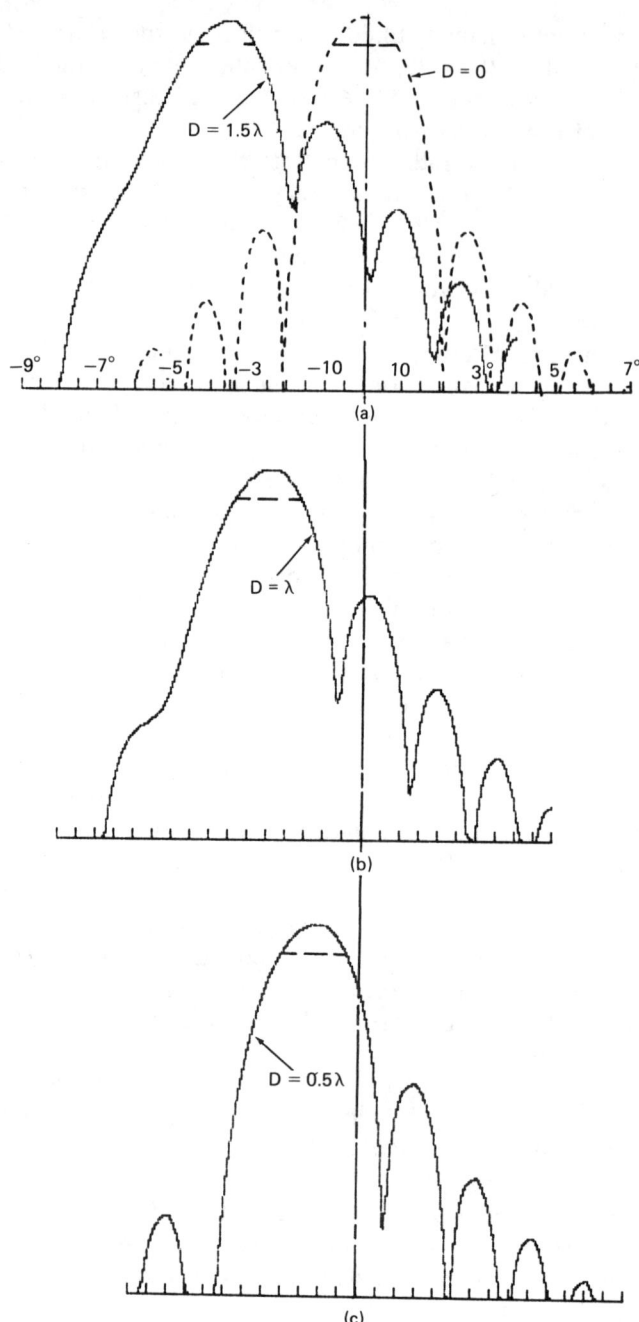

Fig. 7.4 Effects of squint.

Actually, the critical factor is not the number of wavelengths off-axis, but rather the number of beam widths off-axis. The pattern deterioration is proportional to the number of beam widths the pattern is squinted and inversely proportional to the "speed" or focal ratio of the reflector. A typical amateur telescope will have a reflector diameter of 6-inches and a focal length of 48-inches, for a focal ratio of 8. If we consider that the wavelength is $0.5E-6$-meters, or $1.97E-3$-inches, then the beam width will be on the order of

$$HPBW = 71*WAVELENGTH/DIAMETER$$

or $2.33E-4$-degrees. If the field of view of the instrument is taken as ±0.5-degrees, then the outside beam is squinted 1287 half-power beam widths off-axis. It may be seen that the deterioration of the field of view increases very rapidly as the paraboloid becomes "faster." As a matter of fact, the number of usable beam widths of squint approximately progresses as the cube of the F number. Thus an $F = 8$-paraboloid can support about 1296 beam widths of squint, whereas an $F = 0.5$-paraboloid begins to deteriorate at less than one.

Now we come to the term *coma*. The word is from the Latin word for "hair." For an on-focus image, a precise paraboloid aimed at a star will yield a bright central disc surrounded by the rings of one or more side lobes. The rings are often called "Airey Rings" after the British Astronomer Royal Sir George Airey, who first explained them. For an off-axis star, the side lobes away from the axis are suppressed while those nearer the axis are enhanced and blend together. As a result, star will appear to have "hair" streaming toward the axis, and thus the term *coma*.

Another type of distortion arises from the fact that the beams are only about 87 percent as squinted as the feed motion. Consequently, when a photograph is taken of a star field, the stars will be in wrong positions, namely, 1/0.87 times farther from the center than they actually are. This distortion diminishes as the focal ratio grows larger.

Thus, we see that although the Newtonian telescope eliminates spherical aberration, it suffers from a restricted usable field of view because of coma as well as an image distortion in which the outer elements are misplaced.

FOCUSING

Another point of interest is the question of focusing. What happens when the feed is moved toward or away from the prime focus? The curves in Fig. 7.5 show solutions for two cases; In Fig. 7.5(a), the feed has been

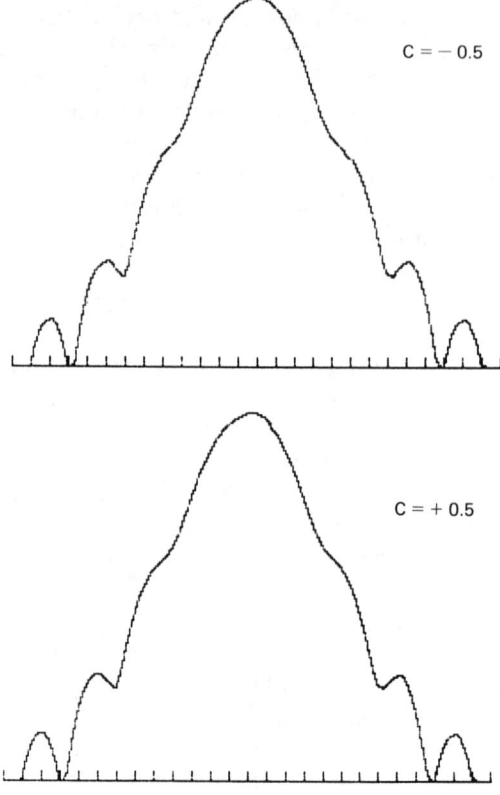

Fig. 7.5 Focusing errors.

moved −0.5-wavelengths away from the prime focus, and in Fig. 7.5(b), +0.5-wavelengths toward it. It may be seen that the patterns are nearly identical. In both, the side lobes have increased and so blended into the main lobe that the main lobe has become "shouldered," with increasingly indistinct side lobes. In the optical telescope, a star image progresses from the appearance of a large diffuse blob to that of a bright disc and ring arrangement and then back to a diffuse blob as it passes through the focussed condition.

In testing a parabolic or paraboloidal antenna, the combination of properties observed can be used for diagnostic purposes. If the side lobes are "shouldered in" and not distinctly separated, a focus error is indicated. If they are asymmetrical, then a squint error is present. If the reflector is "fast," a relatively small squint error can play havoc with the side lobes.

When making such a test, it is also worthwhile to test for inadvertent squint. If the antenna pattern mount (pedestal) is equipped for polar motion, one can do so very simply by taking a pattern cut, "plunging" (reversing the antenna on the polar axis), and then repeating the pattern cut. Superimposing the patterns with absolute angle scales aligned will show up any squint as a misalignment of the beams; these will be separated by twice the squint angle.

A circular paraboloid can be easily checked for mechanical alignment with the polar axis by rotating the dish on the polar axis and testing the rim for run-out. If the reflector outline is not circular, a circle can be drawn on the surface relative to the mechanical vertex and the run-out measured with respect to the circle. A cut paraboloid without such a vertex presents a more challenging mechanical problem. From an electrical viewpoint, the feed position that yields symmetrical, well-defined side lobes can be considered to represent a nonsquinted condition.

PHASE ERRORS

Up to this point, we have considered antennas in a mathematically precise way. The parabolas and paraboloids have been ideal, that is, mathematically perfect, entities. Of course, in the real world, antennas contain a number of flaws. The backing frame warps when it is removed from the welding fixture, the spun aluminum springs back somewhat, the molded fiberglass sags slightly, and so on.

Reflectors must be fabricated with precision, of course, but with how much? The closer the requirement is to perfection, the more expensive and difficult the fabrication becomes. Is there some way to specify antenna tolerances so that "enough" precision is provided without making the reflector too costly? Before answering, it would be worthwhile to take a look at the effects of small errors on an antenna.

In Program 7.3, we have adjusted the squinted pattern calculation to include a random phase error. Line 392 uses the RND function of BASIC to introduce the effect of an adjustable error of this kind. The terms 1.739*RND(4) will yield a string of random numbers between 0 and 1.739, with an RMS value very close to 1. The Apple will give a new random number string for any positive integer in the parentheses. The expression shown gives a string of phase shifts with an RMS value of one-eighth wavelength (PI-4 radians). The two patterns of Fig. 7.6 are nonsquinted patterns deformed by random phase errors. In Fig. 7.6(a), the nonsquinted base pattern with no phase errors (broken-line curve) has been included for comparison. Note that the difference between the curves of 7.6(a) and 7.6(b) arises solely from the fact that they were

```
150 PI = 3.141592654
160  PRINT "INPUT YMAX"
165  PRINT "IN WAVELENGTHS"
170  INPUT YM
180  PRINT "INPUT FOCAL LENGTH"
185  PRINT "IN WAVELENGTHS"
190  INPUT F
200  PRINT "INPUT SQUINT DISPLACEMENT"
205  PRINT "IN WAVELENGTHS"
210  INPUT D
215  PRINT "INPUT FOCAL DISPLACEMENT"
220  PRINT "IN WAVELENGTHS"
230  INPUT C
240  IF C = 0 THEN C = .000001
250 Y = YM
260 AD =  ATN (D / F)
265  REM  NOMINAL SQUINT ANGLE
270  PRINT "INPUT STARTING ANGLE"
275  PRINT "IN DEGREES"
280  INPUT A0
285 A1 = A0 * PI / 180
288  GOT  2020
290  REM  ====================================================
295  REM     CALC. SQUINT PHASE SHIFT
298  REM  ====================================================
300 N = 4 * YM
305  DIM Z(N,5)
310  FOR I = 0 TO N
320 Y = YM - I / 2
330 X = (Y * Y) / (4 * F)
340 FQ = (F + X)
350 AC =  ATN (4 * F * (Y - D) / (4 * (F * F) - (Y * Y) - C))
360 QS = FQ *  COS (AC)
370 SF = FQ *  SIN (AC):Z(I,0) = SF
380 RQ =  SQR ((Y - D) ^ 2 + (F - X - C) ^ 2)
390 PH = (RQ + QS - (2 * F)) * 2 * PI:Z(I,1) = PH
391 Z(I,1) = Z(I,1) + (PI * 1.739 *  RND (4) / 4)
394  REM  A QUARTER WAVE RMS ERROR IS ADDED
400 AA = (Y / YM) * PI / 2.05
410 Z(I,2) =  COS (AA)
420  NEXT I
430  REM  Z(I,2) IS THE AMPLITUDE
```

Program 7.3

OFF-AXIS FEEDS AND PHASE ERRORS 103

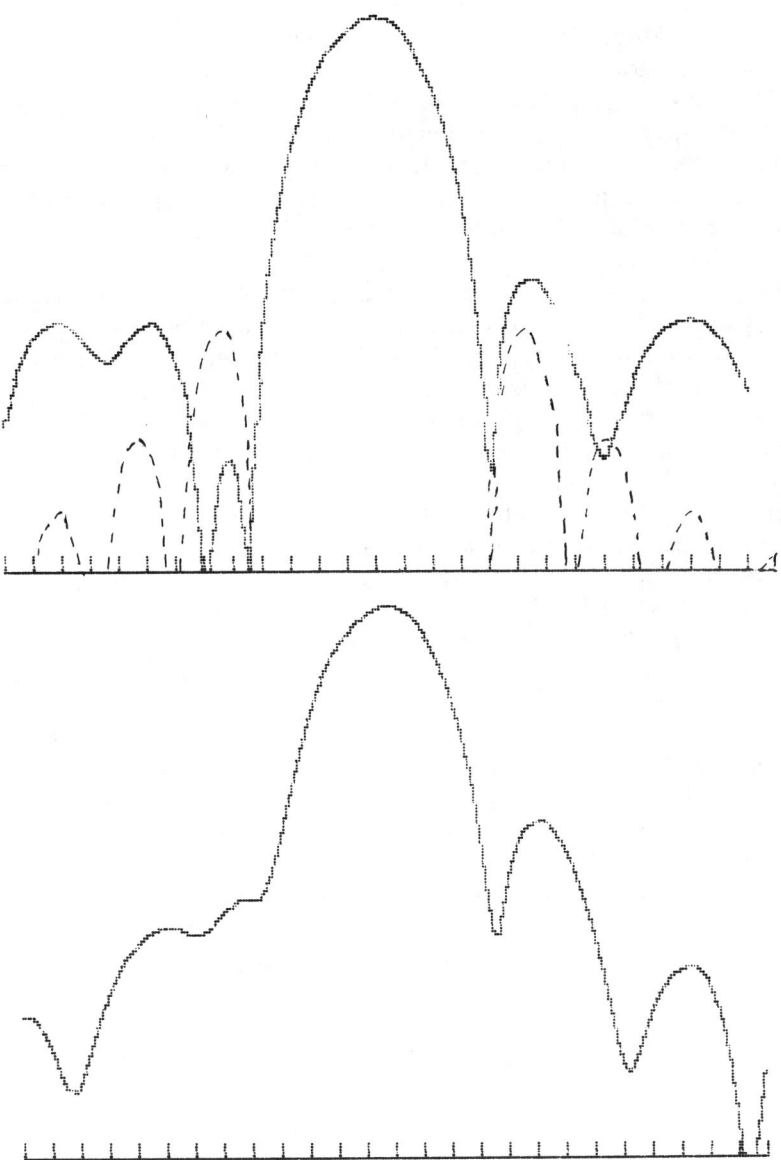

Fig. 7.6 Nonsquinted patterns deformed by random phase errors. (RMS phase error = $\lambda/8$ for both solid-line curves.)

affected by different random errors. They represent successive runs of the self-same program!

There are several other points worthy of note about these patterns. First of all, since there is no long-term order in the errors, there is no detectable steering of the beam. It would not be impossible for the errors to line up in a smooth ramp across the aperture, although the likelihood becomes vanishingly small. As a result, we see that the main beam is essentially unaltered.

On the other hand, we do see that the side-lobe structure has been sorely distorted. It is skewed and the nulls have been filled-in. Compared to the skewing of the side lobes caused by squint, the side-lobe structure is very disordered. This is more or less a dead giveaway that the patterns have been perturbed by phase errors. Whether a geometric collimator, a lens, or an array is involved, a chaotic side-lobe structure indicates that significant phase errors are present in the system.

In Fig. 7.7 we see the effects of reducing the phase error by a factor of 2. In Fig. 7.7(a), the reduction was accomplished by making the RMS phase error taper from one-eighth wavelength at the ends to zero in the center. In Fig. 7.7(b), the error has been reduced to one-sixteenth wavelength overall. There is not much to choose from between the patterns. Although a considerable improvement is evident in the latter, its random distortion is still significant.

Such manipulations apply equally well to an array or a geometric collimator. In a paraboloid, however, the mechanical tolerance will be roughly half as great as the phase error tolerance because of reflection. In other words, for a given amount of phase error in wavelengths, the reflector precision will have to be roughly half that value.

Now, to revert to our original question of the degree of reflector accuracy required, the answer is determined largely by the desired side-lobe level. If we assign a value to the peak voltage of a random side-lobe pattern, experience yields the result:

$$EE = 1.25*ZRMS$$

where

EE = the fractional side-lobe voltage
$Zrms$ = the RMS phase error, in wavelengths

For example, if $Zrms$ = 0.125 wavelength, then EE = 0.0156-V (compared to a normalized 1-V at the peak of the beam). Similarly, a $Zrms$ of 0.0625 wavelength would yield an EE of 0.0078-V. This value of

OFF-AXIS FEEDS AND PHASE ERRORS 105

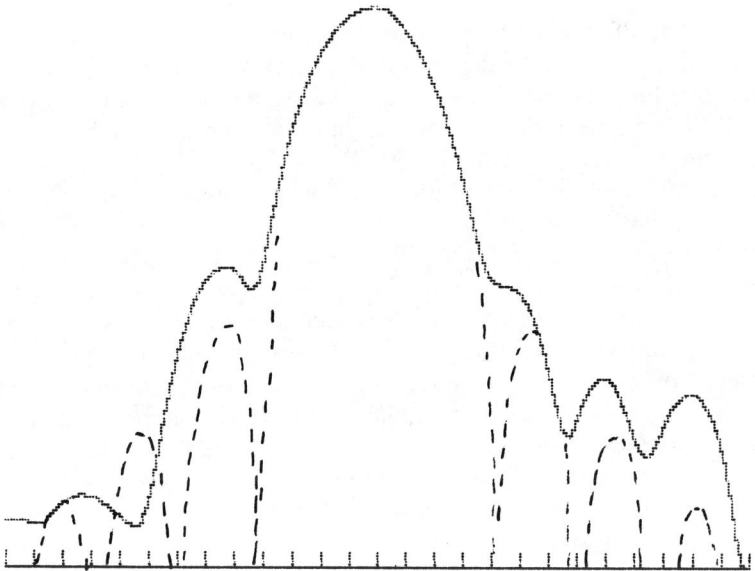

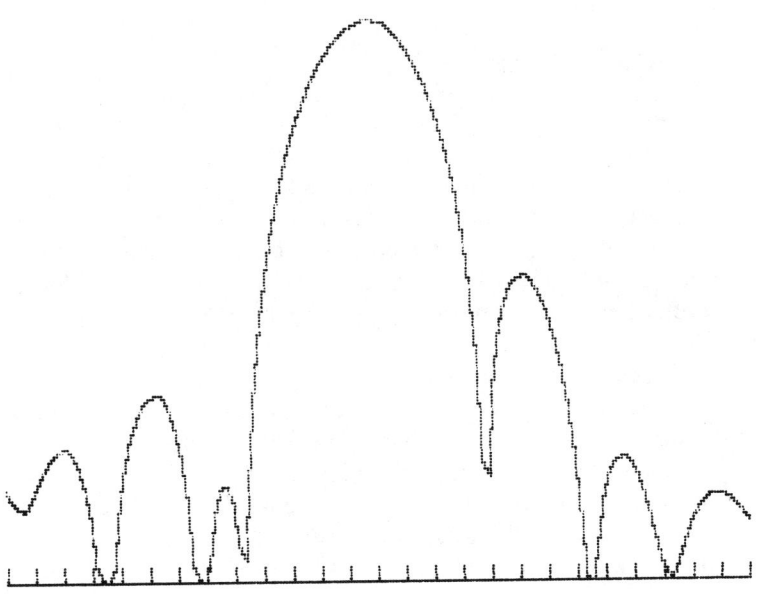

Fig. 7.7 Effects of reducing the phase error by a factor of 2.

EE is −42.15-dB below the peak of the beam. Therefore, a distribution designed to give −42-dB side lobes might deteriorate to give −36-dB side lobes. For the one-eighth-wavelength case, the error side lobes are −36-dB. Therefore, a distribution designed for −36-dB side lobes would give something on the order of −30-dB side lobes.

In another example, a distribution designed to yield −27-dB side lobes would have design side lobes of 0.04467-V, to which we add 0.0156-V (worst case) for a total of 0.06027-V, or −24.4-dB for a one-eighth wavelength RMS error.

As a rule of thumb, this writer has tended to design the distribution for side lobes about 5-dB lower than the specification and then provide tolerance for a 3-dB degradation. Such antennas will usually make it through acceptance testing with minimum effort. This approach optimizes test costs at some slight expense in reflector size and material. If size is critical, as in an airborne installation, the tolerances can be tightened and the distribution designed closer to the side-lobe limit.

In acceptance testing, it is usually possible to find a "best" focus that will minimize the effects of reflector error. This is a time-consuming process, however, and tends to be expensive as well as slow. Two weeks spent during a northern winter laboriously searching for the best focus so that a side lobe might be reduced by 1-dB should suffice to convince anyone that an investment in tighter tolerances is worthwhile!

STACKED-BEAM ANTENNAS

Certain types of systems, particularly radar and surveillance systems, are required to provide antenna coverage of a substantial span of angles and at the same time provide the ability to resolve target position accurately. One solution to this problem is a stacked-beam antenna.

Figure 7.8 shows an ensemble of stacked-beam antenna patterns. A model of a paraboloid with multiple feeds is shown in Fig. 7.9. Because of its numerous feeds, this reflector is exceptionally slow ($F = 1$ to avoid coma distortion).

Two significant points have been flagged in Fig. 7.8: the first and second crossover. The first is usually designed to be as high as possible, since it marks the lowest sensitivity point of the pattern. The second should be somewhat higher than the highest side lobe.

Suppose that the antenna is receiving a signal that is strongest in beam C and equal in beams B and D. This would tell us that the target was nearly aligned with the axis of C. If the signals were equal in C and D, it would tell us that the target was aligned with the crossover. By careful interpolation, it is possible to determine the position to about a tenth of a beam width.

OFF-AXIS FEEDS AND PHASE ERRORS 107

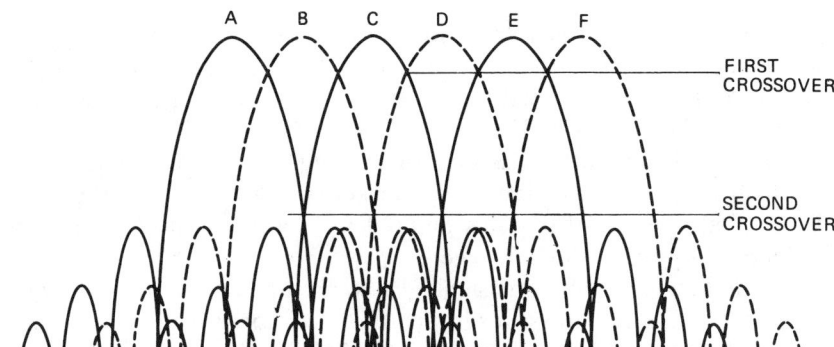

Fig. 7.8 Stacked-beam antenna patterns.

Fig. 7.9 Paraboloid with multiple feeds. Note long focal length and flatness of dish.

In passive countermeasures operations, the stacked-beam antenna permits the use of multiple receivers. In a radar system, it is possible to use a separate radar for each beam. In both cases, the "power" or effectiveness of the system is thereby increased.

A stacked-beam antenna is sometimes used on a search radar as well, in which case the beam is usually stacked in elevation and the power distribution between the beams tailored to "cosecant square" the distribution and optimize the coverage.

The biggest problem in designing a stacked-beam antenna is that of attaining the desired crossover levels. It is desirable for the side lobes to be well below the second crossover level, a condition that implies the attainment of a low level of edge illumination. Furthermore, it is desirable for the dish to have a long focal length and be relatively slow. These factors combine to require a large and relatively directive feed. Conversely, the attainment of high first and second crossovers implies a fairly close spacing of the feeds.

The attainable performance of a stacked-beam antenna is usually a compromise between the attainable side-lobe levels and crossover levels. In the next chapter we shall see how these factors affect monopulse operation as well.

8
Tracking Antennas

The requirement that an antenna be capable of tracking or following a target is a common one in a number of applications. The very first radars were intended to detect and locate enemy aircraft and determine their altitude. The next wave of radars was intended to track aircraft targets precisely. As an interim measure, the sets were arranged to point searchlights so that antiaircraft guns could be fired at night using optical aiming. In the wave that followed (for land-based equipment), the SCR-584 led the way, providing automatic AA pointing and range and "lead" calculations. These sets provided the breakthrough that brought about the defeat of the V 1 (buzzbomb) missile attack on Britain. Their pinpoint accuracy on such unsophisticated targets made the latter ineffective as a weapon.

THE CONICAL SCAN

Nearly all of the early automatic tracking schemes relied upon some form of lobe-switching, of which the conical scan is a subset. The basic form of a conical-scan pattern, defined by Program 8.1, is shown in Fig. 8.1.

Suppose that a paraboloid is so arranged that the feed is offset slightly and can be rotated about its axis. The squint angle will then swept through a cone with each rotation of the feed. A target slightly to the right of the centerline will give a stronger response whenever the feed is left of the latter (when the beam is squinting right). This condition will obtain for all angles up to spurious crossover A, at which point the sense will reverse until spurious crossover B is reached. If the equipment is able to segregate returns into left/right and up/down, it may be seen that the information for an effective servo-tracking system is present, at least for a limited span of angles. When the right echoes are stronger than the left, the antenna is driven to the right, etc. Although the term *echoes* is used in this instance, it can be seen that this tracking scheme would work equally well in a passive system such as a satellite terminal.

Figure 8.2(a) show a model of the feed used on the TPL radar and the SCR-584. Both of these sets used 7/8-in. air-insulated coax line. The dipole was mounted in front of a reflector plate, the latter serving to concentrate the energy from the primary feed onto the dish. Because the

EXPLORING ANTENNAS AND TRANSMISSION LINES

```
150 PI = 3.141592654
200   HGR2
210   HCOLOR= 3
220   POKE - 12524,0
230   POKE - 12525,64
240   POKE - 12529,255
250 U = - 3 * PI
280   REM ******************************************
285   REM    THE PATTERN CALCULATION
290   REM ******************************************
300 U = U + (PI / 46)
310 U = U - 1:W = U + 1
312   IF U = 0 THEN U = .0001
314   IF W = 0 THEN W = .0001
320 Y1 = ((SIN (U)) / U) ^ 2
330 Y3 = 95 * (LOG (Y1)) / 2.30259
350 Y5 =   ABS (Y3)
360 Y2 = ((SIN (W)) / W) ^ 2
370 Y4 = 95 * (LOG (Y2)) / 2.30259
380 Y6 =   ABS (Y4)
390   IF Y5 < 0 THEN Y5 = 0
400   IF Y6 < 0 THEN Y6 = 0
410   IF Y5 > 190 THEN Y5 = 190
420   IF Y6 > 190 THEN Y6 = 190
430   IF X = 0 THEN X0 = X:Y0 = Y5: GOTO 450
440   HPLOT X0,Y0 TO X,Y5
450   HPLOT X,Y6
460 X0 = X:Y0 = Y5
470 X = X + 1
480   IF X < 278 GOTO 300
2000  STOP
```

Program 8.1

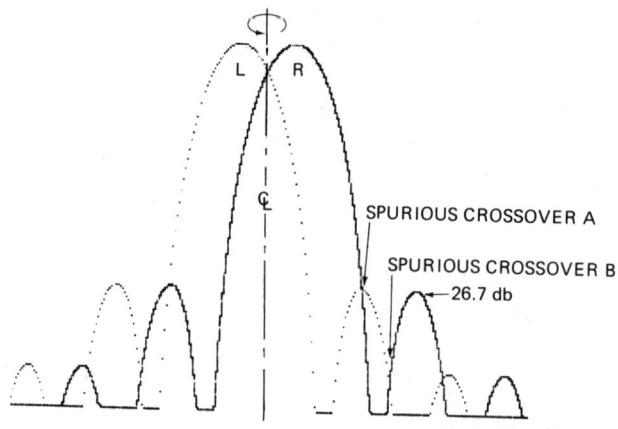

Fig. 8.1 Basic form of conical scan pattern.

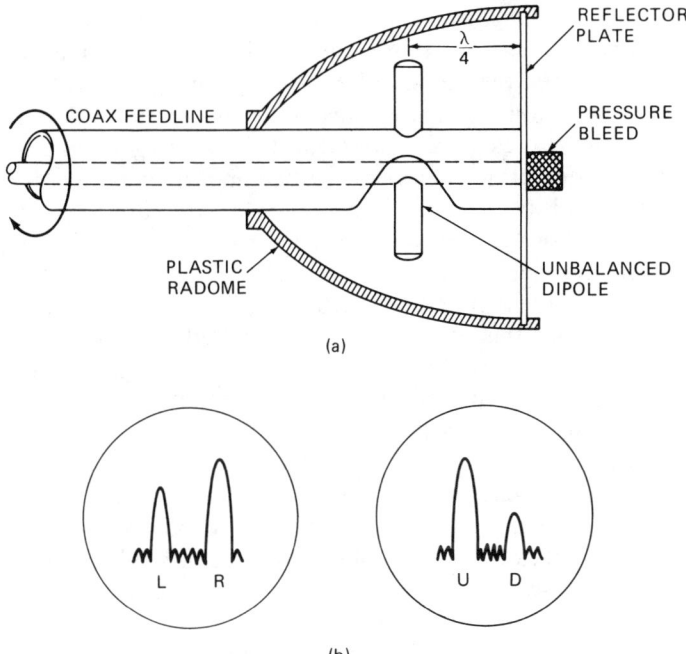

Fig. 8.2 (a) The SCR-584 feed, and (b) "M" scope detail.

dipole was unbalanced (a concept we have not yet treated), the phase center (or effective center) was not on the axis of the pipe; as a result, the antenna squinted in a conical fashion when the feed pipe was rotated. A plastic radome held the air pressure and kept the weather off of the dipole. A pressure bleed was supplied to bleed the pressurized dry air used to keep the transmission line dry.

An "M" scope detail is presented in Fig. 8.2(b). The TPL was an early, small S-Band (approximately 3-GHz) tracker used to point searchlights for night firing. It was not an automatic tracking unit since it was designed as a "quick response" weapon for the air war over Britain. Instead of automatic tracking, a "point" operator and a "train" operator turned handwheels to position the 6-ft dish, taking their information from the "M" scopes.

The M scopes split the target returns into left/right and up/down presentations. As a response to the data in Fig. 8.2(b), the point operator should run the antenna to the *right*, and the train operator should run it *up* in order to equalize the traces.

Yet a third operator manually controlled a short-time slot centered on the target called a *range gate*. The range gate served to pass a narrow bundle of data to the M scopes for only a brief period of time—a precise interval that followed the initial pulse (or "main bang") corresponding to the target range. The range gating was especially necessary for low-flying targets in order to present a clear picture of the target position to the point and train operators. If ground clutter and other targets had come through, it would have been impossible for the operators to track intelligently because the returns from all these targets would have piled on top of one another.

This conical scan was simple and effective, but it had its faults, one of them being the fact that the polarization rotated and the second, that there was a time lapse between the left and right readings. At microwave frequencies, the return pattern from most airplanes looks like a cockleburr. Furthermore, a propeller-driven airplane always shows some propeller modulation. The result is that the left and right returns were usually equal only when the axis pointed toward the plane. Over a given scan, they might be quite different. This disparity is symbolized by the noise of the two lobes in the M scope shown. As a result of this noise, a conical-scan radar would typically "jitter" about the target, and the antenna would tend to shake a bit while tracking.

Referring back to Fig. 8.1, it is not difficult to see that the tracking sensitivity of the system will be approximately maximum when the cone makes the patterns cross at about the -3-dB point. A wider cone would lose sensitivity on-axis, and a narrower cone would produce less distinction between the left and right returns.

The conical scan fell into disuse for some time only to reappear in a new guise. Many of the modern *"adaptive" radars* perform simultaneous track-and-scan operations by means of an electronically scanned array. In these sets, the control pauses briefly during each azimuth sweep to do one conical scan about each known target and obtain a precise position update. Performed electronically, the scan is so fast that most of the jitter of the earlier mechanical scans has dissappeared.

MONOPULSE

In an attempt to get rid of the jitter of con-scan tracking, it was reasoned that a system that derived its azimuth and elevation error signals independently for each pulse would be as jitterfree as possible. In the early 1950s, programs were initiated to develop *monopulse systems* that would do just that. Several different types evolved. We shall treat the simplest and most straightforward first.

AMPLITUDE MONOPULSE

Suppose that we were to equip a paraboloid with a pair of feeds in such a way that one would produce the broken-line and the other the solid-line pattern of Fig. 8.1. Now, if we were to obtain the *coherent* sum and difference of the two feeds, as in Program 8.2, we would obtain the curves shown in Fig. 8.3. This pattern is shown as a voltage rather than a log or decibel pattern in order to illustrate the reversal of sign (algebraic sign, that is) on opposite sides of the centerline—an important concept to get across.

The fact that there is an algebraic sign change or phase reversal when one crosses the centerline makes it possible to create a simple servo tracker. From the construction lines we see that, if we assign an arbitrary value of 1-V to the peak of the sigma beam, the slope of the delta beam at the zero crossing would be about 2-volts/beamwidth. (The designation of the sum pattern as sigma and of the difference pattern as delta is common.) This is larger than that typical of amplitude monopulse patterns. The usefulness of this criterion for comparing monopulse systems will be treated later.

```
150 PI = 3.141592654
200  HGR2
210  HCOLOR= 3
220  POKE  - 12524,0
230  POKE  - 12525,64
240  POKE  - 12529,255
250 U =  - 3 * PI
280  REM  *******************************************
285  REM     THE PATTERN CALCULATION
290  REM  *******************************************
300 U = U + (PI / 46)
310 V = U - 1:W = U + 1
312  IF V = 0 THEN V = .0001
314  IF W = 0 THEN W = .0001
320 Y1 = (( SIN (V)) / V) ^ 2
330 Y2 = (( SIN (W)) / W) ^ 2
340 Y3 = 95 - (67 * (Y1 + Y2))
350 Y4 = 95 - (67 * (Y1 - Y2))
360  IF X = 0 GOTO 380
370  HPLOT X0,Y0 TO X,Y4
375  IF N = 1 THEN N = 0: GOTO 390
380  HPLOT X,Y3
385 N = 1
390 X0 = X:Y0 = Y4
470 X = X + 1
480  IF X < 278 GOTO 300
2000   STOP
```
                                                Program 8.2

114 EXPLORING ANTENNAS AND TRANSMISSION LINES

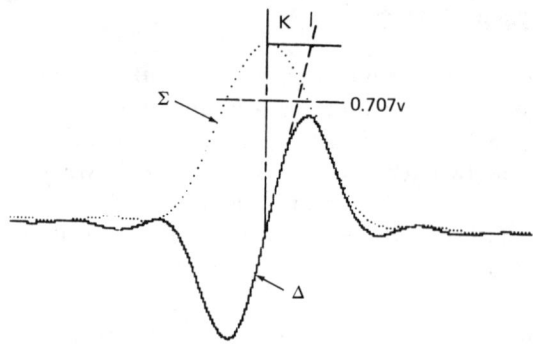

Fig. 8.3 Amplitude monopulse pattern.

Since it is difficult to see side lobes in voltage patterns and since antenna patterns are typically measured with a bolometer or other power sensor at microwave frequencies, the more common presentation of such patterns is that shown in Fig. 8.4, the result of Program 8.3.

In this log presentation, the sign change at the axis crossing has been supressed by the manipulation. We see that the side lobes have actually improved slightly over those in the original pattern and that the steepness of the crossover slope has been exaggerated. Useful tracking information is presented inside the first side-lobe null of the delta pattern. It is not simply by chance that the first side lobe is not the highest. This result was brought about by the nulls of the original single patterns.

The basic pattern first appears in lines 320 through 360 of Program 8.1, these being employed a time-saving measure. The pattern is representative of the relatively low side-lobe distributions to be used in applications such as this one. It approximates the results of the "roof" distribution and has the advantage of being relatively fast running for manipulations in which a number of configurations must be investigated.

FEED CROWDING

Unfortunately, when one proceeds to design an actual amplitude monopulse reflector or stacked-beam reflector, compromises become necessary. Since it can be shown that feed crowding is a little less severe in slower dishes, let us choose a reflector with a focal ratio of 0.5 for our example.

For a dish diameter is 40-wavelengths, the focal length would be 20-wavelengths. For the time being, let us assume that the secondary half-power beam width is 0.025-radian (1/40), or 1.43-degrees. Since each feed

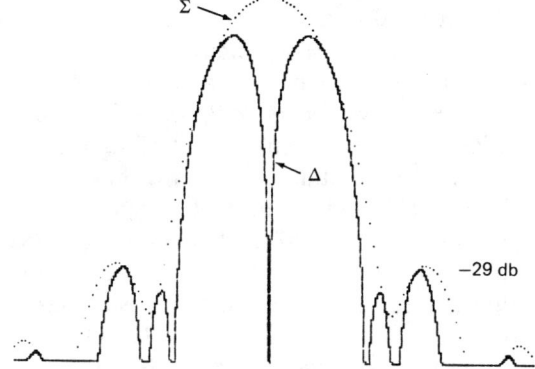

Fig. 8.4 Pattern resulting from Program 8.3.

```
150  PI = 3.141592654
200  HGR2
210  HCOLOR= 3
220  POKE  - 12524,0
230  POKE  - 12525,64
240  POKE  - 12529,255
250  U =  - 3 * PI
280  REM  ******************************************
285  REM      THE PATTERN CALCULATION
290  REM  ******************************************
300  U = U + (PI / 46)
310  V = U - 1:W = U + 1
312   IF V = 0 THEN V = .0001
314   IF W = 0 THEN W = .0001
320  Y1 = (( SIN (V)) / V) ^ 2
330  Y2 = (( SIN (W)) / W) ^ 2
340  Y3 =  ABS (Y1 + Y2)
342  Y4 =  ABS (Y1 - Y2)
344  Y3 =  - 95 * ( LOG (Y3)) / 2.30259
346  Y4 =  - 95 * ( LOG (Y4)) / 2.30259
347  Y3 = Y3 + 15:Y4 = Y4 + 15
348   IF Y4 > 190 THEN Y4 = 190
349   IF Y3 > 190 THEN Y3 = 190
360   IF X = 0 GOTO 375
370  HPLOT X0,Y0 TO X,Y4
375   IF N = 1 THEN N = 0: GOTO 390
378   IF Y3 < YM THEN YM = Y3
379   IF Y3 < 0 THEN Y3 = 0
380  HPLOT X,Y3
385  N = 1
390  X0 = X:Y0 = Y4
470  X = X + 1
480   IF X < 278 GOTO 300
2000  STOP
```

Program 8.3

116 EXPLORING ANTENNAS AND TRANSMISSION LINES

mould have to be squinted by about a half-power beam width, the feed width would be 2*20*0.025, or 1-wavelength.

Now, typically, one can obtain from a waveguide feed horn of one-wavelength width a half-power beam width of 0.88-radians in the E-plane (where the distribution is uniform) and 1.15-radians in the H-plane (where a cosine distribution obtains). A dish with a focal ratio of 0.5 will subtend an angle of 0.9273-radian, or 53.13-degrees.

We can roughly estimate the edge illumination by assuming that the feedhorn patterns are similar in form to the cosine function but different in width. Therefore, since the cosine function passes through the 0.707-V point (half-power point) at 45-degrees, or 0.785-radian, a pattern with a 0.88-radian half-power beam width will be 0.88/0.785, or 1.12 times, wider, and a pattern with a half-power beam width of 1.15-radians will be 1.465 times wider. At an angle of 53.13-degrees, these patterns will have an edge strength of 0.676-V and 0.806-V, respectively.

At this point, an iterative design process is called for. An illumination with an edge strength of 0.676-V will yield a secondary half-power beam width on the order of 0.92-radian/A (A = aperture width, in wavelengths) and side lobes of -17-dB. An edge strength of 0.806-volts will yield about 0.92-radian/A and -16-dB side lobes. This will require a recalculation of the feed width, which turns out to be 0.97-wavelength in the E-plane and 0.92-wavelength in the H-plane. Recalculation also yields 0.695-V and 0.835-V, respectively, for the edge strength.

The two secondary patterns can be simulated by a $(SIN(U))/U$ pattern raised to some other power than 2. For the flatter illumination in the H-plane, an exponent of 1.212 will yield -16-dB side lobes in the individual horn patterns, and for the steeper illumination in the E-plane, an exponent of 1.296 will yield -17.1-dB side lobes.

Figure 8.5 shows the resulting antenna patterns. These are fairly representative of amplitude monopulse patterns for a circular reflector. It may be seen that the side lobes are slightly improved over those for the single horn. Another characteristic quite common in practical monopulse patterns is that their delta side lobes are higher than their sigma side lobes and their delta lobes are wider than their sigma main lobe. The peaks of the delta lobes are also considerably more than 3-dB down. Measured at the -14-dB point, the error slope of the delta pattern measures approximately 1.32-V/beam width.

Our use of E-plane and H-plane terminology has advanced a bit beyond our present discussion. These terms will be treated in detail in later sections on the individual radiator and the waveguide horn. For now, the E-plane can be interpreted to mean parallel with the E vector, and the H-plane to mean perpendicular to the E vector.

```
320 YA = ( SIN (U)) / U
321  IF YA < 0 THEN S =  - 1: GOTO 323
322 S = 1
323 Y1 = S * ABS (YA) ^ 1.296
330 YB = ( SIN (W)) / W
331  IF YB < 0 THEN S =  - 1: GOTO 333
332 S = 1
333 Y2 = S * ABS (YB) ^ 1.296
340 Y3 =  ABS (Y1 + Y2)
342 Y4 =  ABS (Y1 - Y2)
```

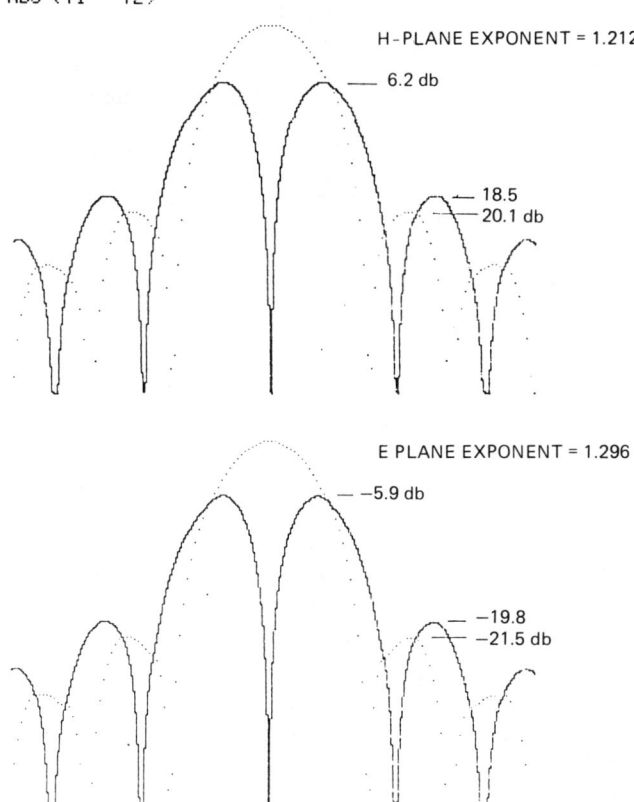

Fig. 8.5 Amplitude monopulse patterns for a circular reflector.

NOISE AND TRACKING

The accuracy of almost any type of tracking is a function of the noise in the system. In a passive tracker, the signal coming from the target to the antenna is subject to Inverse-Square Law spreading, that is, the received power is proportional to the reciprocal of the range squared. Doubling the range would therefore decrease the signal power by a factor of 4, or

118 EXPLORING ANTENNAS AND TRANSMISSION LINES

6-dB. For either of the patterns in Fig. 8.5, the target would be detectable in the sigma beam at twice the range at which it could be tracked.

In a radar situation, the Inverse-Square Law spreading takes place twice, once on the path to the target and again on the return path of the echo. The power is therefore a function of the reciprocal of the range to the fourth power. Doubling the range will thus decrease the signal by 12-dB.

Now let us suppose that we have a target with a signal-to-noise ratio of 20-dB. Assigning an arbitrary 1-V value to the peak of the beam, the noise voltage would be 0.1-V. The tracking error caused by noise would be $\frac{0.1V}{1.32V/\text{beam width}}$, or 0.076-beam width. A 40-dB signal-to-noise ratio would provide tracking with half of the error.

Significant tracking accuracy in most monopulse systems depends upon a substantial signal-to-noise ratio.

FEED ARRANGEMENTS

A variety of arrangements are employed to feed amplitude monopulse systems. Figure 8.6 shows two of the principal ones. The upper arrangement, Fig. 8.6(a), is usually termed a "square four-horn" monopulse feed. The lower, Fig. 8.6(b), is usually termed a "diamond" monopulse array. In the square arrangement, the horns are usually polarized parallel to one of the sides of the square. In the diamond array, the polarization is usually along one of the diagonals of the horns. A diagonally polarized horn generates patterns that are nearly identical in the E- and H-planes, at least through the section used to illuminate the dish.

The diamond arrangement usually yields delta patterns of lesser quality, but it requires less plumbing in the passive antenna components.

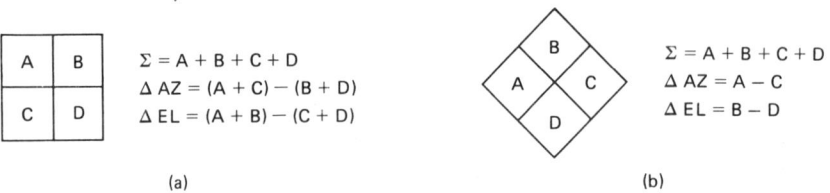

Fig. 8.6 Four-horn amplitude monopulse: (a) Square, and (b) diamond.

PHASE MONOPULSE

Whereas the amplitude monopulse employs two squinted beams from the same aperture (for two-dimensional tracking), the phase monopulse employs two separate apertures that may or may not be squinted. In general, the beams from the individual apertures in a phase monopulse are completely congruent. Young's interferometer (see Fig. 4.4) is, in essence, a kind of phase monopulse.

Figure 8.7 shows a phase monopulse. A paraboloidal reflector has been

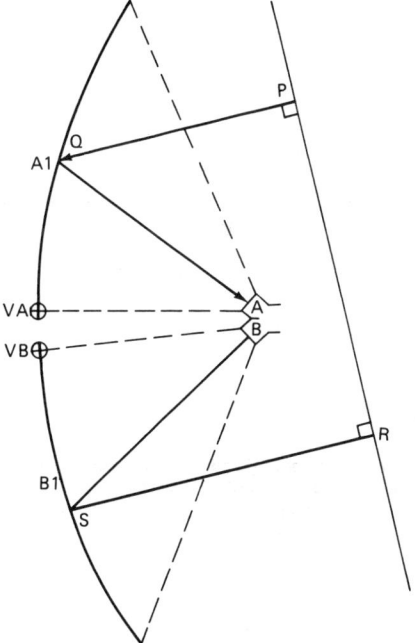

Fig. 8.7 Phase monopulse.

cut along a diameter through the vertex and the two halves slightly separated so that a feed horn can be placed at each of the two separated foci to produce two antennas with nonsquinted beams. A tilted wavefront comes in from a remote point as shown. It can be seen that *PQA* is smaller than *RSB*. If the patterns from *A* and *B* are identical and overlapping, the only difference between the signals will be caused by the time-of-flight difference; this can be translated into phase difference at any given frequency.

```
200 N = 64: REM   NUMBER OF ARRAY ELEMENTS
210 PI = 3.14159
220 DL = .313: REM  ELEMENT SPACING IN WAVELENGTHS
230  DIM X(N,1)
235 AH = PI / (2 * N)
240  FOR I = 1 TO N
250 X(I,0) =  COS (AH * I) *  SQR ( SIN (AH * I))
260  NEXT I
294 ED = 0
295 E0 = 0
296  REM )))))))))))))))))))))))))))))))))))))))
297  REM     THIS COMPLETES APERTURE DISTRIBUTION CALCULATION
298  REM &&&&&&&&&&&&&&&&&&&&&&&&&&&&&&&&&&&&&&&&
300  FOR J = 0 TO 138
310 A0 = (PI / 3600) * J
320 AI = 2 * PI * DL *  SIN (A0)
330  FOR I = 1 TO N
335 AJ = AI * I
340 E0 = E0 + X(I,0) * 2 *  COS (AJ)
342 ED = ED + X(I,0) * 2 *  SIN (AJ)
350  NEXT I
352  REM ##########################################
354  REM     THIS COMPLETES THE SUMMATION
356  REM     FOR ONE ANGLE
358  REM ##########################################
360  IF J = 0 THEN EN = E0: GOTO 380
370  GOTO 430
380  HGR2
390  HCOLOR= 3
400  POKE  - 12524,0
410  POKE  - 12525,64
420  POKE  - 12529,255
422  REM ##########################################
424  REM     THE GRAPHICS AND PRINTER SETUP
426  REM %%%%%%%%%%%%%%%%%%%%%%%%%%%%%%%%%%%%%%%%
428  IF E0 = 0 THEN E0 = .0001
429  IF ED = 0 THEN ED = .0001
430 YA =  ABS (EN / E0)
431 YC =  ABS (EN / ED)
432 YB = 20 * ( LOG (YA)) / 2.30259
433 YD = 20 * ( LOG (YC)) / 2.30259
434 Y = 4.75 * YB
435 YE = 4.75 * YD
436  IF Y > 191 THEN Y = 191
437  IF YE > 191 THEN YE = 191
450  IF J = 0 THEN X0 = X:Y0 = Y
460  HPLOT X0,Y0 TO X,Y
461  HPLOT X,YE
470 X0 = X:Y0 = Y
475 X = X + 2
480 E0 = 0
489 ED = 0
490  NEXT J
```

Program 8.4

```
200 N = 64: REM   NUMBER OF ARRAY ELEMENTS
210 PI = 3.14159
220 DL = .313: REM   ELEMENT SPACING IN WAVELENGTHS
230  DIM X(N,1)
235 AH = PI / (2 * N)
240  FOR I = 1 TO N
250 X(I,0) =  COS (AH * I) *  SQR ( SIN (AH * I))
260  NEXT I
266  GOTO 380
295 E0 = 0
296  REM )))))))))))))))))))))))))))))))))))))
297  REM      THIS COMPLETES APERTURE DISTRIBUTION CALCULATION
298  REM &&&&&&&&&&&&&&&&&&&&&&&&&&&&&&&&&&&&&&&
```

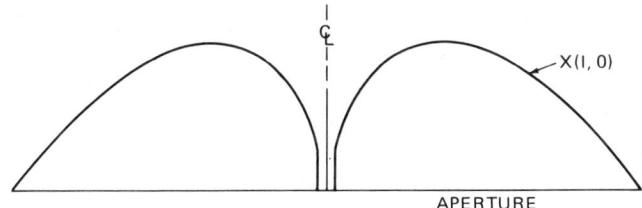

Fig. 8.8 Distribution for line 250 of Program 8.4.

Program 8.4 is very similar to Program 5.2 except for the fact that the aperture distribution is somewhat different and the computation is set up to account for the out-of-phase (SIN) terms in the vector pairs from symmetrical locations (see line 342) as well as the in-phase (COS) terms. Lines 492 through 610 are identical to those in Program 5.2 and have not been reproduced here. The computation is performed at the phase front and a squint allowance is not made.

Obviously, the amplitude distribution on the separate reflectors must be somewhat different from that which would be obtained if a single horn were illuminating the paraboloid. The distribution is determined by line 250 of the program. Figure 8.8 shows the distribution for the stated line. It may be seen that the distribution has a hole punched in the center; this corresponds to the missing strip between the reflectors. Also, the descent is steeper adjacent to the centerline than it is at the dish edge. It is quite common to trim the periphery of the two dish halves into a "jellybean" contour as viewed from the focal axis. When this trimming is integrated into the distribution, it tends to round it.

The resulting antenna pattern is shown in Fig. 8.9. Compared to the amplitude monopulse patterns of Fig. 8.5, the delta pattern lobes are much higher, and thus the phase monopulse could track with about 3-dB

122 EXPLORING ANTENNAS AND TRANSMISSION LINES

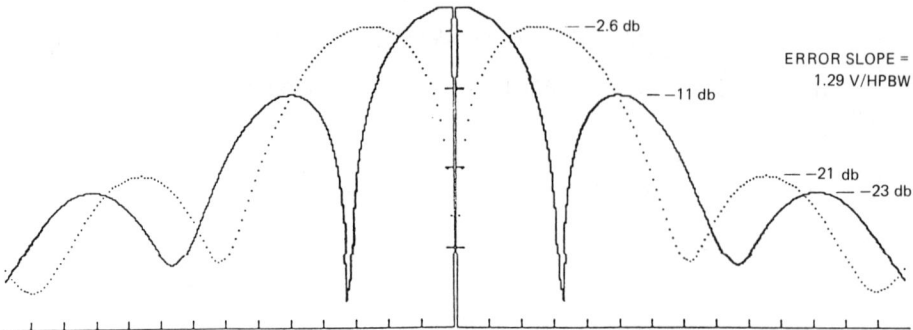

Fig. 8.9 Pattern resulting from Program 8.4.

less signal-to-noise ratio. The delta lobes are much wider, however, as a result, considerably more ground clutter could be picked up by a ground-based tracker at low angles. For an airborne fire-control system, of course, this factor would be insignificant. Note that the sum side lobes are much worse than those in an amplitude monopulse.

As a matter of fact, the secondary pattern behaves exactly as if there were a severe case of aperture blockage, a performance that ceases to be surprising if one examines the distribution. Compared to that of the amplitude monopulse, the error slope is slightly less in terms of volts/beam width. The beam width is about 84-percent of the amplitude monopulse beam width, however, and the absolute slope is thus actually steeper than that of the amplitude monopulse.

An obvious question is whether this distribution is the optimum one for the antenna. Might there be others that would be superior? Figures 8.10 and 8.11 show the results of the printed distributions.

The result for a distribution that works out to be nearly a double cosine is shown in Fig. 8.10(a). The dead spot in the center remains, but the distribution rises from the center at the same rate that it falls toward the periphery. Because the centroids of the distribution are away from the center, the distribution has more "leverage" than it would in a conventional single-peak-in-the-center type, and the sum mode signal shows a narrower main lobe, although there are very severe aperture blocking type effects, including increases in the odd-order side lobes and a supression of the even-order side lobes. On the other hand, the distribution is smoother and more symmetrical when one considers the individual *A* and *B* apertures. The apparent result is a considerable improvement of the delta side lobes.

The distribution in Fig. 8.10(b) occurs in the opposite direction. Here it has become even steeper in the center of the combined apertures thereby

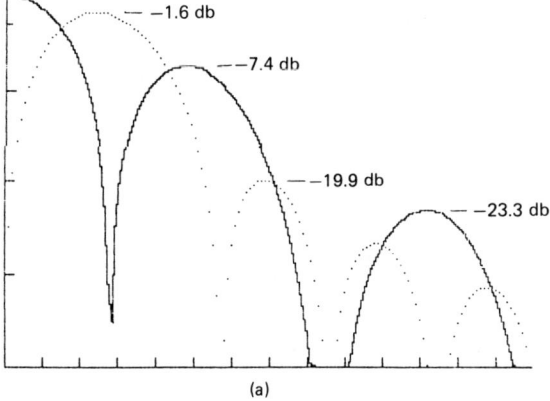
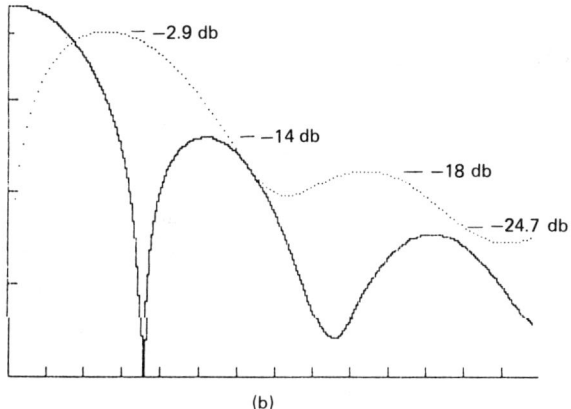

Fig. 8.10 Patterns for various distributions.

reducing the aperture blocking response and improving the side lobes in the sigma lobe at the expense of the delta mode.

The pattern in Fig. 8.11 shows the essential pattern that would be obtained if the reflector were illuminated with a single-cosine distribution covering both apertures—in other words, as if it were running as a single dish. The narrowing of the sigma and delta lobes are shown by the short, heavy broken lines. The narrowing of the half-power beam width previously described is easily extrapolated. Since the central strip is still dead, the aperture-blocking effects brought about by the supression of the even-order side lobes are still visible.

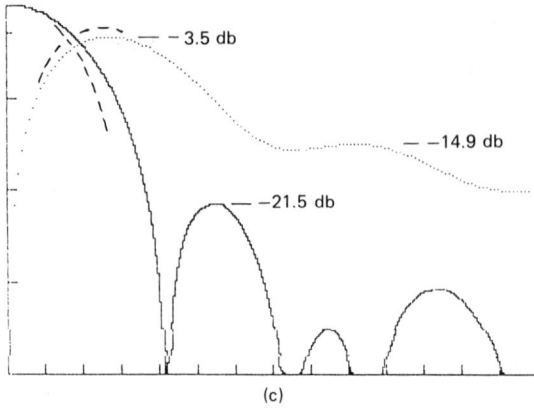

Fig. 8.11

AMPLITUDE VERSUS PHASE MONOPULSE

The information presented here gives a fair comparison between the advantages of amplitude and phase monopulse. The latter clearly has a 3-dB advantage in terms of tracking targets with low S/N, but this is paid for in terms of inferior side lobes and susceptibility to off-axis interference in the delta channel. For a dish of the same overall size, there is little difference in error slope and only a modest difference in gain, with the net advantage going to the phase monopulse.

COMBINATION MONOPULSE

The two separate dishes of the phase monopulse can obviously be twisted slightly with respect to one another in the plane normal to the phase monopulse. The resulting arrangement has the properties of an amplitude monopulse in one plane and of the phase monopulse in the other. Furthermore, since the entire dish is turned, there is no penalty for errors of the squint type, and the feed can be tailored for very low side lobes in the amplitude monopulse plane. This is a rather common configuration. Because of its ability to optimize the distribution in the amplitude monopulse plane, patterns comparable to those in Fig. 8.4 are achievable without the sequelae of feed crowding. In a ground-based system, the amplitude monopulse usually operates in the elevation plane because of the advantages of low side lobes at low angles in reducing ground clutter. The less desirable side lobes of the phase monopulse are placed in the azimuth plane where they are less likely to pick up ground clutter interference.

CASSAGRANIAN SYSTEMS

Nearly all large satellite ground antennas are monopulse. Even a geosynchronous satellite moves slightly in a diurnal figure eight. It is not necessary to track the satellite if the ground antenna beam width is greater than 1-degree, but it becomes necessary if the beam width is smaller.

Most of these antennas are Cassagranian in design because of the advantage provided in spill-over-induced temperature. A Cassagranian antenna equipped with a four-horn feed will exhibit relatively good amplitude monopulse performance. The cassagranian design does not lend itself to phase monopulse.

ARRAYS

It is possible to excite antenna arrays to operate in either the phase or amplitude monopulse modes. Most fire-control radars on modern fighter aircraft are flat monopulse arrays. The techniques of exciting such arrays will be discussed later.

9
Shaped Beam Antennas

Certain applications call for an antenna with some special shape. For example, the patterns of radio and TV stations in coastal cities are usually shaped to minimize the amount of transmission over the water. After all, one cannot sell much coffee or shaving cream to seagulls. The optimum strategy is to tailor the radiation from the antenna to match the population distribution. Even in the middle of the Plains States, a station may be located at one edge of the population distribution so that some tailoring would be desirable. Probably something like a third of present broadcast stations have non-circular radiation patterns.

THE CARDIOID ANTENNA

In the AM broadcast band, running from 550-kHz (545-meters) to 1550-kHz (193.5-meters), antennas with apertures of many wavelengths are usually not feasible because of practical considerations of property costs, material costs, and taxes. Antennas for this band are usually confined to arrays of a few elements spaced over a range of a few wavelengths at most. One directive antenna pattern that can be achieved with an arbitrarily small aperture is the cardioid (heart-shaped) pattern. No matter how small the available space, a cardioid pattern can be constructed using just two elements.

In Fig. 5.1, we saw that two in-phase radiators can be made to generate patterns ranging from nearly circular to very oblong, finally becoming a figure-of-eight at half-wave spacing. It is not difficult to visualize that this property could be used to tailor the distribution to an oblong population distribution of nearly any sort.

A very common problem for stations at the high end of the broadcast band is to prevent interference with another station operating on the same channel or an adjacent channel. To achieve this, it is often sufficient simply to punch a null into the radiation pattern, an end that can be accomplished with a simple two-element array. Suppose that we excite two closely spaced elements in what is more or less the reverse of the end-fire condition described in Fig. 5.5; that is, we phase them so that they cancel in one direction along the axis of the array. In all other directions they will add to some extent, and the pattern will exhibit the desired null.

Program 9.1 achieves this result. Note lines 250 and 260. When $A = 0$, $P0$ and PH cancel, and the minus sign in line 270 assures the null, provided that $E1 = E2$. Figure 9.1 shows that the resulting pattern is zero in the "down" direction and that the overall pattern is heart-shaped.

The shape of the cardioid stays remarkably constant for values of DL ranging from the infinitesimal to 0.2-wavelengths. As the value approaches 0.25, the cardioid commences to become slightly "squshed," but its general shape does not really dissappear until $DL = 0.4$. Note that the patterns of Fig. 9.2 have been scaled down by a factor of 2 since normalizing when $A = PI$ does not work beyond 0.25-wavelengths.

Although this antenna pattern can be generated with an infinitesimally small array, certain disadvantages accrue from making the array any smaller than necessary. If we test the value of the normalizing voltage, $E0$, we find that when $DL = 0.001$, $E0 = 1.579E - 4$, and when $DL = 0.1$, $E0 = 1.382$. In other words, at the tiny spacing, the voltages cancelled not only in the null direction, they very nearly cancelled in all directions! Only when the elements are reasonably separated does the field start to add significantly. This does not really constitute the gain of the array. When the elements are closely spaced, we have to consider such matters as mutual impedance. It stands to reason that an antenna that requires large currents to achieve effects that almost completely cancel cannot be very efficient.

One major use for the electrically tiny cardioid is for radio direction finding with medium frequencies. The direction to a distant transmitter is unambiguously indicated by rotating an antenna with a cardioid pattern until the signal is nulled. Since the atmospheric noise or "static" is much greater in the MF band than the self-generated noise in the receiver, a high efficiency is not required. This matter will be treated in detail later.

On the other hand, when a null is to be punched into the pattern of a broadcast antenna to minimize interference, the efficiency becomes very important, and spacings between 0.25 and 0.4-wavelengths are used, with preference given to the widest spacing that will give suitable coverage in the other directions.

The cardioid is frequently selected whenever small size compared to wavelength is important or only a single null is required.

Aerial navigation is a field that has made extensive use of special antenna patterns since the earliest days. The old A/N beacon modulated the antenna pattern to give dot-dash (A) and dash-dot (N) patterns to the "beams." Similarly, the VOR (Very high frequency OmniRange) operates by tailoring the antenna beam. The basic pattern is a cardioid.

```
150 PI = 3.14159265
160 HGR2
170 HCOLOR= 3
180 POKE  - 12524,0
190 POKE  - 12525,64
200 POKE  - 12529,255
210 REM ::::::::::::::::::::::::::::::::::::::::::::::
220 REM     THE PATTERN CALCULATION
230 REM ::::::::::::::::::::::::::::::::::::::::::::::
240 DL = .001
245 H = PI
250 P0 = 2 * PI * DL
260 PH = 2 * PI * DL *  COS (A)
270 ER = 1 -   COS (PH - P0)
280 EI =  SIN (PH - P0)
290 EP = ((ER * ER) + (EI * EI))
300  IF A = PI THEN E0 = EP
310 EN = (EP / E0) * 95
320 Y = 96 + EN *  COS (A)
330 XL = 140 - EN *  SIN (A)
340 XR = 140 + EN *  SIN (A)
350  IF A = PI THEN Y0 = Y:X0 = XL:X1 = XR
360  HPLOT X0,Y0 TO XL,Y
370  HPLOT X1,Y0 TO XR,Y
380 X0 = XL:X1 = XR:Y0 = Y
390 A = A - PI / 120
400  IF A > 0 GOTO 260
410  HPLOT 0,96 TO 5,96
420  HPLOT 274,96 TO 279,96
1000  STOP
```

Program 9.1

Suppose that an airplane radio is tuned to a constant amplitude transmitter feeding a rotating cardiod pattern. The received signal will be modulated with one cycle for every rotation of the antenna, and the signal will be zero at those instants when the cardiod null is pointed at the airplane. Now, if a signal is placed on the carrier to indicate instants when the null is pointed at magnetic north, the airplane can determine its bearing from the station by noting the time between the north reference and the null.

SHAPED BEAM ANTENNAS 129

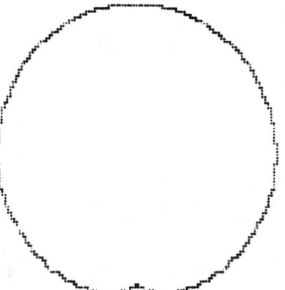

Fig. 9.1 A cardioid pattern.

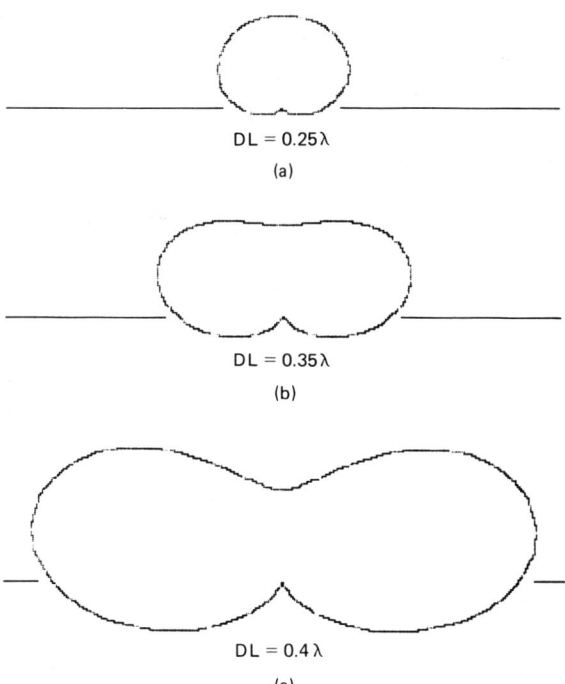

Fig. 9.2 Cardioid patterns for various values of DL.

In actual practice, ten ripples are superimposed on the main pattern of the VOR cardioid. A second "fine" reference is also broadcast to go with these and yield a tenfold increase in accuracy.

THE COSECANT-SQUARED PATTERN

Among all shaped antenna patterns, the cosecant-squared pattern surely represents the queen of the art. A brief discussion might be in order to explain why.

In radar work, the Inverse-Square-Law actually gets in two cracks at the designer. The original signal suffers from Inverse-Square-Law spreading on the way from the set to the target and again on the return from the target to the set. A doubling of the range requires an increase in effectiveness of 12.04-dB. A decent radar receiver seldom departs from a perfect noise figure by any value nearly as large as 12-dB, and an increase of 12-dB in transmitter power is usually a formidable task (e.g., 1-megawatt to 16-megawatts!). Only in the antenna is additional gain easy to come by. Significantly, since the antenna enters into the range equation twice, a 12-dB overall improvement can be obtained by a 6-dB increase in antenna gain. Without further belaboring the point, it is most important to optimize antenna performance.

A search radar should be able to offer coverage similar to that shown in the range-height diagram of Fig. 9.3. The peculiar scales of the chart distort the physical coverage into something more nearly approaching the realities of the job at hand. Our search set would probably have to be able to detect airplanes of a given reflecting area to a range of 100-miles. The highest airplane it is likely to encounter, moreover, is almost sure to be flying at an altitude less than 90,000-feet. With a 100-mile slant range, a vertical angle of 7.7-degrees will encompass such an altitude.

At any angle greater than 7.7-degrees, we would like to restrict the power so that large areas higher than 90,000-feet will not be filled. If we neglect the curvature of the earth, the slant range to a target is equal to the altitude divided by the sine of the angle. Since the power required of the radar varies as the reciprocal of the range to the fourth power, the antenna pattern should radiate a power that is proportional to the reciprocal of the sine to the fourth power. Since the power is proportional to the square of the voltage, moreover, the voltage pattern above the half-power point shown in the figure should vary as the reciprocal of the sine of the elevation angle squared. Since the reciprocal of the sine is the cosecant, this is called a *cosecant-squared pattern*.

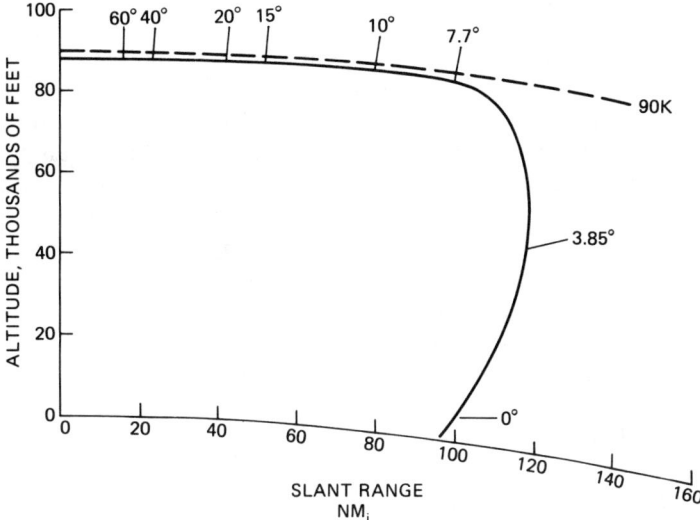

Fig. 9.3 Range-height diagram for cosecant-squared coverage.

This pattern optimizes the distribution of the radar energy in the sense that it supplies an echo that is constant at a constant given altitude for all angles above the nose of the beam without regard to range. Such a distribution represents the least energy that will guarantee detection of the target to the rated altitude. Most microwave surface-based search radars have a cosecant-squared antenna as do most airborne mapping radars.

The concept of the cosecant-squared antenna was developed during WW II, and cosecant-squared reflector designs were first developed in the early 1950s.

If half of a paraboloid is deformed into a shape equivalent to a segment of a beer barrel, an approximation of a cosecant-squared pattern can be obtained. An apocryphal tale says that the first X-band CSC$\wedge$2 dish was created as a result of a small paraboloid and feed falling off the roof of one of the buildings at the MIT Rad Lab. It landed on its edge and was severely deformed. When the deformity was measured, the pattern, proved to be CSC$\wedge$2!

The first precision CSC$\wedge$2 search radar dishes were developed during the early 1950s.

PHYSICAL REALIZATION

In addition to deforming a reflector from a true paraboloid, a variety of other techniques can be used to produce a CSC∧2 beam. The AN/CPS−6 and −6B succeeded in producing one with a paraboloid and an array of feed horns. Figure 7.4(a) shows that a squinted feed can produce an approximation of a CSC∧2 pattern. If the feed were slightly defocused in order to fill in the nulls, the approximation would be even better. The pattern can also be achieved with an array antenna. As a matter of fact, certain very sophisticated, electronically scanned arrays can shift gears into a CSC∧2 mode for search operation. Although it is not within the scope of this text to consider the detailed computation of a reflector shape for such an antenna, we can nevertheless analyze the phase and amplitude distribution that are required to obtain the CSC∧2 pattern.

It seems fair to ask how one can determine the phase and amplitude distribution required to obtain some arbitrary antenna pattern. Up to this point, we have simply postulated arrays or aperture distributions and then investigated the results. The present approach will be the inverse: The desired antenna pattern is known, and we have to find out how to obtain it. The answer to this problem is the Fourier Analysis.

Jean Baptiste Fourier (1768−1830) was a French military engineer and mathematician. In a master work that is widely applicable today, Fourier showed that any single-valued function can be synthesized by a train of sine and cosine functions, that is,

$$Y = F(X) = A1*COS(W) + jB1*SIN(W) + A2*COS(2*W) + \\ jB2*SIN(2*W)... + AN*SIN(N*W) + jBN*SIN(N*W) \quad (9.1)$$

in which the As and Bs are amplitudes of the cosine and sine components, and W, if the function is repetitive, represents the radian frequency of repetition ($W = 2\pi*F$). The sequence is theoretically infinite in length, but for practical applications, it can usually be synthesized with a modest number of terms. If the function is not repetitive, the analysis is still valid, but the repetition frequency will be essentially zero and all frequencies will be present, not just multiples of W.

Now, if we substitute *EP* for *Y* and angle-in-space for *X*, we will find that the resulting expression is identical with the expression for the array antenna patterns that we have been using all along. For reasons of symmetry, we were able to drop the sine terms in certain cases, but in general cases, such as the endfire array of Program 5.3 or the squinted pattern of Program 7.2, all terms were retained. When we consider the

SHAPED BEAM ANTENNAS 133

diversity of antenna patterns that we have already seen, it is not difficult to believe that nearly any shape is capable of being synthesized with such a series.

The reverse process is remarkably similar. The process performed by a digital computer is termed a *discrete fourier transform*, or DFT. This operation is essentially a correlation performed with sine waves. To begin with, we take one cycle of a wave (or pattern), divide it into N segments, and measure the height of each segment, $C1, C2...CN$, as shown in Fig. 9.4.

In doing a discrete correlation, one takes the two functions to be correlated and divides them up into discrete samples. Let us call one function R and the second S. One then multiplies the samples and sums, as follows:

$$R1 \times S1 = T1$$
$$R2 \times S2 = T2$$
$$\vdots$$
$$RN \times SN = \underline{TN}$$
$$\text{Total} = TT$$

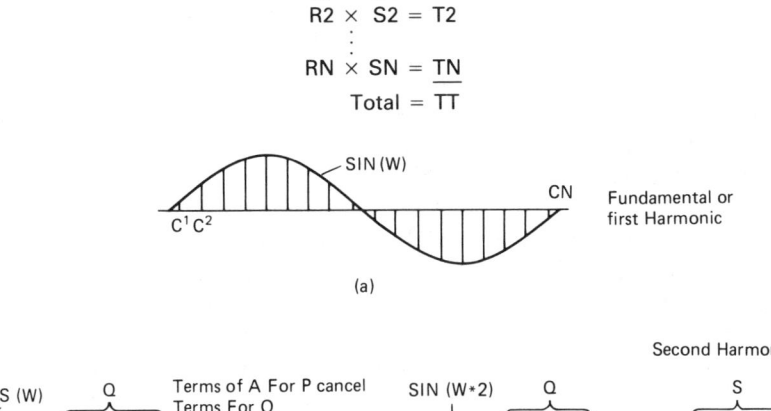

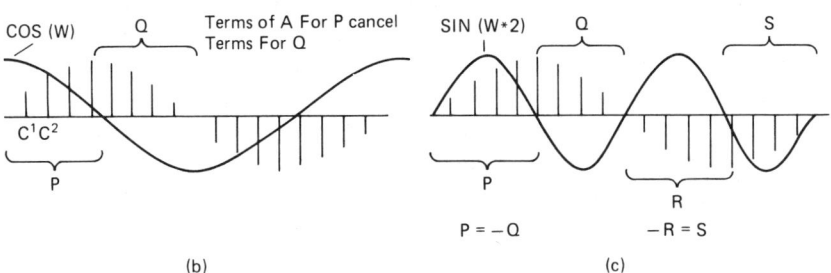

Fig. 9.4 The Discrete Fourier Transform (DFT).

If the functions R and S are identical and completely overlaid, T_T will be maximum. If they are dissimilar or not aligned, however, T_T will be very small.

In the example of Fig. 9.4, we will attempt to obtain the Fourier transform of a trivially simple case, the wave SIN(W).

To determine coefficient B_1, we run through the following series, multiplying the ηth coefficient by the sine of W/N:

$$B_\eta = \sum_0^N C_\eta * SIN(\eta * W/N)$$

To determine coefficient A_1 we run through the cosine series:

$$A_\eta = \sum_0^N C_\eta * COS(\eta * W/N)$$

It is apparent that the summation for B_1 will have a value, whereas in the summation for A_1, the sum of the terms will be zero, as shown in Fig. 9.4(b). Moreover, the product for B_2 will be zero since the sum of the terms will cancel, as shown in Fig. 9.4(c). As a matter of fact, all remaining A_s and B_s will be zero since the example was selected for a perfect correlation with B_1. In a nontrivial analysis, of course, a number of the A_s and B_s will have nonzero values.

This is a computation-intensive operation. Each A requires N trig look-ups, N multiplications, and N additions, as does each B. Furthermore, since the analysis will support any number of A_s and B_s up to N, each manipulation may have to be performed $2*N*N$ times. If $N = 1024$, that means over a million look-ups, multiplications, and divisions. (This figure is given as over a million and not two million because the last half of the terms are symmetrical and may be deduced from the first half.)

Even on the largest and fastest computers, algorithms that increase run time by a power greater than 2 of the number of terms will run slowly and are thus considered to be computationally intensive. It is up to mathematicians to try to find algorithms that will run faster. In 1965, in an historic article in Mathematical Computations, "An algorithm for the Machine Computation of Complex Fourier Series," Cooley and Tukey introduced the FFT, or Fast Fourier Transform.

THE FFT/IFT

The Fast Fourier Transform is used in signal processing to take a waveshape in the time domain and decompose it into a series of harmonically related sine and cosine waves. The Inverse Fourier Transform (IFT) sums a series of harmonically related waves to obtain the waveform.

In computing array patterns, we have actually been doing an Inverse Fourier Transform in summing the contributions from a harmonically spaced array of radiators (DL, $2DL$, $3DL$, etc.) to obtain the shape of the pattern in space. The Discrete Fourier Transform (DFT) will then yield the spacing, amplitude, and phasing for the array of radiators required to produce a given pattern.

The Fast Fourier Transform yields the same result as a Discrete Fourier Transform. It simply runs faster.

Basically, the FFT/IFT reduces the computation by a stratagem in which two FFTs of half the terms are assumed to have only half as many operations as the single original. In the extreme, this reduces to a group of two-term FFT's that must be combined in some fashion. (A more detailed comparison of the DFT and the FFT/IFT is to be found in the author's *Talking Computers and Telecommunications*.[1])

The FFT/IFT is widely used in speech and signal processing to manipulate between time and frequency domains; for example, a given waveshape can be analyzed into a spectrum by FFT. A filter of known phase and amplitude characteristics can then be applied to shape the spectrum. The IFT will then show the waveshape as modified by the filtering. In antenna work, the DFT has not been entirely replaced by the FFT for a variety of reasons. For one thing, the DFT may be worked with any number of terms whereas the FFT must have a binary number of terms. (NOTE: Although bases other than 2 can be implemented (otherwise, division by 2 would yield a fractional radiator), such a discussion would exceed the purpose of this text.)

The IFT is rarely used in place of the ordinary Fourier series for pattern calculation because of this binary (or other base) restriction. Because of its symmetry and freedom from having to include zero-strength terms to "pad" N to a binary number, the straight Fourier series may run faster.

Program 9.2 represents an FFT/IFT edited to run on an Apple II+ equipped with a Silentype printer. It can be used to obtain the amplitude and phase coefficients for an antenna pattern, but note the sequence in lines 32 to 34. As M raises N to a value adequate to do a good job on any worthwhile pattern, it becomes necessary to enter a very large number of terms by hand. Accordingly, the program has been rewritten in Program 9.3 to allow the CSC$\wedge$2 pattern to be entered analytically.

---

[1]John A. Kuecken, *Talking Computers and Telecommunications* (New York: Van Nostrand Reinhold, 1983).

```
6    PRINT "FOR PRINT ENTER 1"
7    PRINT "FOR PLOT ENTER 2"
8    PRINT "FOR BOTH ENTER 3"
9    INPUT PR
11   PRINT "     ENTER FWD OR REV (F/R)    "
13   INPUT AN$
14   IF AN$ = "F" THEN D = 0: GOTO 20
15   IF AN$ = "R" THEN D = 1: GOTO 20
16   PRINT "IT HAS TO BE F OR R DUMMY!"
17   GOTO 11
20   PRINT "ENTER M"
21   INPUT M
22   N = 2 ^ M
23   PRINT "N=",N
24   DIM X(N,2)
25   IF D = 0 THEN  PRINT "INPUT TIME DOMAIN DATA"
26   IF D = 1 THEN  PRINT "INPUT FREQ DOMAIN DATA
30   FOR I = 1 TO N
31   PRINT "##############################"
32   PRINT "INPUT X'("I",0) X("I","1")"
33   INPUT X(I,0),X(I,1)
34   NEXT I
35   IF PR = 2 THEN   GOTO 40
36   IF PR = 1 THEN   GOTO 39
37   IF PR = 3 THEN   GOTO 39
38   GOTO 40
39   PR# 1
40   PRINT "=============================="
41   IF D = 0 THEN  PRINT "TIME DOMAIN DATA IS"
42   IF D = 1 THEN  PRINT "FREQUENCY DOMAIN DATA IS"
45   PRINT "POINT","REAL          IMAGINARY"
50   FOR I = 1 TO N
55   PRINT I,X(I,0),X(I,1)
60   NEXT I
65   PR# 0
70   GOSUB 500
75   IF PR = 2 THEN   GOTO 100
80   IF PR = 1 THEN   GOTO 95
85   IF PR = 3 THEN   GOTO 95
90   GOTO 100
95   PR# 1
100  PRINT "/////////////////////////"
110  IF D = 0 THEN  PRINT "NOW IN FREQ DOMAIN"
111  IF D = 1 THEN  PRINT "NOW IN THE TIME DOMAIN"
112  PRINT "XFRMD DATA IS"
114  PRINT "POINT","REAL          IMAGINARY"
116  FOR I = 1 TO N
118  PRINT I,X(I,0),X(I,1)
120  NEXT I
122  PRINT "=============================="
125  PR# 0:
126  IF PR = 2 THEN   GOTO 1100
127  IF PR = 3 THEN   GOTO 1100
128  END
```

Program 9.2

```
9 PR = 3
13 AN$ = "F"
14  IF AN$ = "F" THEN D = 0: GOTO 20
15  IF AN$ = "R" THEN D = 1: GOTO 20
20 M = 8
22 N = 2 ^ M
23  PRINT "N=",N
24  DIM X(N,2)
25  GOTO 2000
35  IF PR = 2 THEN   GOTO 40
36  IF PR = 1 THEN   GOTO 39
37  IF PR = 3 THEN   GOTO 39
38  GOTO 40
39  REM
40  PRINT "============================="
41  IF D = 0 THEN   PRINT "TIME DOMAIN DATA IS"
42  IF D = 1 THEN   PRINT "FREQUENCY DOMAIN DATA IS"
45  PRINT "POINT","REAL      IMAGINARY"
50  FOR I = 1 TO N
51  IF X(I,0) < 1E - 3 GOTO 60
55  PRINT I,X(I,0),X(I,1)
60  NEXT I
65  PR# 0
70  GOSUB 500
75  IF PR = 2 THEN   GOTO 100
80  IF PR = 1 THEN   GOTO 95
85  IF PR = 3 THEN   GOTO 95
90  GOTO 100
95  PR# 0
100  PRINT "///////////////////////////"
110  IF D = 0 THEN   PRINT "NOW IN FREQ DOMAIN
111  IF D = 1 THEN   PRINT "NOW IN THE TIME DOMAIN"
112  PRINT "XFRMD DATA IS"
114  PRINT "POINT","REAL      IMAGINARY"
116  FOR I = 1 TO N
118  PRINT I,X(I,0),X(I,1)
120  NEXT I
122  IF D = 0 GOTO 3000
125  GOTO 1100
128  END


500  REM     :::::::::::::::::::::::::::::
501  REM     FFT/IFT SUBROUTINE
502  REM     :::::::::::::::::::::::::::::
550 N = 2 ^ M
552  REM     ******DO BIT SHUFFLE******
570 N2 = N / 2
580 N1 = N - 1
590 J = 1
600  FOR I = 1 TO N1
```

Program 9.3

```
610  IF I > = J THEN   GOTO 680
630  T1 = X(J,0)
640  T2 = X(J,1)
650  X(J,0) = X(I,0)
655  X(J,1) = X(I,1)
660  X(I,0) = T1
670  X(I,1) = T2
680  K = N2
690  IF K > = J THEN 730
700  J = J - K
710  K = K / 2
720  GOTO 690
730  J = J + K
740  NEXT I
750  REM ******END OF SHUFFLE***********
760  S1 = - 1
770  IF D = 0 THEN   GOTO 790
780  S1 = 1
790  PI = 3.1415926
800  FOR L = 1 TO M
810  L1 = 2 ^ L
820  L2 = L1 / 2
830  U1 = 1
840  U2 = 0
850  W1 = COS (PI / L2)
860  W2 = S1 * SIN (PI / L2)
870  FOR J = 1 TO L2
880  FOR I = J TO N STEP L1
890  I1 = I + L2
895  REM *******DO BUTTERFLY*************
900  V1 = (X(I1,0) * U1 - X(I1,1) * U2)
910  V2 = (X(I1,1) * U1 + X(I1,0) * U2)
920  X(I1,0) = X(I,0) - V1
930  X(I1,1) = X(I,1) - V2
940  X(I,0) = X(I,0) + V1
950  X(I,1) = X(I,1) + V2
960  NEXT I
970  REM ******DO TWIDL FACTOR**********
975  U3 = U1
976  U4 = U2
980  U1 = (U3 * W1 - U4 * W2)
990  U2 = (U4 * W1 + U3 * W2)
1000  NEXT J
1010  NEXT L
1020  IF D = 1 THEN   GOTO 1060
1030  FOR I = 1 TO N
1040  X(I,0) = X(I,0) / N
1045  X(I,1) = X(I,1) / N
1050  NEXT I
1060  RETURN
1075  PRINT "#############################"
1080  PRINT "  START OF PLOT ROUTINE"
1085  PRINT "PLOTS BLACK ON WHITE ON HGR2"
1090  PRINT ">>>>>>>>>><<<<<<<<<<<<<<<"
1100  DIM Y(N,1)
```

Program 9.3 (*continued*)

SHAPED BEAM ANTENNAS

```
1110 X% = 4 * 558 / N
1115 M4 = - 430
1120  REM   CALC. HORIZONTAL SCALE
1130  HGR2
1140  HCOLOR= 3
1150  POKE  - 12524,0
1160  POKE  - 12525,64
1170  POKE  - 12529,255
1180  REM   SETS TO PLOT B/W ON HGR2
1190  IF D = 1 THEN  GOTO 1400
1200  FOR I = 1 TO N
1210 H = I
1220 Y(H,0) =  SQR (X(I,0) ^ 2 + X(I,1) ^ 2)
1230  IF Y(H,0) > Y1 THEN Y1 = Y(H,0)
1240  IF Y(H,0) < Y2 THEN Y2 = Y(H,0)
1250  NEXT I
1260 N4 = 190 / (Y1 - Y2)
1270  REM   THIS SETS VERTICAL SCALE
1280 M4 = 0
1290 M5 = (N / 2) + 1
1300  FOR H = 1 TO M5
1310 Y% = 190 - (Y(H,0) * N4)
1320  HPLOT M4,190 TO M4,Y%
1330 M4 = M4 + X%
1340  NEXT H
1345 M4 = M4 - X%
1350  HPLOT 0,190 TO M4,190
1360  END
1400  FOR I = 1 TO N
1410 H = I
1420 Y(H,0) =  SQR (X(I,0) ^ 2 + X(I,1) ^ 2)
1430  IF X(I,0) < 0 THEN Y(H,0) = Y(H,0) *  - 1
1440  IF Y(H,0) > Y1 THEN Y1 = Y(H,0)
1450  IF Y(H,0) < Y2 THEN Y2 = Y(H,0)
1455  NEXT I
1460 N4 = 190 / (Y1 - Y2)
1470  REM   CALC. VERTICAL SCALE
1480 N5 = N4 * Y2
1490  REM   LOCATE AXIS
1495  FOR H = 1 TO N
1500 Y% = 190 + N5 - (Y(H,0) * N4)
1510  IF H = 1 THEN M6 = 0:N6 = Y%
1515  IF M4 < 0 OR M6 < 0 GOTO 1530
1520  HPLOT M6,N6 TO M4,Y%
1530 M6 = M4
1540 M4 = M4 + (X% / 2)
1550 N6 = Y%
1560  NEXT H
1570 Y% = 190 + N5
1575 M4 = M4 - (X% / 2)
1580  HPLOT 0,Y% TO M4,Y%
1590  END
```

Program 9.3 (*continued*)

140 EXPLORING ANTENNAS AND TRANSMISSION LINES

```
1950 REM  *******************************************
1960 REM       THE COSECANT SQUARED PATTERN
1970 REM  :::::::::::::::::::::::::::::::::::::::::::
2000 PI = 3.1415926
2010 A0 = 2 * PI / N
2020 FOR I = 1 TO N
2030 U = (I * A0 * 19.93) - (N * A0 * 19.93 / 2) - 1.79256
2035 IF I > 3 * N / 4 GOTO 35
2040 IF U > 1.39256 GOTO 2130
2050 IF U < - PI THEN X(I,0) = 0:X(I,1) = 0: GOTO 2100
2055 IF U = 0 THEN U = .000001
2060 X(I,0) = ( SIN (U)) / U
2070 PRINT U,I,X(I,0)
2100 NEXT I
2120 GOTO 35
2130 U1 = (U / 19.93) + .069872556
2295 IF U1 = 0 THEN U1 = .000001
2300 X(I,0) = .101002144 / SIN (U1)
2310 X(I,1) = 0
2890 GOTO 2100
2900 REM  *******************************************
2910 REM       THIS ROUTINE FILTERS
2920 REM       THE FREQUENCY DOMAIN DATA
2930 REM  :::::::::::::::::::::::::::::::::::::::::::
3000 D = 1
3010 FOR I = 18 TO 238
3020 X(I,0) = 0
3030 X(I,1) = 0
3040 NEXT I
3050 GOTO 70
```

Program 9.3 (*continued*)

Program 9.2 is incomplete as presented. Lines 500 through 1060 of Program 9.3 should be added verbatim, and lines 1075 through 1590 should be added with the following changes:

>1110 X% = 558/N
>1115 DELETE
>1515 DELETE

Lines 1115 and 1515 were deleted to render a more pleasing scale for the unique characteristics of the CSC/\2 pattern.

In the general case program, it is necessary to display the entire window for the time domain data (angle-in-space domain for antenna work).

Program 9.3 dispenses with some of the menu-driven stuff in the original. It is set into the print-and-plot mode in line 9. Manual loading of the antenna pattern is eliminated since this operation is performed by the suboutine at line 2000 as initiated by line 25. The routine from lines 2030 to 2070 loads the pattern data for a $(\text{SIN}(X))/X$ between the below-the-horizon axis crossing and the upper 3-dB point (approximately). Line 2030 places the lower 3-dB point on the horizon.

Line 2300 blends a $1/(\sin(u))$ or cosecant curve into the nose at the last point before the upper 3-dB point.

The "time domain" and "frequency domain" labels have been retained from the original to strengthen the tie-in of the two programs. We shall deal with the routine at line 3000 later.

A printout of the amplitude distribution is shown in Fig. 9.5. The period of the program is considered to be covered in 360-degrees of elevation. Since M was specified to be 8, $N = 256$, and the pattern is thus divided into 256 readings. Readings from 0 to 124 are zero, and reading 128 is assumed to be the horizon. Readings 134 and over represent the cosecant curve of amplitude although the radar-based CSC$\wedge$2 title is retained.

Running the program yields the coefficients shown in Fig. 9.6 and 9.7. A check of Fig. 9.7 will suffice to prove that the distribution is symmetrical except for the base term. The plot at the bottom of Fig. 9.7 shows the magnitude of the coefficients from 0 through 128. It may be seen that the distribution falls rapidly to relatively insignificant values; as a matter of fact, much of the listing has been deleted for that reason.

In terms of physical significance, the real listings correspond to the A_s and the imaginary listings to the B_s. In the time or angle domain, the significance of the imaginary terms eludes the writer. They were included in the original FFT/IFT program to provide the symmetry needed to solve an IFT in which imaginary terms do have a real physical significance. You will never load an antenna pattern that has imaginary terms, but the program may return some after it has been processed through the IFT. These represent phase errors caused by truncation of some of the terms.

The successive terms may be considered to be either aperture samples or array elements spaced a half wavelength apart. The question arises of how many of these terms must be included to simulate the pattern. All 256? Some fraction thereof? The routine at line 3000 was added to investigate this question. This line sets some portion of the lower terms to zero and sends the program back into the IFT routine. To see the distribution plotted, delete line 122 and make the same corrections to the plot routine as those listed for the general-purpose program.

```
TIME DOMAIN DATA IS
POINT         REAL          IMAGINARY
125          .0971528004         0
126          .293838241          0
127          .505949244          0
128          .706725679          0   ←— 0° ELEVATION
129          .869420854          0
130          .971644101          0
131          .999065077          0
132          .947809351          0
133          .825108962          0   UPPER 3 DB
134          .638351308          0   ←— +0.1384 RAD
135          .590787307          0       +7.92°
136          .517719759          0
137          .460983766          0
138          .415680467          0          Σ X(I,0)=25.3487
139          .378692476          0          Σ X(I,0)/N=0.09902
140          .3479416            0
141          .321989143          0
142          .29960753           0
143          .280643256          0
144          .263931246          0
145          .249239432          0
146          .236231813          0
147          .224643172          0
148          .21426138           0
149          .204914761          0
150          .196462898          0
151          .18878984           0
152          .181799004          0
153          .175409306          0
154          .169552179          0
155          .16416926           0
156          .159210568          0
157          .15463306           0
158          .150399476          0
159          .146477413          0
160          .142838561          0
161          .13945809           0
162          .13631414           0
163          .133387394          0
164          .130660736          0
165          .128118948          0
166          .125748466          0
167          .123537171          0
168          .121474211          0
169          .119549848          0
170          .117755329          0
171          .116082778          0
172          .114525093          0
173          .11307587           0
174          .111729326          0
175          .110480242          0
176          .109323901          0
177          .10825605           0
178          .107272849          0
179          .106370845          0
180          .105546907          0
```

Fig. 9.5 Elevation amplitude distribution ($\Sigma X(I,0) = 25.3487$ and $\Sigma X(I,0)/N = 0.09902$).

SHAPED BEAM ANTENNAS 143

```
////////////////////////////
NOW IN FREQ DOMAIN
XFRMD DATA IS
POINT      REAL           IMAGINARY
1          .0677682562    0
2          -.057022951    .0230667352
3          .0367091082    -.0284060425
4          -.0259467652   .0202411118
5          .0266848515    -.0149574804
6          -.0268984726   .0173805335
7          .0215258185    -.0190402311
8          -.0170496575   .0155521073
9          .0173706088    -.0123738877
10         -.0178147643   .0131908761
11         .0149233394    -.0140491442
12         -.0121054622   .0117823002
13         .0122860086    -9.43053617E-03
14         -.0127309884   9.75881114E-03
15         .0108199824    -.0103243599
16         -8.75235743E-03 8.66325428E-03
17         8.85522223E-03 -6.78916242E-03
18         -9.27642869E-03 6.92513131E-03
19         7.88714399E-03 -7.37423509E-03
20         -6.24795794E-03 6.10426073E-03
21         6.30372305E-03 -4.56416286E-03
22         -6.70515844E-03 4.6269711E-03
23         5.64471545E-03 -5.0393063E-03
24         -4.28865508E-03 4.05897992E-03
25         4.31841037E-03 -2.77739757E-03
26         -4.71016608E-03 2.82342764E-03
27         3.88294754E-03 -3.23820954E-03
28         -2.73483412E-03 2.48784152E-03
29         2.75468004E-03 -1.41639551E-03
30         -3.14716989E-03 1.47242357E-03
31         2.50012176E-03 -1.9074811E-03
32         -1.51677114E-03 1.34492376E-03
33         1.53866834E-03 -4.46947712E-04
34         -1.94031877E-03 5.23095965E-04
35         1.44059755E-03 -9.83354161E-04
36         -5.94214133E-04 5.74466081E-04
37         6.25996491E-04 1.81225739E-04
38         -1.04212885E-03 -8.49835073E-05
39         6.66591516E-04 -3.97296878E-04
40         6.3339145E-05   1.12181863E-04
41         -1.79581911E-05 5.29276671E-04
42         -4.14438997E-04 -4.19050452E-04
43         1.44369396E-04 -7.72819205E-05
44         4.8644872E-04   -1.10914544E-04
45         -4.27627834E-04 6.63592747E-04
46         -1.93770233E-05 -5.48630493E-04
47         -1.62097359E-04 4.84546204E-05
48         7.09919977E-04 -1.6358507E-04
49         -6.4110883E-04  6.50071422E-04
50         1.84061608E-04 -5.40495348E-04
51         -2.93016411E-04 4.69115935E-05
52         7.734162E-04    -1.09421925E-04
53         -7.00492696E-04 5.48914456E-04
54         2.40022688E-04 -4.54017152E-04
55         -2.91870281E-04 -2.38195784E-05
56         7.19795907E-04 -2.66725473E-06
```

Fig. 9.6 Distribution coefficients yielded by Program 9.3.

221	6.25997448E-04	-1.81225678E-04
222	-5.94215533E-04	-5.74466167E-04
223	1.4405988E-03	9.83354487E-04
224	-1.94032068E-03	-5.23096309E-04
225	1.53866833E-03	4.46947777E-04
226	-1.51677268E-03	-1.34492337E-03
227	2.50012263E-03	1.90748094E-03
228	-3.14717107E-03	-1.47242384E-03
229	2.75468075E-03	1.41639563E-03
230	-2.73483525E-03	-2.48784175E-03
231	3.88294824E-03	3.23820992E-03
232	-4.7101672E-03	-2.82342839E-03
233	4.31841062E-03	2.77739789E-03
234	-4.28865605E-03	-4.05898051E-03
235	5.64471592E-03	5.03930902E-03
236	-6.70515927E-03	-4.62697227E-03
237	6.30372342E-03	4.56416361E-03
238	-6.24795864E-03	-6.1042618E-03
239	7.88714432E-03	7.37423631E-03
240	-9.27642958E-03	-6.92513295E-03
241	8.85522177E-03	6.78916299E-03
242	-8.75235798E-03	-8.66325527E-03
243	.0108199821	.0103243609
244	-.0127309883	-9.75891297E-03
245	.0122860082	9.43053741E-03
246	-.0121054619	-.0117823018
247	.0149233384	.014049146
248	-.0178147637	-.0131908787
249	.0173706076	.0123738892
250	-.0170496566	-.0155521091
251	.0215258165	.0190402331
252	-.0268934708	-.017380537
253	.026684848	.0149574832
254	-.0259467631	-.0202411144
255	.0367091039	.0284060462
256	-.057022947	-.023066743

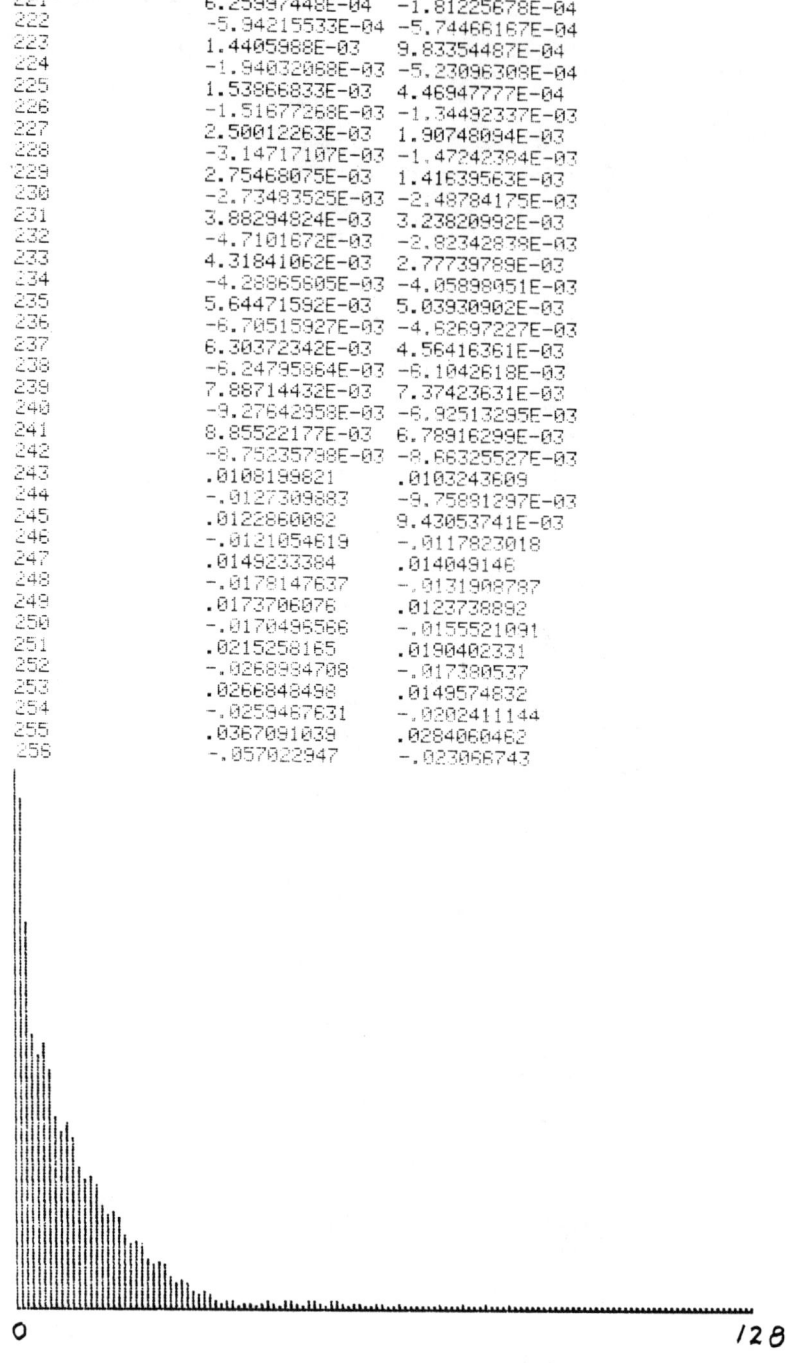

Fig. 9.7 Distribution coefficients yielded by Program 9.3.

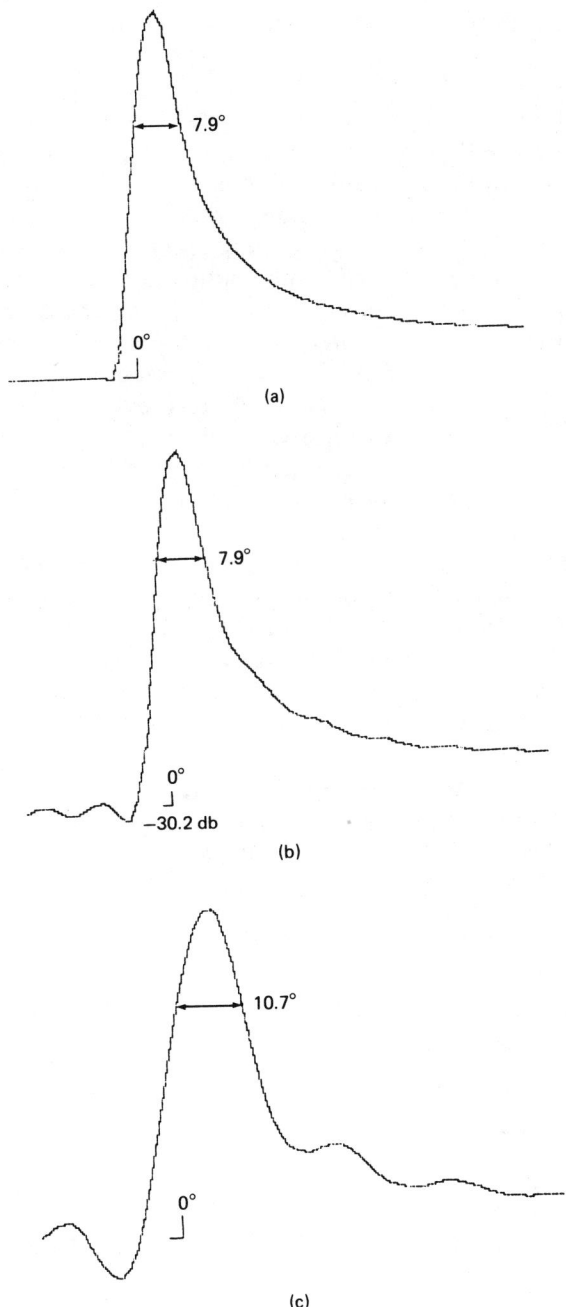

Fig. 9.8 The cosecant-squared pattern (a) for all 256 terms, (b) terms 32 to 224, and (c) terms 18 to 238.

Figure 9.8 shows the results with all 256 terms, with the 32 end terms, and with 18 of the end terms. Figure 9.8(a) shows the originally desired CSC$\wedge$2 pattern with little error. At 32 terms [Fig. 9.8(b)], the errors are just beginning to be noticeable. Cutting back to 18 terms [Fig. 9.8(c)] clearly brings about some deterioration in the main beam width as well as substantial ripples in the CSC$\wedge$2 portion and high side lobes. Since at 32 terms, or 16-wavelengths, the beam width will be just under 8-degrees, the aperture is not terribly efficient considering that a cosine-squared distribution will give an 8-degree beam width at 9-wavelengths.

One clue is the sharp corner below the horizon on the original pattern specification. Nature abhors sharp corners on antenna patterns and causes us to pay for them with many high-frequency terms. If a modest side lobe is allowed, the requirement for high-frequency terms might be reduced, and a smaller aperture might yield a satisfactory CSC$\wedge$2 curve and nose width.

Although the program runs fairly slowly, requiring about 10-minutes, an indication of its effectiveness will help put this drawback in perspective. In 1952 when the writer was involved in the development of the MPS-10 radar, a recent math grad sat down at her desk every morning for nine months and punched at a mechanical Friden calculator all day long. She was solving essentially the same problem that the Apple completed in 10-minutes!

The Fourier analysis is a powerful tool for developing antennas with arbitrary pattern shapes. The computing ability of a small personal computer places this powerful design tool within the grasp of nearly everyone.

10
Unfilled Arrays

In our previous treatment of arrays, we have largely confined ourselves to "filled" arrays, that is, arrangements in which an antenna is placed every half wavelength or so. However, a large class of antennas, used most often in radio astronomy, are not so constructed. This chapter is intended to provide a brief background in such antennas.

For typical radar and communications applications, the designer usually wishes to collect as much energy as possible; this, of course, means as much collecting area as possible. For radar, directivity is useful for resolution, but for communications, it is not valuable in itself. If one could collect a large signal without creating too narrow an antenna pattern, it would tend to make life simpler. Now and then one would like a narrow beam in order to discriminate against interference, but a narrow beam is more often just a nuisance that requires sturdy antenna supports and precise hardware to keep the antenna aimed.

Astronomy often presents the opposite requirement. Here, signal gathering is not a severe problem, but the resolution of closely spaced sources is a matter of the first importance. For large arrays in particular, the prime requirement is for resolution rather than signal-gathering power. Accordingly, the aperture of many astronomical instruments is filled with only 5-percent or less of the normal complement of elements.

THE SIMPLE INTERFEROMETER

We noted in Chap. 4 that Young's interferometer experiment is much easier to perform with the aid of a laser. In an experiment conducted by the author, the spirit lamp was replaced with an Aerotech LSR-2R HeNe Laser with a wavelength of 6.328E − 9-meters. The beam was so tiny that it proved necessary to spoil it slightly with a small, double-convex lens of −48-mm focal length in order to make it illuminate both pinholes evenly. At a range of 0.96-meters from the pinhole pair, a set of 8 fringes (these were quite striking and had good contrast) were spread across a more or less elliptical spot 6.2-mm across. This result indicated that there were approximately 1247-wavelengths between the pinholes, or 0.031-inches. Individually, the two pinholes showed more or less oblong spots that were 6.8 × 5.5-mm on target for one and 10.4 × 5.5-mm on target for the

```
200 N = 256
210 PI = 3.14159
220 DL = .313: REM  ELEMENT SPACING IN WAVELENGTHS
230  DIM X(N,1)
235 AH = PI / (2 * N)
240  FOR I = 1 TO 192:X(I,0) = 0: NEXT
242  FOR I = 193 TO N
244 X(I,0) =  SIN ((PI / 64) * (I - 192))
250  PRINT I,X(I,0)
260  NEXT I
295 E0 = 1
296  REM ))))))))))))))))))))))))))))))))))))))))
297  REM     THIS COMPLETES APERTURE DISTRIBUTION CALCULATION
298  REM &&&&&&&&&&&&&&&&&&&&&&&&&&&&&&&&&&&&&&
300  FOR J = 0 TO 138
310 A0 = (PI / 3600) * J
320 AI = 2 * PI * DL *  SIN (A0)
330  FOR I = 1 TO N
332  IF X(I,0) = 0 GOTO 350
335 AJ = AI * I
340 E0 = E0 + X(I,0) * 2 *  COS (AJ)
350  NEXT I
352  REM ##############################################
354  REM     THIS COMPLETES THE SUMMATION
356  REM       FOR ONE ANGLE
358  REM ##############################################
360  IF J = 0 THEN EN = E0: GOTO 380
370  GOTO 430
380  HGR2
390  HCOLOR= 3
400  POKE  - 12524,0
410  POKE  - 12525,64
420  POKE  - 12529,255
422  REM ##############################################
424  REM     THE GRAPHICS AND PRINTER SETUP
426  REM %%%%%%%%%%%%%%%%%%%%%%%%%%%%%%%%%%%%%%%%%
430 YA =  ABS (EN / E0)
432 YB = 20 * ( LOG (YA)) / 2.30259
434 Y = 4.75 * YB
436  IF Y > 191 THEN Y = 191
450  IF J = 0 THEN X0 = X:Y0 = Y
460  HPLOT X0,Y0 TO X,Y
470 X0 = X:Y0 = Y
475 X = X + 2
480 E0 = 1
490  NEXT J
492  REM $$$$$$$$$$$$$$$$$$$$$$$$$$$$$$$$$$$$$$$$$$$
494  REM     ADD THE SCALES
496  REM $$$$$$$$$$$$$$$$$$$$$$$$$$$$$$$$$$$$$$$$$$$$
500  HPLOT 0,1 TO 0,191
510  HPLOT 0,191 TO 279,191
530  HPLOT M,186 TO M,191
540 M = M + 20
560  IF M < 280 THEN  GOTO 530
570  HPLOT 0,14 TO 3,14
580  HPLOT 0,48 TO 5,48
590  HPLOT 0,95 TO 5,95
600  HPLOT 0,143 TO 5,143
610  STOP
```

Program 10.1

other, indicating that the smallest pinhole dimension was 292-wavelengths (0.000146-inches) and the largest, 401-wavelengths (0.000254-inches). The holes were pricked through a calling card to be as close to one another as possible. Side lobes from the individual apertures proved to be ragged but nevertheless very obvious when the spots were viewed one at a time.

There are several points of interest in this experiment. For one thing, it manifested the principle of pattern multiplication. It is obviously impossible to obtain fringes unless both apertures illuminate the same area. Second, the effects of refraction were remarkably evident even though the apertures were quite large in terms of wavelength compared to typical radio practice.

THE TWO-ELEMENT INTERFEROMETER

Program 10.1 is intended to synthesize the radio frequency equivalent of Young's experiment. It employs two apertures of 20-wavelength diameter separated by 120-wavelengths on the centerlines. The routine in lines 240 through 260 sets up an array with an aperture distribution consisting of 2 half sine wave beginning at element 192 and running to element 256. Elements 0 to 192 have zero excitation. Since the array is considered to be symmetrical about zero, it consists of 512-elements spaced 0.313-wavelengths aparts. The sine distribution of the excited "wings" of the aperture gives them a well-behaved radiation pattern with approximately −23-dB side lobes. This pattern is shown as a broken curve in Fig. 10.1(a).

Because of the wide separation between the two active apertures, the array presents a multilobed structure with very narrow lobes. As may be seen from the figure, this lobe structure lies effectively below the envelope of the individual apertures. In Fig. 10.1(b), the lobe structure is magnified 4 times to give a clearer picture of its details.

One mode of operation of a radio telescope is to hold it fixed and let stars drift through the field as governed by the rotation of the earth. It may be seen that an interferometer would be capable of resolving the position of a radio star much more precisely than individual apertures could because of the narrowness of the lobes in its overall structure. At the same time, using two small dishes or apertures would be a great deal less expensive than using one filled aperture of the same size. In effect, this antenna consists of only 128 out of a possible 512 radiators, for a 25-percent "fill factor."

Of course, it could be difficult to determine which of the narrow lobes the radio star was occupying. If the spacing between the apertures could

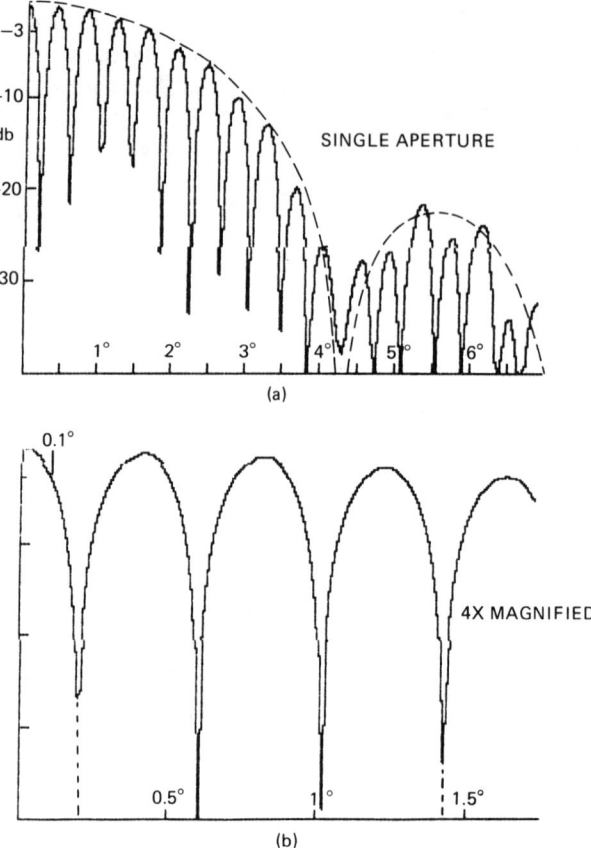

Fig. 10.1 Two-element interferometer, with $D = 20\lambda$ and $S = 140\lambda$.

be changed in order to make a second, somewhat different interferometer pattern available, the precise lobe could be distinguished since only one of them would match. Since the siderial rotation of the earth would place the same star in the same location 86,164.09-seconds later, an interferometer pass could be taken and recorded with an accurate time base. The spacing between antennas could then be changed and a second recording made the next day. The two points that tallied with a time lapse of 86,164.09-seconds could be interpreted as indicating the lobe of the system that the star was occupying. If the altered spacing was chosen correctly, no other lobe would coincide.

Data gathering on this relaxed scale is of relatively little value in the radar or communications fields but is quite suitable for astronomy, substituting, as it does, ingenuity for additional budget allocation. No matter how much one has to spend, an interferometer arrangement will always provide greater precision in locating a source than a single dish.

THE UNFILLED ARRAY

Although the two-element interferometer represents a neat way to locate a single source precisely, there are times when several sources are so close to one another, or when even a single source is so extended, that they span the gap between adjacent lobes on a wide interferometer. It seems logical to ask whether it would be possible to add a few antennas to obtain the very narrow beams desired rather than go to the cost of building a full-aperture antenna.

Program 10.2 is designed to treat an interferometer array with an arbitrary number of elements and an arbitrary spacing. Lines 310 through 325 represent an evaluation of an aperture 20-wavelengths in diameter and with a distribution designed to yield −23-db side lobes. Each element of the array is assumed to be an antenna of this sort. The program produces a pattern very similar to the envelope of Fig. 10.1, but it is much faster running.

Figure 10.2(a) shows the result for a thirteen-element array of 20-wavelength dishes. The dishes are spaced uniformly, 74.3-wavelengths apart, and cover a span of 892-wavelengths between the centerlines of the outer dishes. Figure 10.2(b) shows this pattern magnified four times to show some of the details.

The individual lobes have a beam width on the order of 0.1-degree, and the lobe spacing is approximately 0.76-degree. In contrast to the two-element interferometer, there is a large low area between the lobes that helps distinguish between closely spaced sources. In Fig. 10.2(a), the outer lobes fall off perceptibly as a result of the effects of the individual dishes, but this falloff does not really become pronounced until about 3.5-degrees off axis and beyond.

These lobes are usually referred to as *grating lobes* in analogy to an optical diffraction grating. They arise precisely because of the regular spacing between the elements. Every 0.77-degrees, the phase shift between adjacent dishes extends to 1-wavelength, and, as a result, the array generates a new major lobe with all elements contributing fully in-phase.

A logical question is whether the spacing can be made uneven in such a way that only a few elements will contribute at any angle except along the axis. Figure 10.3 shows such an arrangement.

This array is the same size as in the previous example, but the spacings

152 EXPLORING ANTENNAS AND TRANSMISSION LINES

```
FOR I= 1  X= 446
FOR I= 1  X= -446
FOR I= 2  X= -372
FOR I= 3  X= -297
FOR I= 4  X= -223
FOR I= 5  X= -149
FOR I= 6  X= -74
FOR I= 7  X= 0
FOR I= 8  X= 74
FOR I= 9  X= 149
FOR I= 10 X= 223
FOR I= 11 X= 297
FOR I= 12 X= 372
FOR I= 13 X= 446
```

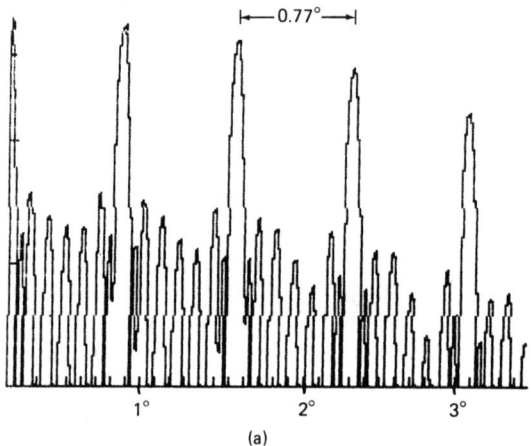

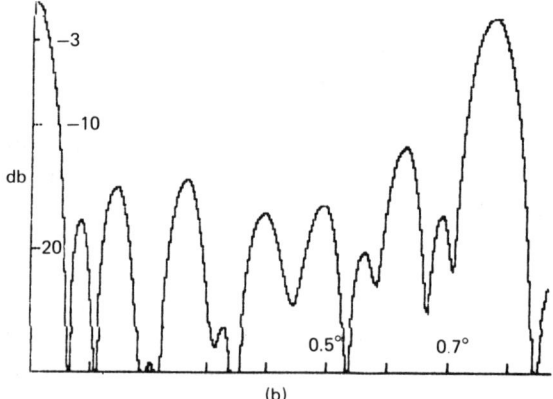

Fig. 10.2 (a) Thirteen-element array with uniform 74.3λ spacing, and (b) this pattern magnified four times.

```
200 N = 12
205 XS = 20
206  PRINT A0
210 PI = 3.1415926
215  DIM X(N,1)
220  FOR I = 1 TO N
230  PRINT "FOR I= ";I;" INPUT X(I,0)"
235  INPUT X(I,0)
240  PR# 1
245  PRINT "FOR I= ";I;" X= ";X(I,0)
250  PR# 0
255  NEXT I
300  FOR J = 0 TO 278
305 A0 = (PI / 14400) * J
307  IF A0 = 0 THEN E1 = 1: GOTO 320
310 A1 = 30.19 * A0
315 E1 = ( SIN (A1) / A1)
320 E1 =  ABS (E1)
325 E2 = (E1) ^ 1.78
327 ER = 1
330  FOR I = 1 TO N
335 A3 = 2 * PI * X(I,0) *  SIN (A0)
340 ER = ER +  COS (A3)
345 EI = EI +  SIN (A3)
350  NEXT I
351 E0 = E2 * ( SQR ((ER * ER) + (EI * EI)))
352 EI = 0:ER = 0
356  REM     FOR ONE ANGLE
358  REM ##########################################
360  IF J = 0 THEN EN = E0: GOTO 380
370  GOTO 430
380  HGR2
390  HCOLOR= 3
400  POKE  - 12524,0
410  POKE  - 12525,64
420  POKE  - 12529,255
422  REM ##########################################
424  REM     THE GRAPHICS AND PRINTER SETUP
426  REM %%%%%%%%%%%%%%%%%%%%%%%%%%%%%%%%%%%%%%%%%%
430 YA =  ABS (EN / E0)
432 YB = 20 * ( LOG (YA)) / 2.30259
434 Y = 6.333 * YB
436  IF Y > 191 THEN Y = 191
450  IF J = 0 THEN X0 = X:Y0 = Y
460  HPLOT X0,Y0 TO X,Y
470 X0 = X:Y0 = Y
475 X = X + 1
480 E0 = 1
490  NEXT J
492  REM $$$$$$$$$$$$$$$$$$$$$$$$$$$$$$$$$$$$$$$$$$
494  REM     ADD THE SCALES
496  REM $$$$$$$$$$$$$$$$$$$$$$$$$$$$$$$$$$$$$$$$$$
500  HPLOT 0,1 TO 0,191
510  HPLOT 0,191 TO 279,191
530  HPLOT M,186 TO M,191
540 M = M + 8
560  IF M < 280 THEN  GOTO 530
570  HPLOT 0,19 TO 3,19
580  HPLOT 0,63 TO 5,63
590  HPLOT 0,127 TO 5,127
610  STOP
```

Program 10.2

```
FOR I= 1 X= -381.33
FOR I= 2 X= -211.47
FOR I= 3 X= -117.27
FOR I= 4 X= -65.0400001
FOR I= 5 X= -36.06
FOR I= 6 X= -20
FOR I= 7 X= 0
FOR I= 8 X= 26.86
FOR I= 9 X= 48.43
FOR I= 10 X= 87.33
FOR I= 11 X= 157.48
FOR I= 12 X= 283.97
FOR I= 13 X= 512
```

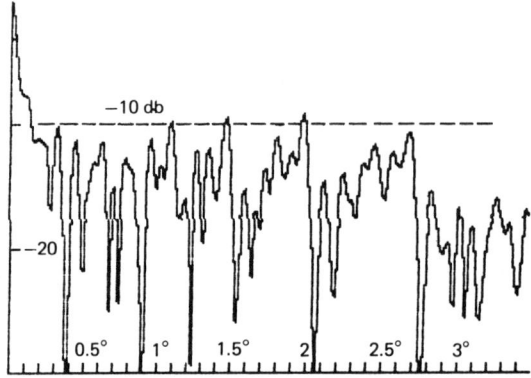

Fig. 10.3 Result of thinning the array.

have been chosen in geometric progression with alternate signs reversed. The widest spacing is 512 − 283.97, or 228.03-wavelengths. The pair of elements at this point have a first grating lobe at 0.251-degree. The most closely spaced pair, at 20-wavelengths, have a first grating lobe at 2.866-degrees. The remainder have their first grating lobes scattered in between. It may be seen that this strategy more or less successfully supresses the grating lobes, but at some expense in the side-lobe level, which rises slightly above −10-dB. If this side-lobe level can be tolerated, the fix is simple and straightforward. The fill factors for this array and also for that in Fig. 10.2 are (20 + 13)/(892 + 20), or 0.285.

The economic advantages of thinning an array to this extent are obvious. The principal limitation is the "lack of contrast" obtained. Suppose that one were looking at a single weak source surrounded by several stronger sources removed at an angle of 1 or 2-degrees. The latter

```
FOR I= 1 X= -251.83
FOR I= 2 X= -166.94
FOR I= 3 X= -106.62
FOR I= 4 X= -60.78
FOR I= 5 X= -26.05
FOR I= 6 X= 0
FOR I= 7 X= 22.04
FOR I= 8 X= 53.01
FOR I= 9 X= 92.52
FOR I= 10 X= 147.09
FOR I= 11 X= 221.03
```

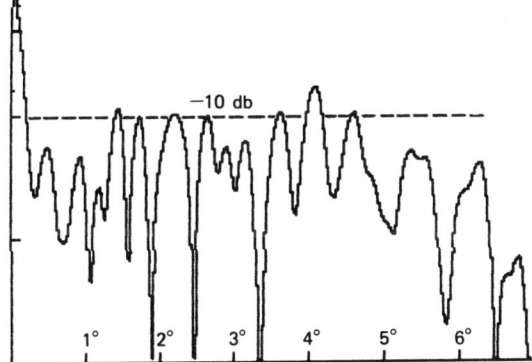

Fig. 10.4 Interleaved grating lobes.

could easily overpower the weak source in the main beam, thereby "blurring the picture."

Placing the elements in geometric progression is obviously only one approach, and perhaps somewhat simple-minded. Might some other approach, perhaps more sophisticated, yield a lower sidelobe background? The distribution of Fig. 10.4 was selected by considering pairs of antennas as individual interferometers and selecting their spacing so that the grating lobes interleaved and the grating lobes from no pair fell upon the grating lobes of any other pair with more than a 25-percent overlap, considering the beam width of 0.1-degree. Even though the 13-elements add up to a somewhat smaller array than that of Fig. 10.3, the side-lobe results are really no better. Because of the smaller span, however, the main lobe is somewhat broader.

It also seems logical to question whether a tapering of the illumination would help. (Note that there is a certain amount of tapering already present as a result of reserving the closer spacings for the center.) The

156 EXPLORING ANTENNAS AND TRANSMISSION LINES

```
FOR I= 1 X= -251.83
FOR I= 2 X= -166.94
FOR I= 3 X= -106.62
FOR I= 4 X= -60.78
FOR I= 5 X= -26.05
FOR I= 6 X= 0
FOR I= 7 X= 22.04
FOR I= 8 X= 53.01
FOR I= 9 X= 92.52
FOR I= 10 X= 147.09
FOR I= 11 X= 221.03

]LIST 330,351

330   FOR I = 1 TO N
335   A3 = 2 * PI * X(I,0) *  SIN (A0)
337   A4 = 11 -  ABS (6 - I)
340   ER = ER + A4 *  COS (A3)
345   EI = EI + A4 *  SIN (A3)
350   NEXT I
351   E0 = E2 * ( SQR ((ER * ER) + (EI * EI)))
```

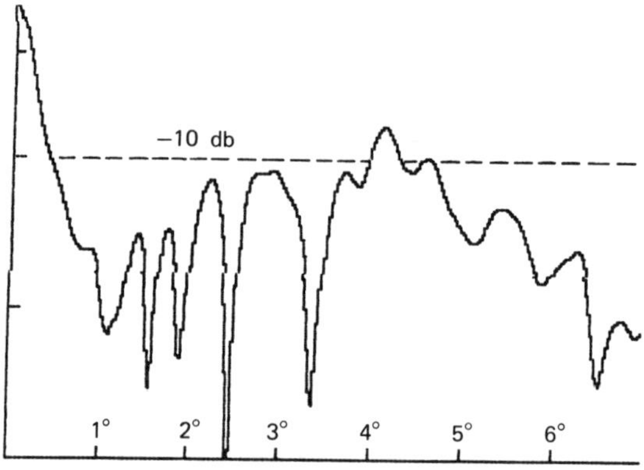

Fig. 10.5 Effect of illumination taper.

result shown in Fig. 10.5 reveals that aperture tapering offers little help. The main beam itself is thickened at the base, and the side lobes are reduced by only a negligible amount. The gaps caused by the missing elements seem to overpower any tapering effects.

Of course, the array exhibits high resolution only in the plane normal to the line of the array. In the other plane, the beam pattern is simply the pattern of a single dish. To overcome the difference, antennas of this kind are frequently constructed in the outline of a cross, a "*T*," or a "*Y*" so that high resolution may be obtained in two dimensions. In modern

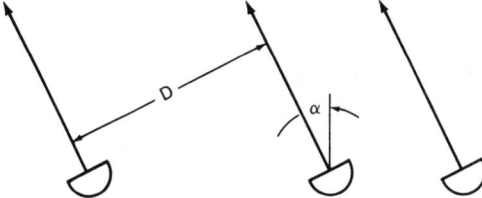

Fig. 10.6 The multi-dish array.

practice, such arrays—for example, the VLA (Very Large Array) telescope—may have dimensions measuring in miles.

The VLA located at Socorro, New Mexico, has three tracks in a "Y" configuration, with the span of the "Y" measuring something like 27-miles! The individual dishes are approximately 10-meters in diameter and equipped with polar mounts. The mount and base run on dual railroad tracks so that the dishes may be moved into different geometries as well as into a central shop for servicing. In addition to its huge size and corresponding high resolution, the arrangement provides a remarkable level of flexibility. It can be spaced and configured to optimize the performance for a specific measurement. In addition, the arrangement permits a "pay as you go" type of operation in which element dishes are added as they are paid for and constructed. Not all of them were needed to put the system into operation.

One point is worthy of note with respect to these arrays. Although the individual dishes are tracked to point in the direction of the main lobe of the array, doing so changes the array geometry as a result of the change in the angle off the axis of the array (see Fig. 10.6). The array can go from a broadside to a near endfire condition. The effect is somewhat different from that of the scanning array used in radar in that the element patterns swing to point at the target.

THE CORRELATION ANTENNA

In the arrays we have considered up to this point, the signals from the individual elements have combined additively—the natural action of a power-splitter feed—and the output has been an algebraic summation of the inputs—the natural operation of passive linear elements. This is not the only operation possible with electrical signals, however. Others become possible if we permit the use of active devices. In particular, we can multiply one signal by another.

Multiplication of one *ac* signal by another can be accomplished in a variety of ways. A double-balanced mixer can be used to multiply two *RF* signals. In fact, the mixing technique was the first to be practically applied. A Hall Effect device will output a current that is the product of an input current and a magnetic field strength and thus the product of two currents. Performing a log-and-antilog operation with Op-Amps will also achieve multiplication. Most recently, it has become possible to multiply *RF* signals with high-speed analog-to-digital converters.

Various digital techniques are now being applied. Perhaps the most sophisticated one involves digitizing the signals for manipulation in a high-speed computer. For astronomical purposes, the operation need not be performed in "real time," and thus extensive computer manipulation is not too onerous a burden.

Suppose, for example, that we were to construct an array of 12 dishes, each 20-wavelengths in diameter, and that these dishes are spaced in a geometric progression ranging from 20 to 512-meters. Next let us suppose that each dish has a receiver running with a single common local oscillator so that the coherence between them is retained. We then combine the outputs of the receivers in such a way that adjacent antennas form a two-element interferometer, for example, *AB, BC, CD, DE,....* For an output, we take the product as follows:

$$AB*BC*CD*...NN = EO.$$

This manipulation is performed in lines 330 through 352 of Program 10.3. The remainder of the program is identical to the previous multi-element program.

The output is shown in Fig. 10.7. Note the change in the vertical scale, which has been compressed to show a range of 50-dB. The improvement in side-lobe level compared to that in the additive arrangement is dramatic, with the near side lobes suppressed to -36-dB and the wider side lobes suppressed even more. The product combination of signals is obviously a powerful discriminant.

In some of the early mixing type of multipliers, the performance achieved was not nearly so dramatic because the multiplication was limited in dynamic range. In the all-digital techniques now being used, particularly those employing a digital computer, suppression of the side-lobes is almost always excellent.

Perhaps the most advanced elaboration of this technique is that of "extended baseline interferometry." In this procedure, two (or more) radio telescopes train on the same object and tape record their data. In order to obtain phase coherence, both must either be synchronized to the same frequency reference or else a time reference must be supplied in the

```
200 N = 12
210 PI = 3.1415926
215 DIM X(N,1)
220 FOR I = 1 TO N
230 PRINT "FOR I= ";I;" INPUT X(I,0)"
235 INPUT X(I,0)
240 PR# 1
245 PRINT "FOR I= ";I;" X= ";X(I,0)
250 PR# 0
255 NEXT I
260 REM &&&&&&&&&&&&&&&&&&&&&&&&&&&&&&&&&&&&&&&&&&
265 REM        START ARRAY CALCULATION
300 FOR J = 0 TO 278
305 A0 = (PI / 57600) * J
307 IF A0 = 0 THEN E1 = 1: GOTO 320
310 A1 = 30.19 * A0
315 E1 = ( SIN (A1) / A1)
320 E1 =  ABS (E1)
325 E2 = (E1) ^ 1.78
327 E0 = 1
330 FOR I = 1 TO N
335 A3 = PI * X(I,0) *  SIN (A0)
340 E3 = E2 *  COS (A3)
350 E0 = E0 * E3
352 NEXT I
356 REM     FOR ONE ANGLE
358 REM ###############################################
360 IF J = 0 THEN EN = E0: GOTO 380
370 GOTO 430
380 HGR2
390 HCOLOR= 3
400 POKE  - 12524,0
410 POKE  - 12525,64
420 POKE  - 12529,255
422 REM ###############################################
424 REM    THE GRAPHICS AND PRINTER SETUP
426 REM %%%%%%%%%%%%%%%%%%%%%%%%%%%%%%%%%%%%%%%%%%%%
430 YA =  ABS (EN / E0)
432 YB = 20 * ( LOG (YA)) / 2.30259
434 Y = 3.8 * YB
436 IF Y > 191 THEN Y = 191
450 IF J = 0 THEN X0 = X:Y0 = Y
460 HPLOT X0,Y0 TO X,Y
470 X0 = X:Y0 = Y
475 X = X + 1
480 E0 = 1
490 NEXT J
492 REM $$$$$$$$$$$$$$$$$$$$$$$$$$$$$$$$$$$$$$$$$$$$$
494 REM    ADD THE SCALES
496 REM $$$$$$$$$$$$$$$$$$$$$$$$$$$$$$$$$$$$$$$$$$$$$
500 HPLOT 0,1 TO 0,191
510 HPLOT 0,191 TO 279,191
530 HPLOT M,186 TO M,191
540 M = M + 32
560 IF M < 280 THEN  GOTO 530
570 HPLOT 0,11 TO 3,11
580 HPLOT 0,38 TO 5,38
590 HPLOT 0,76 TO 5,76
595 HPLOT 0,114 TO 5,114
600 HPLOT 0,152 TO 5,152
610 STOP
```

Program 10.3

160 EXPLORING ANTENNAS AND TRANSMISSION LINES

```
FOR I= 1 X= 20
FOR I= 2 X= 26.86
FOR I= 3 X= 36.06
FOR I= 4 X= 48.43
FOR I= 5 X= 65.0400001
FOR I= 6 X= 87.33
FOR I= 7 X= 117.27
FOR I= 8 X= 157.48
FOR I= 9 X= 211.47
FOR I= 10 X= 283.97
FOR I= 11 X= 381.33
FOR I= 12 X= 512
```

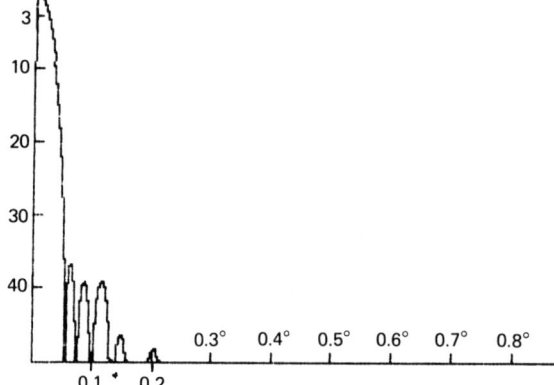

Fig. 10.7 The correlation pattern.

tape record. The records may then be coherently combined to provide an interferometer with a base line that may ultimately approach the diameter of the earth. Variations of this technique have produced radio maps with a resolution that exceeds that of the finest optical telescopes.

11
The Telegraphers Equation

The properties of light and radio waves can be explained in a variety of ways. Light can be considered to be a particle, a wave, or an electrical phenomenon. Each description has applications where it fits the observed facts the best. Beginning with Chap. 3 we have been considering light as a wave because this description best accords with the diffraction properties of antennas. In the next few chapters, we shall be considering light as an electrical phenomenon since this provides the easiest explanation for the physical phenomena actually observed. In this section, we shall deal with electrical circuits and discuss lumped and distributed circuit elements.

A circuit is said to consist of lumped elements if it is made up of resistors, batteries, condensers, and coils, all of which are visible and can be assigned a fixed location in space. A flashlight or a toaster is an example of a lumped element circuit. The distributed element circuit takes a little more explaining.

The invention of the telegraph in 1844 provided the impetus for the construction of the first long electrical lines. By the 1850s, a large number of fairly long telegraph lines had been constructed, and it was becoming apparent that a very long electrical circuit did not behave in the same manner as a short one. As noted in Chap. 1, the first trans-Atlantic telegraph cable was laid in 1858 by Cyrus Field. It failed electrically in a few weeks, but it managed to operate long enough to let the builders know that there was something radically different about that cable compared to a short telegraph line. Because of the mirror galvanometer invented by William Thompson, the sensitivity had not been a problem at all since a large deflection could be obtained. The problem came from the fact that at any rate faster than a few words per minute, the characters became hopelessly jumbled and the information could not be decoded. If the cable was to be a financial success, some way would have to be devised to speed up the transmission.

Thompson was appointed Engineer-in-Charge of a reorganized and refinanced cable company in an attempt to make the cable fast enough to be a financial success. His effort probably represents the first example of applying mathematical analysis to solve an electrical engineering problem prior to experimentation.

Thompson, following ideas suggested by Karl Friedrich Gauss, had been doing work on heat flow and transfer. He applied some of the same differential equations to the flow of electricity in long lines heavily loaded with capacitance. In 1861, Western Union completed the transcontinental telegraph line, solving the transmission problem by using repeaters to break up the line into a series of shorter lines. Since repeaters could not be powered and serviced upon the ocean floor, this solution was not applicable to trans-Atlantic transmission. The solution eventually discovered by Thompson was embodied in what has come to be known as Kelvin's Telegraphers Equation. The cable of 1866 was an immediate technical and financial success. Thompson was knighted Lord William Thompson Kelvin by Queen Victoria, and his work was to strongly influence James Clerk Maxwell.

Imagine a cable 3000-miles (4.83E6-meters) long. Furthermore, suppose for the sake of argument that the electricity traveled on the cable at half the speed of light, or 1.5E8-meters per second. When the telegraph key closed in Boston, it would take 0.0322-seconds for the signal to reach England. A fast telegrapher might open the key again if he were sending a dot. Obviously, something must have established the ratio of voltage-to-current in the line for at least the length of time it took the signal to get to the end of the line, and probably twice that much since the information would have to get back to the battery somehow.

Consider the circuit of Fig. 11.1 showing a transmission line of infinite length and uniform construction. A small portion of the line (ΔX) will experience some resistance and inductance in series. The resistance, of course, comes from the fact that all wires offer some resistance. The inductance comes from the fact that a current in the wire will establish a magnetic field in space. Delta X will also experience some shunt conductance, or leakage, and some shunt capacitance. The conductance stems from the fact that insulation between wires cannot be perfect. The capacitance stems from the fact that a potential difference between wires will establish an electric field in space. Since the line is uniform, all segments will have the same equivalent circuit. The resistance is measured in ohms, the inductance in henries, the conductance in mhos, and the capacitance in farads. The usual practice is to state these quantities "per meter."

The instant that key SW is closed, a current rushes into the line and attempts to charge the capacitor. One section must feed another, however, and the second section cannot experience a voltage until the first capacitor has acquired a charge.

Equations 11.1 and 11.2 show the change in voltage and the change in current across ΔX. In order to make the problem a little bit simpler, let

THE TELEGRAPHERS EQUATION

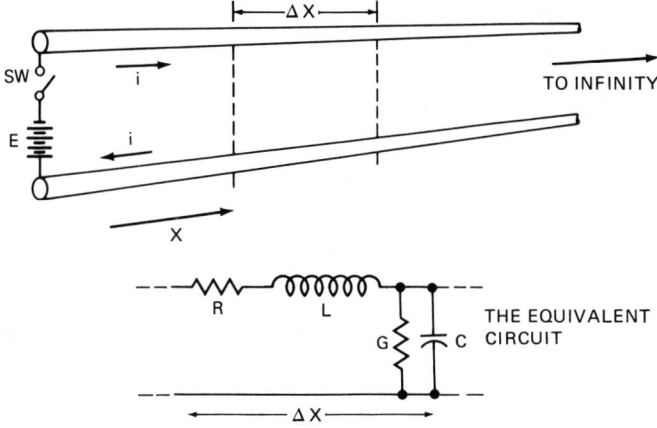

Fig. 11.1 The transmission line.

us assume for the moment that R and G are both zero. These volues produce Eqs. 11.3 and 11.4.

$$\frac{\delta E}{\delta X} = -\left(Ri + L\frac{di}{dt}\right) \text{ volts/meter} \quad (11.1)$$

$$\frac{\delta i}{\delta x} = -\left(GE + C\frac{dE}{dt}\right) \text{ amperes/meter} \quad (11.2)$$

For $R = 0$ and $G = 0$,

$$\frac{\delta E}{\delta x} = -L\frac{di}{dt} \quad (11.3)$$

$$\frac{\delta i}{\delta x} = -C\frac{dE}{dt} \quad (11.4)$$

Differentiating Eq. 11.3 with respect to t and Eq. 11.4 with respect to x gives

$$\frac{\delta^2 E}{\delta x \delta t} = -L\frac{d^2 i}{dt^2} \quad (11.5)$$

$$\frac{\delta^2 i}{\delta x^2} = -C\frac{d^2 E}{dt^2} \quad (11.6)$$

Rearranging Eqs. 11.5 and 11.6, we obtain

$$\frac{1}{C}\frac{\delta^2 i}{\delta x^2} = L\frac{\delta^2 i}{\delta t^2} \tag{11.7}$$

or

$$\frac{\delta^2 i}{\delta x^2} = LC\frac{\delta^2 i}{\delta t^2} \tag{11.8}$$

If we had reversed the order of differentiation, we would have obtained

$$\frac{\delta^2 E}{\delta x^2} = LC\frac{\delta^2 E}{\delta t^2} \tag{11.9}$$

It can be shown rigorously, although it should be more or less intuitively apparent, that the velocity at which the wave propagates is inversely proportional to the amount of inductance and also inversely proportional to the amount of capacitance. The larger that L is, the slower the current will rise, and the larger that C is, the slower the voltage will rise. Therefore,

$$LC = \frac{1}{V^2} \tag{11.10}$$

Substituting gives

$$\frac{\delta^2 i}{\delta x^2} = \frac{1}{V^2}\frac{\delta^2 i}{\delta t^2} \tag{11.11}$$

$$\frac{\delta^2 E}{\delta x^2} = \frac{1}{V^2}\frac{\delta^2 E}{\delta t^2} \tag{11.12}$$

Equations 11.11 and 11.12 are termed the *travelling wave equations*. They are second-degree differential equations with a general solution of the form

$$I = I_f\left\{t - \frac{x}{v}\right\} + I_r\left\{t + \frac{x}{v}\right\} \tag{11.13}$$

$$E = E_f\left\{t - \frac{x}{v}\right\} + E_r\left\{t + \frac{x}{v}\right\} \tag{11.14}$$

THE TELEGRAPHERS EQUATION 165

in which the braces imply a functional relationship, the specific function having not yet been specified.

The negative sign in the expressions is noteworthy. Given the coordinates of Fig. 11.1, it implies that whatever may happen to E_f or I_f will happen later the farther to the right we go since the x/v term will tends to cancel t; in other words, the "younger" they get. The opposite is true of E_r and I_r. The farther to the right we go, the "older" they get. It follows that the f subscripts represent forward-going waves—that is, from the source to the end of the line—and the r subscripts represent backward-going waves—that is, traveling right to left in Fig. 11.1.

These equations always seem to give a bit of trouble when one tries to visualize what they represent; at least, they have troubled the present writer. As an aid to visualization, Program 11.1 has been included. This will plot the traveling waves on your screen. If you question whether E_f goes right-to-left as it is supposed to do, simply set E_r equal to zero. The plot of Figure 11.2 is a bit of a cheat since it represents two screenfuls of data. The top half was run first, and then T was set equal to 144 and the program rerun. A bit of artistic splicing followed. The presentation was deliberately designed to yield a pseudo-three-dimensional effect. It is best seen by looking at it from a grazing angle. If one thinks of the display as

```
160   HGR2
170   HCOLOR= 3
180   POKE  - 12524,0
190   POKE  - 12525,64
200   POKE  - 12529,255
205 M = 20
210 TF = (T - X) / 20
220 TR = (T - 279 + X) / 20
230 EF =  EXP ( - 1 * (TF * TF))
240 ER =  EXP ( - 1 * (TR * TR))
250 E = 18 * (EF + ER)
260 Y = M - E
270   HPLOT X,Y
280 X = X + 1
290   IF X < 280 GOTO 210
320 T = T + 8
330 X = 0
340 M = M + 10
350   IF M < 191 GOTO 210
3000   END
```

Program 11.1

Fig. 11.2 Plot of the traveling waves.

showing two straight-fronted waves that cross about in the center, the illusion will be heightened.

In treating waves mathematically, they are often considered to be sinusoidal whereas nature imposes no such limitation. As a matter of fact, many naturally occurring waves are distinctly nonsinusoidal. Gravity waves in water approach a sinusoidal shape only in very deep water, and

Fig. 11.3 Destructive addition of two waves.

even then only approximately. The wave in the illustration represents a Gauss Error Function as plotted by lines 230 and 240. This is a nice function to illustrate since it is within 1 percent of being one cycle of a sinewave, but it is uniquely defined from negative to positive infinite time. As we shall see shortly, a pair of forward and reverse sinewaves produce a more sophisticated pattern. The singular occurrence of the

168 EXPLORING ANTENNAS AND TRANSMISSION LINES

wave in the Gauss Error Function makes the pattern much easier to comprehend.

Note that in the center of the illustration, the two oppositely directed waves pass through one another and add but are not altered by this crossing and go their separate ways as before. The phenomenon can be easily seen on water when two boat wakes cross one another, adding and cancelling as they pass but emerging unaltered.

The reverse wave need not have the same sign as the forward wave, as shown in Fig. 11.3, which was run with the algebraic sign of E_r reversed. In this case, the two waves add destructively, producing a complete cancellation in the center of the illustration. It is not hard to imagine how the waves of Fig. 11.2 could cross one another, add, and emerge unscathed, but the idea of the waves completely disappearing and then reconstitution themselves nothing is a little harder to swallow. The phenomenon is worth a little more explanation.

For convenience, let

$$\left\{ t - \frac{x}{v} \right\} = \left\{ a_f \right\} \tag{11.15}$$

$$\left\{ t - \frac{x}{v} \right\} = \left\{ a_r \right\} \tag{11.16}$$

Then, returning to Eqs. 11.4 and 11.5, we may write

$$\frac{\partial}{\partial x}(E_f + E_r) = -L\frac{\partial}{\partial t}(I_f + I_r) \tag{11.17}$$

$$\frac{\partial}{\partial x}(I_f + I_r) = -C\frac{\partial}{\partial t}(E_f + E_r) \tag{11.18}$$

But if can be shown that

$$\frac{\partial E_f}{\partial x} = -\frac{1}{V}\frac{dE_f}{da_f} \tag{11.19}$$

$$\frac{\partial E_r}{\partial x} = \frac{1}{V}\frac{dE_r}{da_r} \tag{11.20}$$

THE TELEGRAPHERS EQUATION

A similar manipulation is possible for I_f and I_r. Then,

$$\frac{1}{V}\left(-\frac{dE_f}{da_f} + \frac{dE_r}{da_r}\right) = -L\left(\frac{dI_f}{da_f} + \frac{dI_r}{da_r}\right) \tag{11.21}$$

$$\frac{1}{V}\left(-\frac{dI_f}{da_f} + \frac{dI_r}{da_r}\right) = -C\left(\frac{dE_f}{da_f} + \frac{dE_r}{da_r}\right) \tag{11.22}$$

Transposing gives

$$\frac{dI_f}{da_f} + \frac{dI_r}{da_r} = \frac{1}{LV}\left(\frac{dE_f}{da_f} - \frac{dE_r}{da_r}\right) \tag{11.23}$$

$$\frac{dI_f}{da_f} - \frac{dI_r}{da_r} = CV\left(\frac{dE_f}{da_f} + \frac{dE_r}{da_r}\right) \tag{11.24}$$

We had noted earlier, however, that $V^2 = 1/LC$ and thus,

$$V = 1/\sqrt{LC} \tag{11.25}$$

Thus,

$$\frac{1}{LV} = \frac{1}{\sqrt{L/C}} \tag{11.26}$$

and

$$CV = \frac{1}{\sqrt{L/C}} \tag{11.27}$$

The radical, $\sqrt{L/C}$, is termed the *characteristic impedance* of the line and is usually designated as Z_0. Adding and subtracting Eqs. 11.23 and 11.24 gives

$$2\frac{dI_f}{da_f} = \frac{2}{Z_0}\left(\frac{dE_f}{da_f}\right) \tag{11.28}$$

$$2\frac{dI_r}{da_r} = \frac{2}{Z_0}\frac{dE_r}{da_r} \tag{11.29}$$

Integrating gives

$$I_f = \frac{E_f}{Z_0} + K_1 \qquad (11.30)$$

$$I_r = \frac{E_r}{Z_0} + K_2 \qquad (11.31)$$

in which the K's are constants or dc terms that can be omitted, thus giving

$$I_f = \frac{E_f}{Z_0} \qquad (11.32)$$

$$I_r = \frac{E_r}{Z_0} \qquad (11.33)$$

The answer to the first of our questions is a bit surprising. The ratio of current-to-voltage in the line at the instant of key closure is determined by Z_0, which is in turn determined by L and C.

Now suppose that we have a uniform line that is infinitely long. Since it would take forever for the current from the sending end to reach infinity, the current would be determined by Z_0 for eternity. Moreover, since we see from Eq. 11.32 that its form is identical to that of Ohm's Law, the dimension of Z_0 has to be in ohms.

Let us now take this infinite line and cut off a 1-meter piece at the sending end. The remaining infinite segment would still behave electrically like a resistor of Z_0-ohms. If we were to terminate the 1-meter section in a resistor of Z_0-ohms, it would be impossible to distinguish the finite line from the infinite one by electrical measurement.

It is noteworthy that our equations also came up with an equation for a reverse wave. Is the reverse wave merely a meaningless mathematical by-product—one of those extra solutions that may be neglected? If not, then where does it come from? Well, in an infinite uniform line, a reverse wave would never occur; at least, it would never arrive. On a finite line terminated at Z_0 ohms, there would also be no reverse wave; its amplitude would be zero. However, in a line terminated at something other than Z_0, there will always be a reverse wave.

Let us consider our long telegraph line again. At the instant the operator closes the key, current races into the line, and a wave begins racing toward the far end with the current equal to the battery voltage divided by Z_0. Now, let us suppose that there is no termination at the far end. When the wave reaches the far end, the current must immediately drop to zero since no current can flow in an open circuit. The sending end, however, does not yet know about this; it is still happily pumping current into the line. What brings the situation back to equilibrium? The answer is the reverse wave. At the instant the wave strikes the open circuit, a reverse wave with the same voltage as the forward wave appears. Since the voltages are identical, the voltage doubles and the currents cancel. The reverse wave races back toward the sending and, doubling the voltage and cancelling the current along the way. Only when this wave arrives does the battery discover the open circuit. Program 11.2 and Fig. 11.4 illustrate this phenomenon.

If the far end of the line had been short-circuited (terminated at 0-ohms), the situation would be somewhat different. A short circuit

```
160  HGR2
170  HCOLOR= 3
180  POKE  - 12524,0
190  POKE  - 12525,64
200  POKE  - 12529,255
205 M = 5
210 TF = (T - X) / 20
220 TR = (T - 560 + X) / 20
230  IF TF > 0 THEN EF = 1: GOTO 240
235 EF = 0
240  IF TR > 0 THEN ER = 1: GOTO 250
245 ER = 0
250 E = 4 * (EF + ER)
260 Y = M - E
270  HPLOT X,M TO X,Y
280 X = X + 1
290  IF X < 280 GOTO 210
320 T = T + 24
330 X = 0
340 M = M + 10
350  IF M < 191 GOTO 210
3000  END
```

Program 11.2

172 EXPLORING ANTENNAS AND TRANSMISSION LINES

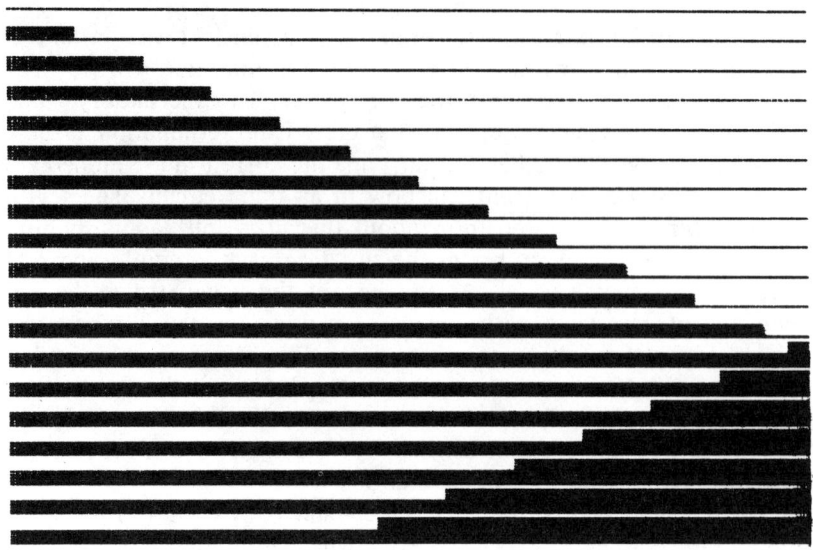

Fig. 11.4 Effect of the reverse wave in the unterminated telegraph line.

would support the current but not the voltage (one cannot develop much voltage across a short circuit). Accordingly, the reverse wave would be of equal magnitude but opposite polarity to the forward wave. It would race back to the sending end, doubling the current and cancelling the voltage along the way. If one considers the display of Fig. 11.4 to be current, then the line is short-circuited at the far end.

The original cable problem becomes a little clearer now. When the telegraph key was closed in Boston, the initial wave would rush to England. If the impedance of the galvanometer in England was much higher than Z_0, a wave of like polarity would race back to Boston. If by that time the key were open again, the wave would again race toward England. The result would be a jumble in which all of the echoes would be interfering with one another. One had to send a character and wait for all of the echoes to die out before sending the next. The correct solution, of course, was to terminate each end of the cable with a load equal to the cable characteristic impedance, thereby eliminating the echoes.

The echo problem becomes significant whenever the time scale of data transmission approaches the propagation time. For 100-MHz (or megabaud) digital data, a 1.5-meter cable will suffice to scramble data if the cable is not properly terminated.

VOLTAGE REFLECTION COEFFICIENT

Of course, all terminations are not as violently disruptive as an open or short circuit. A termination closer to Z_0 than 0 or infinity will have less reflection. The *voltage reflection coefficent* is defined as follows:

$$\Gamma = ER/EF \tag{11.34}$$

This can also be expressed by

$$\Gamma = (Z1 - Z0)/(Z1 + Z0) \tag{11.35}$$

in which gamma is a number less than unity. When the load impedance Z_1 is equal to Z_0, the voltage reflection coefficient goes to zero. In this case, the line is described as "matched."

In the next chapter we shall see that the voltage reflection coefficient may not only go from $+1$ to -1 but can also be a complex or imaginary number.

TIME DOMAIN REFLECTOMETERS

In recent years a class of instruments called *time domain reflectometers* (TDRs) have come into widespread use. These instruments essentially operate by placing a stepped voltage wave on the line and monitoring the echoes. Any change in L or C changes the Z_0 of the line and therefore presents a mismatched condition that will be responsible for a reverse wave. Pinched spots in the cable, cable fittings, etc., can all be detected quickly and painlessly with a TDR. This could be a horrendous task using older techniques, particularly when there was more than one discontinuity on the line.

Figure 11.5 shows the basic elements of a TDR and a simulated display. If we were checking a very long line and were not interested in resolving fine details, we could use a battery, a resistor, and a telegraph key, and we could do the monitoring with any old oscilloscope. However, let us suppose that we wished to be able to resolve discontinuities that were 3-cm apart. Light will travel 0.03-meters in 0.03/3E8-seconds, or 1E − 10-seconds, which is a pretty short time. The signal makes a round trip, but nonetheless the rise time for the pulse and the rise time on the oscilloscope must be on the order of 1E − 10-sec. In terms of frequency response, the oscilloscope would have to show a cutoff frequency in excess of 12.5-GHz! For this reason, the step generator actually keys

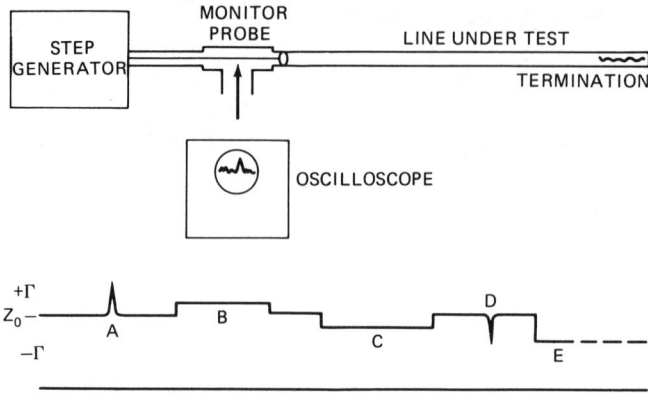

Fig. 11.5 Basic elements of the TDR and simulated display.

repetitive square waves with a very fast rise time, and a sampling oscilloscope is used.

In the display of Fig. 11.5, A represents a point with a little too much inductance, as in a connector in which the undersized segment of the pin is not fully swallowed. Since L is in the numerator of Z_0, this condition brings about an increase in Z_0, and the trace deflects upward. B and C represent line segments with a Z_0 that is greater than and less than the nominal Z_0. D is a point with excess capacitance, as in a pinched point or a poorly installed connector. E is a termination with the resistor smaller than Z_0, and the trace continues indefinitely at this height after the termination.

Particularly for cables installed in airplanes or ships where access to them is usually very difficult, the TDR is a great boon to the serviceman since he can quickly check out and locate any fault.

ATTENUATION

At the start of this discussion we stated that we would neglect the effects of R and G. If the values are small enough, we can often do this, but if they are not negligible, it is slightly different. Up to this point, we have not mentioned frequency, an omission that may have been a bit surprising since the equivalent circuit of Fig. 11.1 is identical to that of a low-pass filter. However, since a real line can be divided into arbitrarily small segments, the L and C both tend toward zero, and the cutoff frequency tends toward infinity. A practical upper limit in frequency for a low-loss

THE STEADY-STATE AC CASE   175

line arises when the dimensions begin to be comparable to a half wavelength.

When R and G are not negligible, it can be shown that the characteristic impedance becomes

$$Z0 = \text{SQR} ((R + jWL)/(G + jWC)) \qquad (11.36)$$

where $W = 2*PI*FREQUENCY$.

If the ratio of R to L is the same as the ratio of G to C, then the phase angles of numerator and denominator cancel, Z_0 remains a pure real number and the properties of the line are independent of frequency. The line attenuates the signal, but it doesn't change the response with change in frequency.

On the other hand if this relationship is not maintained, the Z_0 of the line does change with frequency, as does the propagational velocity. A line whose propagational velocity changes with frequency is said to be "dispersive." A pulse like that of a telegraph key closure contains a wide spectrum of frequencies. If they travel at different velocities, the pulse gets "smeared out" on the line and loses its sharp rise.

The early trans-Atlantic cable consisted of a single wire insulated with gutta percha and cloth wrapping and with steel armor on the exterior. The latter, along with the ocean, served as a return conductor. The leakage conductivity of the insulation was rather higher than that of a good modern insulation would be, and the capacitance was also quite large. To bring the inductance into line, Oliver Heaviside suggested that inductive "loaders" be periodically added. This reduced the dispersion of the cable and improved the rise time of the pulses so that high speed telegraph became possible. Modern telephone cables are frequently loaded with inductance to flatten the dispersion.

In the remainder of this section we shall be treating techniques of dealing with transmission lines and the problems of impedance matching.

12
The Steady-State AC Case

In Chap. 11 we treated the development of the telegraphers' equation and developed the concepts of wave equations, characteristic impedance, and propagation velocity. The signals considered were simple pulses like telegraph characters. In modern computer usage as employed in Ethernet and other networks, this is termed "baseband video."

In typical radio work, the cable is usually used for transporting some relatively narrow band of *ac*. In contrast to video applications, the bandwidth of the signal is usually narrow compared to the carrier frequency. In some applications like cable TV, however, a great many such signals may be present simultaneously and can propagate freely without interfering with one another.

Considering a single sine-wave frequency, the wave equations can be written (in computerese) as follows:

$$EX = EF*EXP(jW*(T - X/V)) + ER*EXP(jW*(T + X/V)) \qquad (12.1)$$

$$IX = (EF/Z0)*EXP(jW*(T - X/V)) - (ER/Z0)*EXP(jW*(T + X/V)) \qquad (12.2)$$

in which *EX* and *IX* represent, respectively, the voltage and current on the line at point *X*, *jW* is the radian frequency; and *T*, *X*, and *V* have the same definitions given in Chap. 11. The EXP(*jW*...) expressions are exponential notations based on Euler's equation as set forth in Eqs. 2.1 and 2.2.

Program 12.1 is a slight rehash of Program 11.1 to accommodate sine waves. Since a sine wave is defined to apply to any instant in time, any one instant will look like any other once things have phased out. In order to show how things develop on the line, lines 225 and 235 were added; they permit the sine waves to begin at a single instant.

Figure 12.1, starting from the top, shows two sine waves rushing toward one another. About a third of the way down, they collide, and an entirely new type of structure begins to build as the waves pass through one another. At the bottom, the standing wave pattern has filled the full width of the figure. Whereas at the top, the forward and backward waves were rushing to the right and to the left, respectively, at the bottom, the waves stand still and simply bounce up and down. The dashed lines represent the loci of points with no motion, or *standing wave nulls*. These

THE STEADY-STATE AC CASE 177

nulls are spaced exactly a half wave apart (for the wavelength in the line). The standing waves themselves arise from the same contrarotating vector situation we investigated in Chap. 4 (see Fig. 4.1).

If we move to the right (increasing X), the vector for the forward wave rotates backward (representing an earlier time), and the vector for the reverse wave rotates forward (representing a later time). In free space, the standing wave patterns may be any length greater than a half wavelength since the waves may approach from any angle. On a transmission line, however, the two waves are always traveling diametrically opposite, thus giving interference patterns with a half-wavelength spacing.

It seems fair to ask how the wave gets past the nulls since these points have been said to have permanently zero voltage. The answer is to be found in Eq. 12.2. You will note that there is a sign difference between the reverse and forward current waves since the direction of current flow is defined oppositely for the two waves. As was shown in Fig. 11.3, reversing the sign of the backward wave will replace a null with a loop, or maximum, and vice versa. If we imagine Fig. 12.1 to represent a voltage standing wave, we would find that the loops of the current standing wave would fall where the voltage exhibits nulls and the nulls of the current

```
160    HGR2
170    HCOLOR= 3
180    POKE  - 12524,0
190    POKE  - 12525,64
200    POKE  - 12529,255
205  M = 10
207  T = 0
210  TF = (T - X) / 20
220  TR = (T - 279 + X) / 20
225    IF TF < 0 THEN EF = 0: GOTO 235
230  EF =  SIN (TF)
235    IF TR < 0 THEN ER = 0: GOTO 250
240  ER =  SIN (TR)
250  E = 9 * (EF + ER)
260  Y = M - E
265    IF Y > 191 GOTO 280
270    HPLOT X,Y
280  X = X + 1
290    IF X < 280 GOTO 210
320  T = T + 8
330  X = 0
340  M = M + 10
350    IF M < 191 GOTO 210
3000   END
```

Program 12.1

Fig. 12.1 Standing waves.

standing wave would replace the loops in the voltage pattern. As a result, the energy is never zero at any point on the line.

It can be said that waves propagate by exchanging potential energy for kinetic energy and then exchanging kinetic energy back for potential energy. This is really true of all waves—waves in water, waves on the string of a musical instrument, sound waves, and light waves. Standing waves can be observed in each of these types. A cup of coffee resting on a vibrating machine will show a particularly striking "bullseye" pattern of standing waves. The peaks represent potential energy analogous to the voltage peaks on an electrical standing wave. Midway between the peaks the motion of the water will be maximized, representing a peak in kinetic energy analogous to the current peaks in an electrical standing wave.

Of course, the situation shown in Fig. 12.1 represents a special case in which the backward wave is equal to the forward wave.

The voltage reflection current can assume any value between $+1$ and -1 and can also be complex and have any phase angle—as determined by the termination. We are used to thinking of an impedance as something physical: a resistor, a capacitor, a motor, etc. In reality, however, an impedance is a ratio of voltage to current. In the standing wave, we see that the ratio of voltage to current varies with the position on the line. Indeed, if the voltage reflection coefficient has an absolute magnitude of unity (regardless of phase angle), we find that the voltage goes to zero at points where the current is finite, thus implying zero impedance (a short circuit). Conversely, at a quarter wave to either side, the voltage is maximum and the current is zero, implying an infinite impedance, or open circuit. The line impedance thus cycles between zero and infinite impedance in every quarter wavelength.

It is important not to confuse this impedance with the characteristic impedance of the line, or Z_0. Years ago, Z_0 was called the *surge impedance* of the line bacause it represents the impedance shown to the first surge of power on the line, before the standing wave has a chance to return. The only time that a transmission line will actually show an impedance exactly equal to Z_0 is when the line is terminated in a resistor in which $R = Z_0$ or an equivalent "matched" load. In this case, the line impedance will be equal to Z_0 everywhere along its length. In all other cases, a nonzero backward wave exists, and the impedance will cycle from a value higher than Z_0 to one lower than Z_0 without ever passing through Z_0.

Although the equations of Program 12.1 were set up to permit us to see the collision of the two waves, radio work is interested mainly in the "steady state condition" rather than the "instantaneous condition," just as in ordinary ac measurements, we compute voltages, currents, and

180 EXPLORING ANTENNAS AND TRANSMISSION LINES

impedances on the average rather than an instantaneous basis, understanding that we have omitted an EXP(jWT). On this basis, we may rewrite Eqs. 12.1 and 12.2 as follows:

$$EX = (EF*EXP(-jWX/V)) + (ER*EXP(jWX/V)) \quad (12.3)$$

$$IX = ((EF/Z0)*EXP(-jWX/V)) - ((ER/Z0)*EXP(jWX/V)) \quad (12.4)$$

Since the voltage cycles along the line, we may define a new parameter, the *voltage standing wave ratio*. Obviously, the voltage will be maximum at those points on the line where the forward and reverse waves add. Conversely, it will be minimum where they cancel. The voltage standing wave ratio is thus repressed as follows:

$$VSWR = (EF + ER)/(EF - ER) \quad (12.5)$$

But since $ER = VR*EF$, in which VR is the absolute magnitude of the voltage reflection coefficient, we thus have

$$VSWR = (1 + VR)/(1 - VR) \quad (12.6)$$

Since the British definition of the VSWR is the inverse of this relationship, British VSWRs are always numbers less than 1, whereas American VSWRs are always numbers greater than 1.

Now from Eq. 11.35 we have

$$ER = \Gamma*EF = ((Z1 - Z0)/(Z1 + Z0))*EF \quad (12.7)$$

Substituting in Eqs. 12.3 and 12.4 gives

$$EX = \frac{EF}{(Z1 + Z0)}*[((Z1 + Z0)*EXP)-jWX/V)) + ((Z1 - Z0)*EXP(jWX/V))] \quad (12.8)$$

$$IX = \frac{EF}{Zo(Z1 + Z0)}*[((Z1 + Z0)* EXP(-jWX/V)) - ((Z1 - Z0)*EXP(jWX/V))] \quad (12.9)$$

$$ZX = \frac{EX}{IX} = Z0*\left[\frac{((Z1 + Z0)*EXP(-jWX/V)) + (Z1 - Z0)*EXP(jWX/V))}{((Z1 + Z0)*EXP(-jWX/V)) - (Z1 - Z0)*EXP(jWX/V))}\right] \quad (12.10)$$

Now if we let $X = -\ell$ (because we wish to measure distance from the termination of the line) and observe that wavelength λ is equal to V/F, then

THE STEADY-STATE AC CASE

$$-\beta\ell = WX/V \quad \text{and} \quad \beta = \frac{2*PI}{\lambda} \qquad (12.11)$$

Thus, Eq. 12.10 becomes

$$ZX = Z0 * \left[\frac{((Z1 + Z0)*EXP(-j\beta\ell)) + ((Z1 - Z0)*EXP(j\beta\ell))}{((Z1 + Z0)*EXP(-j\beta\ell)) - ((Z1 - Z0)*EXP(j\beta\ell))}\right] \qquad (12.12)$$

However, from Euler's Equations we have

$$Z1(\varepsilon^{j\beta\ell} + \varepsilon^{-j\beta\ell}) = 2*Z1*COS(\beta\ell) \qquad (12.13)$$

$$Z0(\varepsilon^{j\beta\ell} - \varepsilon^{-j\beta\ell}) = 2*j*Z0*SIN(\beta\ell) \qquad (12.14)$$

Thus,

$$ZX = Z0 * \left[\frac{(2*Z1*COS(\beta\ell)) + (2*j*Z0*SIN(\beta\ell))}{(2*Z0*COS(\beta\ell)) + (2*j*Z1*SIN(\beta\ell))}\right] \qquad (12.15)$$

or,

$$ZX = Z0 * \left[\frac{(Z1*COS(\beta\ell)) + (j*Z0*SIN(\beta\ell))}{(Z0*COS(\beta\ell)) + (j*Z1*SIN(\beta\ell))}\right] \qquad (12.16)$$

Equation 12.16 is the transmission line equation for a lossless, uniform transmission line. Note that part of the expression is imaginary and that Z_1 can be real, complex, or imaginary, as can ZX. We shall next treat a few special cases that are quite illustrative.

LINE TERMINATED IN A SHORT CIRCUIT

For the shorted line, $Z_1 = 0$, the upper left and lower right terms in the expression are zero. Therefore,

$$ZX = Z0 \left(\frac{0 + (j*Z0*SIN(\beta\ell))}{Z0*COS(\beta\ell)) + 0}\right) \qquad (12.17)$$

or,

$$ZX = j*Z0*TAN(\beta\ell) \qquad (12.18)$$

182 EXPLORING ANTENNAS AND TRANSMISSION LINES

THE OPEN-CIRCUITED LINE

In the open-circuited line, Z_1 is infinite, and we thus have to do some manipulating first. If we divide each of the terms by Z_1, then the terms containing Z_1 become 1 and the terms containing Z_0 become zero. We thus have

$$ZX = Z0 * \left[\frac{(COS(\beta \ell)) + (j*(Z0/Z1)*SIN(\beta \ell))}{((Z0/Z1)*COS(\beta \ell)) + (j*SIN(\beta \ell))} \right] \quad (12.19)$$

$$ZX = Z0 * \left[\frac{COS(\beta \ell)}{j*SIN(\beta \ell)} \right] \quad (12.20)$$

$$ZX = -j*Z0*CTN(\beta \ell) \quad (12.21)$$

Now there is something interesting about these results. When we consider the trigonometric identity

$$TAN(A) = -CTN(A + PI/2) \quad (12.22)$$

We see that the result for the open circuit and the result for the short circuit are identical except for the position involved. Immediately at the short circuit, the impedance is zero, and it becomes so again a half wave away, whereas the zero impedance for the open circuit falls a quarter wavelength away and three quarters of a wavelength from the termination. At a single frequency, if the length of the line were unknown, it would therefore be impossible to determine whether a given line was terminated in an open or a short circuit. For situations in which more than one frequency can be used, however, it will be possible to determine the electrical length of the line from the impedance behavior at the input end.

Program 12.2 was written to solve Eq. 12.16. In order to illustrate the circulating nature of the line input impedance, Lines 660, 670, and 680 were added; these generate the tables of Figs. 12.2 and 12.3. For more normal usage, they may be deleted. The program section from 330 to 400 calculates real and imaginary parts of the numerator, sums the terms, and then performs the conversion to polar notation. Lines 450 through 510 do the same for the denominator. Lines 530 through 570 perform the division and conversion back to rectangular coordinates. It does not take a great deal of imagination to see that this is a substantial operation when performed with a sliderule, pencil and paper.

Figure 12.2 shows the results for a short-circuited termination with the line length incremented in one-sixteenth wave steps. The program has

```
145  HOME
150  PI = 3.1415926
160  Z0 = 50
170  PRINT "INPUT REAL PART OF LOAD"
180  INPUT ZR
185  IF ZR = 0 THEN ZR = .000001
190  PRINT "INPUT IMAGINARY PART OF LOAD"
200  INPUT ZI
205  IF ZI = 0 THEN ZI = .000001
210  PRINT "INPUT LINELENGTH IN WAVELENGTHS"
220  INPUT BL
230  PR# 1
250  PRINT " LINE OF ";Z0;" OHMS, WITH AN ELECTRICAL LENGTH OF ";BL;" WAVE
     LENGTHS"
260  IF ZI < 0 GOTO 290
270  PRINT "TERMINATED IN A LOAD OF ";ZR;" +J ";ZI;" OHMS."
280  GOTO 300
290  PRINT "TERMINATED IN A LOAD OF ";ZR;" -J "; - ZI;" OHMS"
300  PR# 0
310  BL = 2 * PI * BL
320  REM  ::::::::::::::::::::::::::::::::::::::::::::::
330  REM      THE CALCULATION
340  N1 = ZR *  COS (BL)
350  N2 = ZI *  COS (BL)
360  N3 = Z0 *  SIN (BL)
370  N4 = N2 + N3
380  N5 =  SQR ((N1 * N1) + (N4 * N4))
390  IF N1 = 0 THEN N1 = .000001
400  N6 =  ATN (N4 / N1)
450  D1 = Z0 *  COS (BL)
460  D2 = ZR *  SIN (BL)
470  D3 =  - ZI *  SIN (BL)
480  D4 = D1 + D3
490  D5 =  SQR ((D4 * D4) + (D2 * D2))
500  IF D4 = 0 THEN D4 = .000001
510  D6 =  ATN (D2 / D4)
530  IF D5 = 0 THEN D5 = .000001
540  Z2 = Z0 * (N5 / D5)
550  Z3 = N6 - D6
560  Z4 = Z2 *  COS (Z3)
570  Z5 = Z2 *  SIN (Z3)
590  PR# 1
600  IF Z5 < 0 GOTO 630
610  PRINT "THE INPUT IMPEDANCE IS ";Z4;" +J ";Z5;" OHMS"
620  GOTO 650
630  PRINT "THE INPUT IMPEDANCE IS ";Z4;" -J "; - Z5;" OHMS"
650  PRINT ""
660  BL = BL + PI / 8
670  PRINT "WITH AN ELECTRICAL LENGTH OF   ";BL / (2 * PI);" WAVELENGTHS"
675  PR# 0
680  IF BL <  = PI GOTO 340
3000 STOP
```

Program 12.2

184 EXPLORING ANTENNAS AND TRANSMISSION LINES

```
RUN
INPUT REAL PART OF LOAD
?0
INPUT IMAGINARY PART OF LOAD
?0
INPUT LINELENGTH IN WAVELENGTHS
?0
 LINE OF 50 OHMS, WITH AN ELECTRICAL LENGTH OF 0 WAVELENGTHS
TERMINATED IN A LOAD OF 1E-06 +J 1E-06 OHMS.
THE INPUT IMPEDANCE IS 9.99999999E-07 +J 1E-06 OHMS

WITH AN ELECTRICAL LENGTH OF  .0625 WAVELENGTHS
THE INPUT IMPEDANCE IS 1.18162304E-06 +J 20.7106785 OHMS

WITH AN ELECTRICAL LENGTH OF  .125 WAVELENGTHS
THE INPUT IMPEDANCE IS 2.01151237E-06 +J 49.9999997 OHMS

WITH AN ELECTRICAL LENGTH OF  .1875 WAVELENGTHS
THE INPUT IMPEDANCE IS 6.88700351E-06 +J 120.710676 OHMS

WITH AN ELECTRICAL LENGTH OF  .25 WAVELENGTHS          ←——— OPEN
THE INPUT IMPEDANCE IS 2.27220631E+09 +J 719440237 OHMS      CIRCUIT

WITH AN ELECTRICAL LENGTH OF  .3125 WAVELENGTHS
THE INPUT IMPEDANCE IS 6.8573422E-06 -J 120.710681 OHMS

WITH AN ELECTRICAL LENGTH OF  .375 WAVELENGTHS
THE INPUT IMPEDANCE IS 2.0022263E-06 -J 50.0000004 OHMS

WITH AN ELECTRICAL LENGTH OF  .4375 WAVELENGTHS
THE INPUT IMPEDANCE IS 1.17653396E-06 -J 20.7106793 OHMS

WITH AN ELECTRICAL LENGTH OF  .5 WAVELENGTHS
THE INPUT IMPEDANCE IS 1E-06 -J 1.66982554E-06 OHMS

WITH AN ELECTRICAL LENGTH OF  .5625 WAVELENGTHS
```

Fig. 12.2 The short-circuited line

been adjusted to keep the numbers within the range of the Apple so that the "short circuit" becomes merely very small and the "open circuit" becomes merely very large. However, it can be seen that the impedance cycles between the open and the short circuit every quarter wavelength, and the whole pattern repeats every half wavelength.

It may be seen that the impedance is essentially a pure reactance except at the short-circuit and open-circuit points. The reactance goes both positive and negative, or capacitive and inductive. Use is frequently made of this characteristic in impedance matching, particularly at very high frequencies and above. A shorted or open line can be used to generate an inductance or capacitance sufficient to match a line.

The data of Fig. 12.3 shows the impedance generated by a 50-ohm line terminated by a 25-ohm resistor. This calculates to a VSWR of 2:1. It may be seen that the impedance takes on resistive values ranging from

```
]RUN
INPUT REAL PART OF LOAD
?25
INPUT IMAGINARY PART OF LOAD
?0
INPUT LINELENGTH IN WAVELENGTHS
?0
 LINE OF 50 OHMS, WITH AN ELECTRICAL LENGTH OF 0 WAVELENGTHS
TERMINATED IN A LOAD OF 25 +J 1E-06 OHMS.
THE INPUT IMPEDANCE IS 24.9999997 +J 9.99970224E-07 OHMS

WITH AN ELECTRICAL LENGTH OF  .0625 WAVELENGTHS
THE INPUT IMPEDANCE IS 28.0846799 +J 14.8941513 OHMS

WITH AN ELECTRICAL LENGTH OF  .125 WAVELENGTHS
THE INPUT IMPEDANCE IS 40.0000003 +J 30.0000002 OHMS

WITH AN ELECTRICAL LENGTH OF  .1875 WAVELENGTHS
THE INPUT IMPEDANCE IS 69.4762968 +J 36.8453699 OHMS

WITH AN ELECTRICAL LENGTH OF  .25 WAVELENGTHS
THE INPUT IMPEDANCE IS 100 +J 0 OHMS

WITH AN ELECTRICAL LENGTH OF  .3125 WAVELENGTHS
THE INPUT IMPEDANCE IS 69.4762968 -J 36.8453699 OHMS

WITH AN ELECTRICAL LENGTH OF  .375 WAVELENGTHS
THE INPUT IMPEDANCE IS 40.0000004 -J 30.0000002 OHMS

WITH AN ELECTRICAL LENGTH OF  .4375 WAVELENGTHS
THE INPUT IMPEDANCE IS 28.0846799 -J 14.8941513 OHMS

WITH AN ELECTRICAL LENGTH OF  .5 WAVELENGTHS
THE INPUT IMPEDANCE IS 24.9999997 -J 1.00232747E-06 OHMS

WITH AN ELECTRICAL LENGTH OF  .5625 WAVELENGTHS
```

Fig. 12.3 Line with 2:1 VSWR.

half the line Z_0 to twice the line Z_0 at a quarter wavelength. As with the more extreme case, the impedance circulates between these values and repeats the pattern every half wavelength.

An interesting conclusion can be drawn from the data. Note that between 0.125 and 0.1875-wavelength and between 0.3125 and 0.375-wavelength the resistive component traverses 50-ohms, and the entire mismatch is reactive. If the line were interrupted at these points and a capacitor or inductor (respectively) of approximately 30-ohms reactance inserted, the mismatch would be canceled and the line would be "matched" thereafter. This question will be treated in the next chapter in somewhat greater detail.

Another property of the mismatched transmission line is that it works like a transformer. This property is frequently put to use in impedance matching. We noted that the 25-ohm load on the 50-ohm Z_0 line was transformed to 100-ohms a quarter wavelength away. Line Z_0 is the geometric mean between the lowest and highest resistances. If we had used a line with a Z_0 of 35.35-ohms, then a quarter wavelength away from the 25-ohm load, we would have obtained $50 + j0$-ohms. This type of matching is frequently put to use in microwave circuits, particularly those assembled in stripline, for which the odd Z_0s are easy to fabricate.

A typical quarter-wave transformer will have a modest bandwidth because segments cease to be a quarter wave when the frequency changes. However, multiple-step transformers can be designed for a bandwidth approaching 50-percent.

The transformer need not always be a quarter-wave long. Suppose that we have a $68 - j26$-ohm load. A 73-ohm line 0.125-wavelengths long will bring the impedance to $50 + j0$-ohms. If the original load had been $68 + j0$, we could have added the $-j26$-ohms in the form of a lumped capacitor or a line stub so that we could use the standard 73-ohm Z_0 line for a transformer.

Graphical and semigraphical solutions to impedance matching problems will be the subject of the next chapter.

13
The Smith Chart

Before the advent of the programmable calculator and the personal computer, the principal tool for solving transmission line problems was the Smith Chart, introduced by P. H. Smith of Bell Labs in 1939. This graphical solution of transmission line problems was a boon to telephone and radio engineers.

For the power utility engineer, the transmission line equations had to be taken into account only when working on long transmission lines, a hundred miles or more, and even then there was usually only one frequency to be considered. In these relatively infrequent cases, finding solutions for the transmission line equations was not too onerous.

For the telephone or radio engineer, on the other hand, the matter was more complex. The telephone man had to deal with a wide range of frequencies. The radio man was faced with the fact that even a few feet of co-axial cable or open pair could be many wavelengths long. Having to solve transmission line equations for each of a large number of frequencies was tedious and time-consuming. The swift and simple graphical solutions made possible by the Smith Chart were a welcome relief. Even today, with programmable calculators and computing facilities a common feature of most antenna and RF laboratories, the Smith Chart holds its position as the preferred method for displaying impedance/frequency plots.

An example of the usefulness of the Smith Chart can be obtained by considering the following questions:

1. With a given impedance or admittance, at what point along the transmission line will a lossless reactance cancel the reflected wave? What size reactance is required?
2. Having measured the impedance-frequency plot of a device at one point on a transmission line, what does the data look like at another point on the line?
3. What is a given impedance when transformed into admittance?

With the Smith Chart, questions 1 and 3 can very nearly be answered by inspection, and question 2 takes only a relatively small amount of calculation.

188 EXPLORING ANTENNAS AND TRANSMISSION LINES

Over the years the basic Smith Chart has been expanded to accommodate a variety of different scales for the purpose of solving different problems. Some of the major ones will be treated in this chapter. To begin with, the Smith Chart consists of a series of constant resistance and reactance circles (or they may be conductance and susceptance circles). These circles are orthogonal (they always cross at right angles), and they display every possible value of resistance and reactance. On a full (not expanded) Smith Chart, it is always possible to plot any impedance. Figure 13.1 shows a Smith Impedance Chart normalized to 50-ohms. Smith Charts can also be obtained in admittance form and normalized to 1-ohm or some other value.

The nature of these curves is illustrated in Fig. 13.2. As indicated by Fig. 13.2(a), all resistance and reactance circles pass through the infinity point ($R/Z_0 = \infty$). The chart can also contain constant VSWR circles, all of which are concentric about the center of the chart. The formula for locating these is given in Fig. 13.2(b). If a fixed impedance is used to terminate the line (here considered lossless), the locus of all impedances that would be seen in traveling along the line will appear as a constant VSWR circle on the chart. At the point where the circle crosses the real axis (the $X = 0$ line in the center), VSWR = R/Z_0. This point makes for an easy definition of the radius of VSWR circle C_R.

Programs 13.1 and 13.2 show a technique for having the Apple draw the resistance and reactance curves. Line 270 of the former establishes the relationship between R in the program (or A, in Fig. 13.2) and the resistance. In Program 13.2, line 250 establishes the relationship between the arc radius R_X (B_R in Fig. 13.2) and the reactance X_A. The curves plotted by Programs 13.1 and 13.2 are shown in Figs. 13.3 and 13.4, respectively.

The point should be made that the lack of Greek letters in the Apple and the lack of lower-case letters in BASIC have made it necessary to depart from traditional notation in many instances. An attempt has been made to choose a substitute notation on a logical basis, but it has not always been possible because of various conflicts.

Although not a feature of the original Smith Chart, polar notation is sometimes useful, especially when working with instruments such as a vector voltmeter or a vector impedance meter. Programs 13.3 and 13.4 plot such curves, which are shown in Figs. 13.5 and 13.6. At first glance, one might be tempted to consider the phase contours meridians, but a bit of study will tell us that they are not. Although the zero-degree and ninety-degree curves are meridians, it is obvious that the 75-degree contour is well spaced from the 90-degree contour, whereas the 75th meridian would lie nearly on top of it (79.5 versus 96.6-percent of the radius).

THE SMITH CHART 189

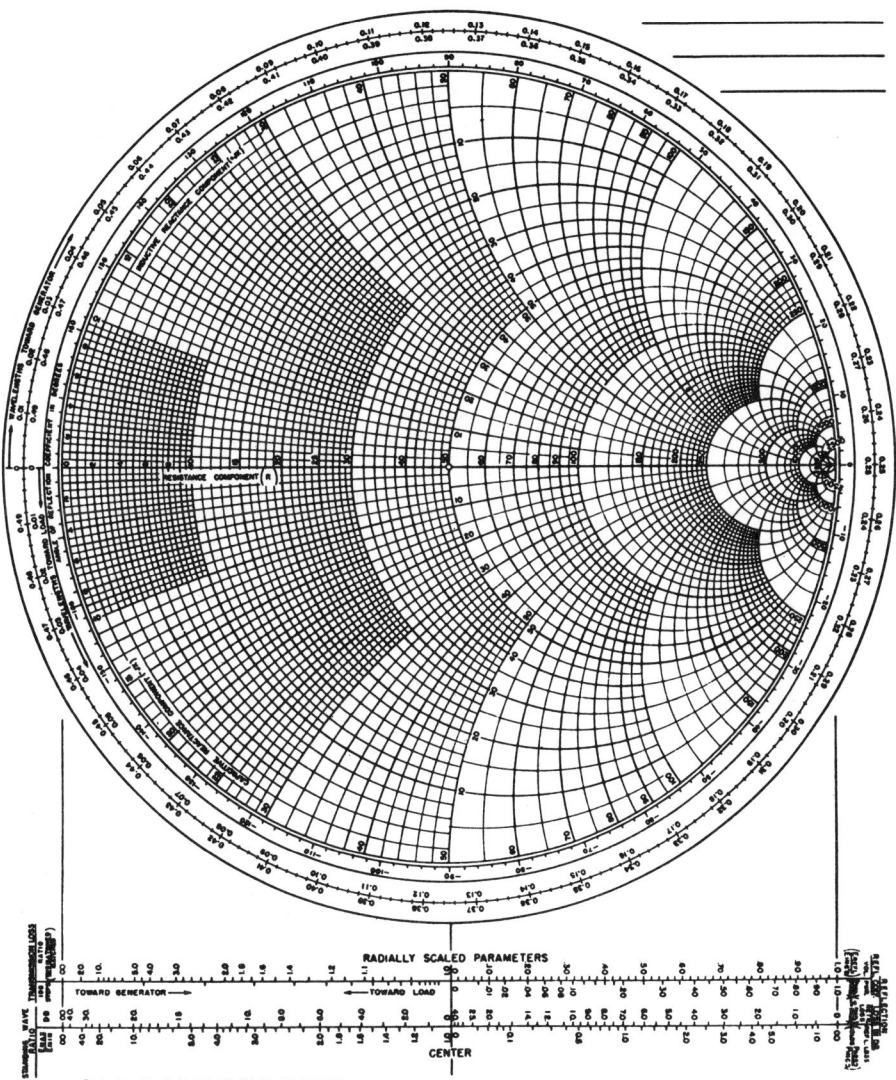

Fig. 13.1 The Smith Chart.

190 EXPLORING ANTENNAS AND TRANSMISSION LINES

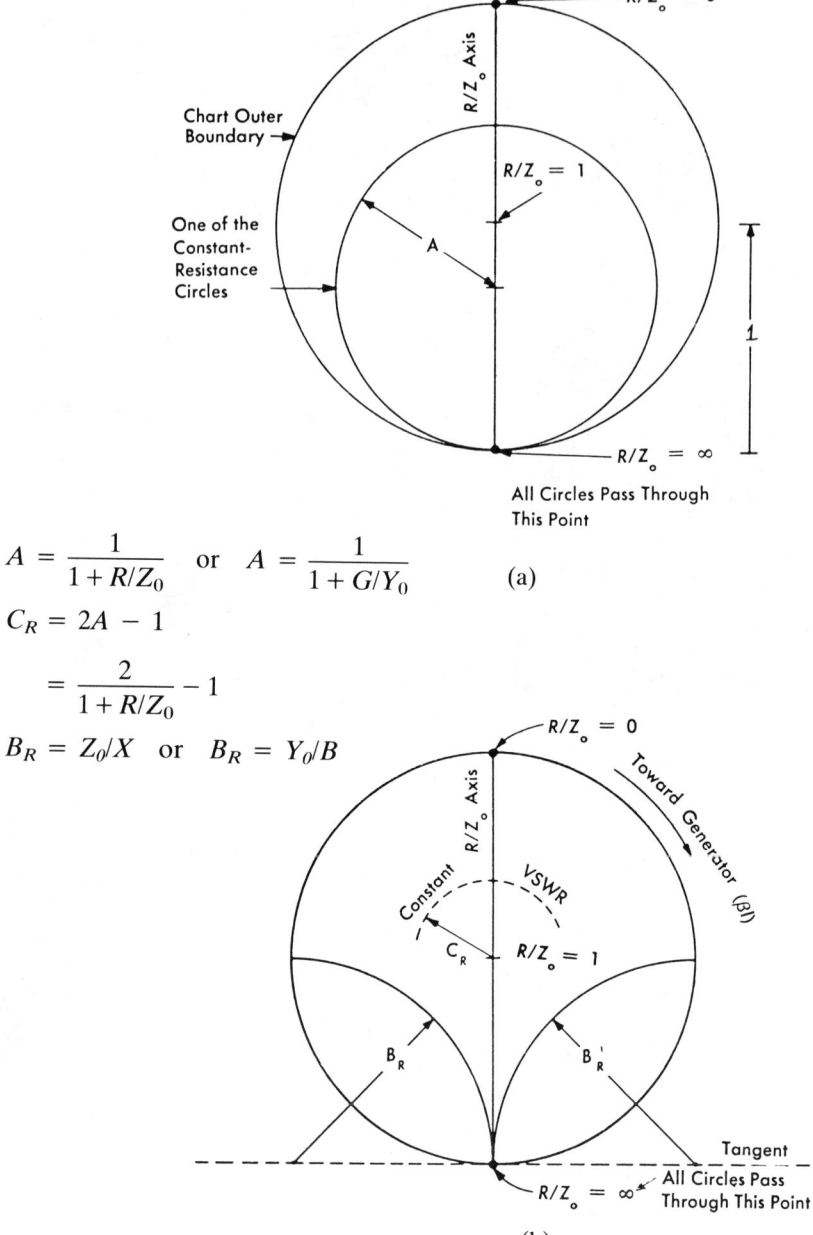

$$A = \frac{1}{1 + R/Z_0} \quad \text{or} \quad A = \frac{1}{1 + G/Y_0} \quad \text{(a)}$$

$$C_R = 2A - 1$$

$$= \frac{2}{1 + R/Z_0} - 1$$

$$B_R = Z_0/X \quad \text{or} \quad B_R = Y_0/B$$

Fig 13.2. Make up of the Smith Chart.

THE SMITH CHART 191

```
135  HGR2
140  HCOLOR= 3
150  POKE  - 12524,0
160  POKE  - 12525,64
190  POKE  - 12529,255
195 PI = 3.14159265
260 A = 0
270 R = 96 / (1 + RE / 50)
275  FOR A = 0 TO 2 * PI STEP PI / 36
276 Y = 192 - R * (1 + SIN (A))
277 X = 140 + R * COS (A)
278  IF Y = 192 THEN Y = 191.9
280  IF A = 0 THEN XA = X:YA = Y
290  HPLOT XA,YA TO X,Y
300  XA = X:YA = Y
310  NEXT A
320  IF RE < 50 THEN RE = RE + 10: GOTO 270
330 RE = RE + 25
340  IF RE < 325 GOTO 270
350  HPLOT 138,94 TO 142,94 TO 142,98 TO 138,98 TO 138,94
3000  STOP
```

Program 13.1

```
135  HGR2
140  HCOLOR= 3
150  POKE  - 12524,0
160  POKE  - 12525,64
190  POKE  - 12529,255
195 PI = 3.14159265
240  FOR A = 0 TO 2 * PI STEP PI / 120
245  IF XA = 0 THEN XA = .000001
250  RX = 96 * 50 / XA
260 Y = 192 - RX * SIN (A)
265  IF Y =  > 192 THEN Y = 191.9
270  XL = 140 - RX + RX * COS (A)
280  XR = 140 + RX - RX * COS (A)
290  DY = Y - 96
300  DX = 140 - XL
310  RA =  SQR ((DY * DY) + (DX * DX))
315  IF RA > 96.1 GOTO 400
330  IF A = 0 THEN Y0 = Y:M = XL:N = XR: NEXT A
340  HPLOT M,Y0 TO XL,Y
350  HPLOT N,Y0 TO XR,Y
360  Y0 = Y:M = XL:N = XR
370  NEXT A
400  IF XA < 50 THEN XA = XA + 10: GOTO 240
410  XA = XA + 25
420  IF XA < 350 GOTO 240
430  HPLOT 140,0 TO 140,191
440  HPLOT 138,94 TO 142,94 TO 142,98 TO 138,98 TO 138,94
3000  STOP
```

Program 13.2

192 EXPLORING ANTENNAS AND TRANSMISSION LINES

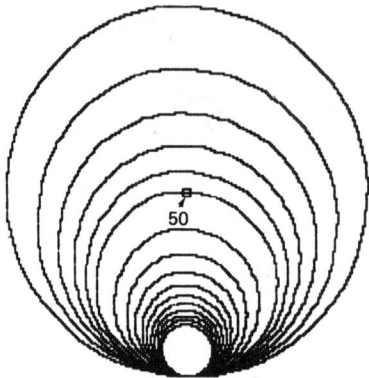

Fig. 13.3 Resistance curves plotted by Program 13.1: 10-Ω steps for $R < 50$ and 25-Ω steps for $R > 50$.

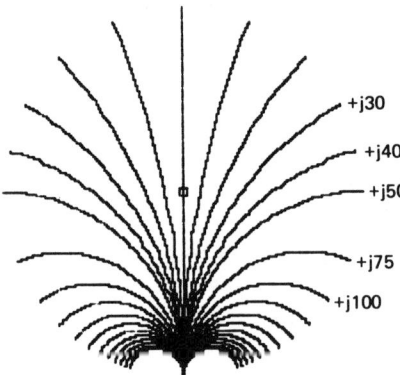

Fig. 13.4 Reactance curves plotted by Program 13.2.

We must also be careful not to confuse the phase angles in these two figures with the angular scale around the periphery of the Smith Chart. They are conventional phase angles in the low-frequency sense of relating to the power factor of the point. The scales around the periphery of the chart, on the other hand, represent transmission-line rotation measured in fractions of a wavelength.

```
135  HGR2
140  HCOLOR= 3
150  POKE  - 12524,0
160  POKE  - 12525,64
190  POKE  - 12529,255
200 RA = 95: REM   RADIUS OF CHART
205 Z0 = 50
210 PI = 3.14159265
215 R = 1
220 TH = PI / 12
230 X = R * TAN (TH)
240 A = RA / (1 + R / Z0)
250   IF X = 0 THEN X = .000001
260 B = RA * Z0 / X
270 A1 =  ATN (A / B)
280 Y = 191 - B * SIN (2 * A1)
290 XL = 140 - B + B * COS (2 * A1)
300 XR = 140 + B - B * COS (2 * A1)
310   IF S = 0 THEN Y0 = Y:X1 = XL:X2 = XR
320   HPLOT X1,Y0 TO XL,Y
330   HPLOT X2,Y0 TO XR,Y
335 R = R * 1.5
350 Y0 = Y:X1 = XL:X2 = XR
352 S = 1
355   IF R < 2000 GOTO 230
360 TH = TH + PI / 12
362 R = 1
370   IF TH < PI / 2 THEN S = 0: GOTO 230
375 TH = 0
380 Y = 96 + 95 * COS (TH)
390 XL = 140 - 95 * SIN (TH)
400 XR = 140 + 95 * SIN (TH)
410   IF TH = 0 THEN Y0 = Y:X1 = XL:X2 = XR
420   HPLOT X1,Y0 TO XL,Y
430   HPLOT X2,Y0 TO XR,Y
440 Y0 = Y:X1 = XL:X2 = XR
450 TH = TH + PI / 36
460   IF TH < (2.10 * PI) GOTO 380
3000   STOP
```

Program 13.3

Traditionally, line length has always been described in wavelengths rather than in degrees or radians. Since the standing wave pattern repeats itself every half wavelength, The circumference of the Smith Chart represents a half wavelength, and the distance between an open circuit ($R/Z_0 = \infty$) and a short circuit ($R/Z_0 = 0$) is a half rotation of the chart, or a quarter wavelength. Direction is described as being "toward the generator" or "toward the load".

```
135  HGR2
140  HCOLOR= 3
150  POKE  - 12524,0
160  POKE  - 12525,64
190  POKE  - 12529,255
200 RA = 95: REM  RADIUS OF CHART
205 Z = 10
210 PI = 3.14159265
215 Z0 = 50
217 DR = .2
220  IF R > Z GOTO 490
230 X =  SQR (Z ^ 2 - R ^ 2)
240 A = RA / (1 + R / Z0)
250  IF X = 0 THEN X = .000001
260 B = RA * Z0 / X
270 A1 =  ATN (A / B)
280 Y = 191 - B * SIN (2 * A1)
290 XL = 140 - B + B * COS (2 * A1)
300 XR = 140 + B - B * COS (2 * A1)
310  IF S = 0 THEN Y0 = Y:X1 = XL:X2 = XR
320  HPLOT X1,Y0 TO XL,Y
330  HPLOT X2,Y0 TO XR,Y
340 R = R + DR
345 S = 1
350 Y0 = Y:X1 = XL:X2 = XR
360  GOTO 220
490 DR = DR * 1.5
495  HPLOT XL,Y TO XR,Y
500  IF Z < 55 THEN Z = Z + 10:R = 0:S = 0: GOTO 220
510 Z = Z + 25:R = 0:S = 0
520  IF Z < 210 GOTO 220
3000  STOP
```

Program 13.4

USING A SMITH CHART

To use a Smith Chart manually, one simply takes the given impedance or admittance of a line and locates it on the chart at the appropriate intersection with its R and jB component. The chart used must be one having the same characteristic impedance or admittance as that of the line. To find the impedance at any other location on the line, one simply rotates this point about the center of the chart (corresponding to a constant VSWR circle) through the appropriate number of wavelengths toward or away from the generator. To illustrate, let us consider a specific example of Question 1 at the beginning of this chapter. Given a load admittance of

THE SMITH CHART 195

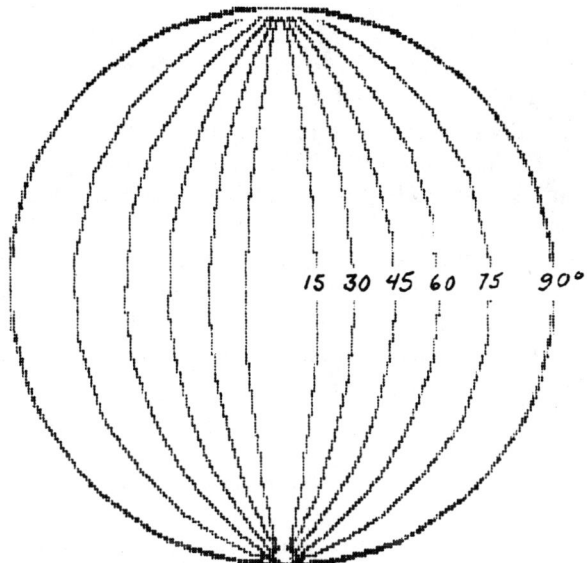

Fig. 13.5 Curves plotted by Program 13.3.

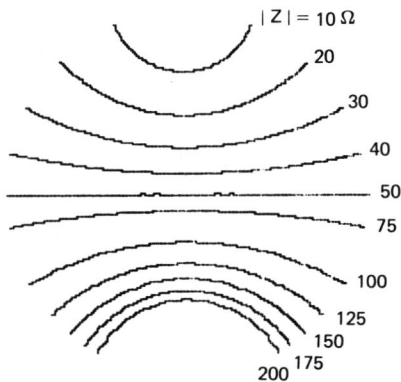

Fig. 13.6 Curves plotted by Program 13.4.

$Y_1 = 10 + j2$-mmhos, at what point on a 50-ohm line will a shunt stub exactly match the load, and what is the shunt-matching element?

To begin with, since we are interested in a shunt-matching element, we will use the admittance chart for 20-mmho. Note that a shunt-matching element is usually easiest to install on a co-axial line since it can be done with standard "Tee" connectors. Conversely, a series-matching element usually requires a special box to carry the outer conductor around the matching element.

In Fig. 13.7, having located Y_1 as shown, we must rotate it from the original plot point at 0.02-wavelengths through 0.134-wavelengths to place it on the 20-mmho conductance curve. At this point, the admittance is $20 + j14.2$-mmho. Obviously, a shunt inductance of $-j14.2$-mmho will perfectly match the load at this point.

The Smith Chart can also be used to determine the required stub length. Presuming that the stub will consist of a length of 50-ohm (20-mmho) line of the same type used for the main transmission line, we start at a short circuit ($Y = $ infinity) and proceed around the periphery of the chart till we arrive at a susceptance of $-j14.2$-mmho—a distance of 0.152-wavelengths. This is the length of 50-ohm stub that will match the load impedance. The net circuit is shown at the bottom of Fig. 13.7.

An important point is to be noted here. Let us suppose that the load impedance remains a constant $10 + j2$-mmho. At a slightly lower frequency, with the same physical placement, the line length in wavelengths between the load and the stub would be shorter, and the conductance would be less than 20-mmho. In addition, the susceptance of the stub would be sufficiently higher to be pushed across the center, or $B = 0$, line. The matching network has a finite bandwidth. In general, the shorter the line length involved, the broader the bandwidth will be. If the load-to-stub length were increased by a half wavelength to 0.634-wavelengths, the same stub would match the load, but the bandwidth would be much smaller.

THE COMPUTER SMITH CHART

Since a personal computer can alleviate the drudgery of solving the transmission line equations, one is tempted to ask why the Smith Chart remains popular. The answer is that it presents data in an organized way that the eye can easily grasp. Most people cannot percieve patterns nearly as well by looking at tables of numbers and as they can be looking at graphic displays. Perception of patterns in data can save a great deal of analytical time, as will be demonstrated shortly.

THE SMITH CHART 197

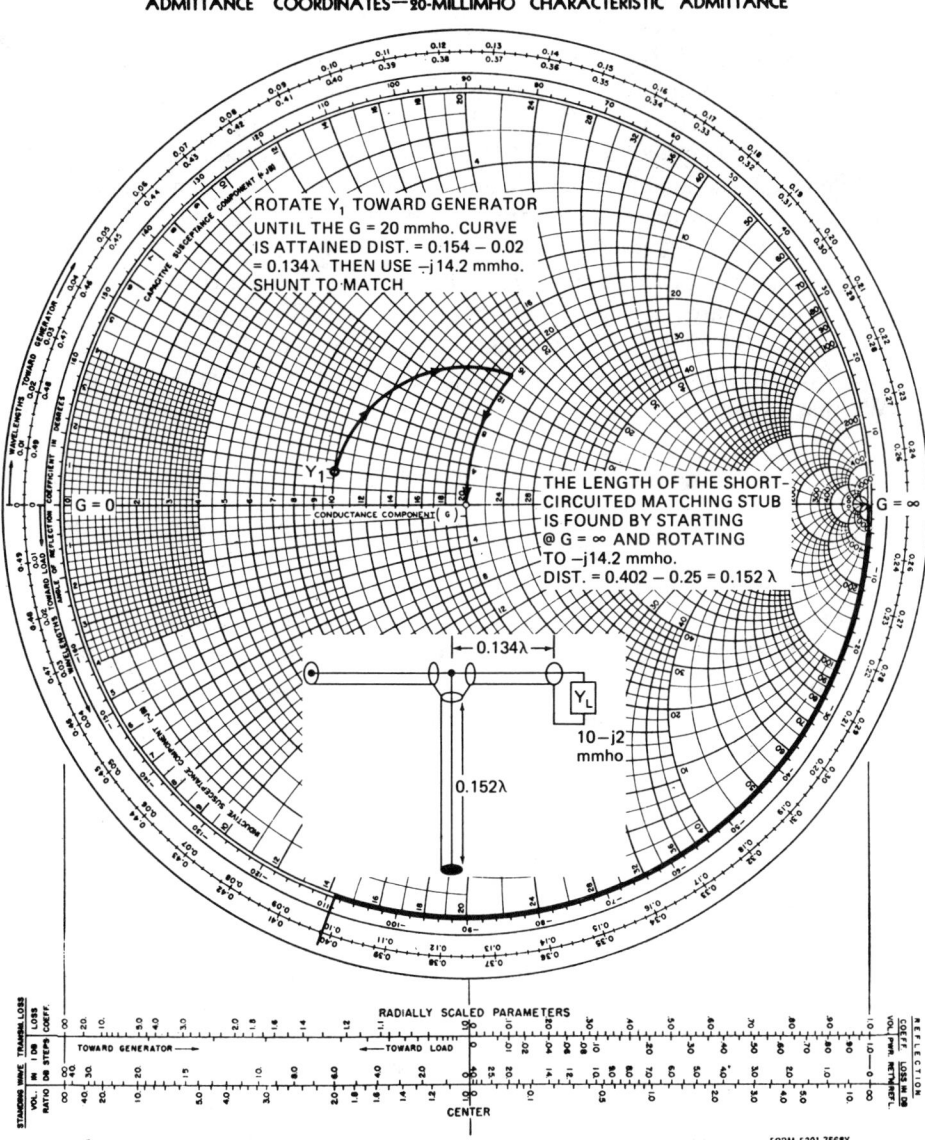

Fig. 13.7 Using the Smith Chart to find a shunt-matching element.

The principal effort involved in plotting Smith Chart impedance (or admittance) data stems from the rather odd coordinates of the Smith Chart as defined by Fig. 13.2. Most plotters require rectangular (Cartesian) coordinates. The problem is enhanced slightly by the coordinate set used by the Apple's HGR-2 screen, in which 0,0 represents the upper left-hand corner and 280,191 the lower right-hand corner. The Apple is unforgiving about straying outside this rectangle. Such a faux pas will stop the program cold and solicit an ILLEGAL QUANTITY ERROR IN XXXX message.

For a given impedance or admittance the, "spot" belongs at the intersection of the arcs swung by B_R and A in Fig. 13.8 (see also Fig. 13.2). Triangles *CDE* and *CDF* are congruent by reason of three equal sides, and since the angle at *E* is a right angle, the angle at *F* must be one also. We thus obtain

$$dY = BR*SIN(2*H1) \tag{13.1}$$

$$dX = BR - BR*COS(2*A1) \tag{13.2}$$

and the coordinates of point *F* are $X_F = 140 + dX$ and $Y_F = 191 + dY$.

NOTE: A variety of schemes will suffice to prevent plotting off the screen. If the chart radius is 95 and the plot starts with point *E* at 140, 191, the plot will usually stay on screen since it will take a truncation error of 0.000001 to trip the illegal quantity response.

Program 13.5 was written to solve one of the most common problems involving the use of the Smith Chart—as exemplified by Question 2 at the beginning of the chapter.

It is often the case, particularly at higher frequencies, to be unable to measure an antenna's impedance at the antenna proper. (You are on the ground and the antenna is in the air, etc.) In order to do anything meaningful about impedance matching, you must know the impedance at the antenna. Conversely, the impedance at the antenna may be known from previous measurements, and you need to determine the impedance that will be present at the end of the cable in the radio room.

In Program 13.5, lines 200 through 530 are run last. They draw a rudimentary Smith chart over the data points previously plotted. Lines 622 through 792 solve the geometry of Fig. 13.8. Lines 800 through 880 correct the line length. Lines 890 through 930 print out the corrected VSWR and the wavelengths from the generator for the corrected data. Note that the latter can be validly and easily transferred to a large-scale (and thus more accurate) Smith Chart. Lines 1000 through 1070 plot the data as a series of points (actually small squares).

The program is fully prompted and asks for each data item as it is required. It first asks for the number of the points you wish to plot (line 622) and then for the Z_0 of the line. The next request is for the *electrical length* of the line in meters (there are 39.37 inches in a meter). The electrical length is the physical length corrected for the velocity of propagation. To make this correction for a teflon co-ax, divide its

```
200 R = 96
205 Y = 96 + R * SIN (A)
210 X = 140 + R * COS (A)
220  IF A = 0 THEN XA = X:YA = Y
230  HPLOT XA,YA TO X,Y
235 XA = X:YA = Y
240 A = A + .02
250  IF A < 6.28 THEN  GOTO 205
255 A = 0
260 R = 48
270 Y = 144 + R * SIN (A)
280 X = 140 + R * COS (A)
290  IF A = 0 THEN XA = X:YA = Y
300  HPLOT XA,YA TO X,Y
310 XA = X:YA = Y
320 A = A + .02
330  IF A < 6.28 THEN  GOTO 270
340  HPLOT 140,190 TO 140,0
350 A = 0
360 R = 96
370 Y = 191 - R * SIN (A)
380 XL = 44 + R * COS (A)
390 XR = 236 - R * COS (A)
400  IF A = 0 THEN X1 = XL:X2 = XR:Y2 = Y
410  HPLOT X1,Y2 TO XL,Y
420  HPLOT X2,Y2 TO XR,Y
430 X1 = XL:X2 = XR:Y2 = Y
440 A = A + .02
450  IF A < 1.57 THEN  GOTO 370
455 :A = 0
460 R = 32
470 Y = 96 + R * SIN (A)
480 X = 140 + R * COS (A)
490  HPLOT X,Y
500 A = A + .2
510  IF A < 6.28 THEN  GOTO 470
530  STOP
```

Program 13.5

```
600 REM ******************************************
610 REM       THIS IS THE DATA ROUTINE
620 REM (((((((((((((((((((((((((((((((((((
621 HOME
622 PRINT "ENTER NUMBER OF POINTS": INPUT N
623 PRINT "ENTER LINE Z0": INPUT Z0
624 PRINT "INPUT LINE ELECTRICAL LENGTH": PRINT "IN METERS": INPUT LE
625 PRINT "ENTER ONE-WAY VOLTAGE ATTENUATION FACTOR": INPUT AN
627 PI = 3.14159265
630 DIM XA(N,3)
633 DIM XB(N,4)
635 FOR I = 1 TO N
640 PRINT "INPUT FREQUENCY"
642 PRINT "RESISTANCE IN OHMS"
644 PRINT "REACTANCE IN OHMS"
650 INPUT XA(I,0),XA(I,1),XA(I,2)
660 NEXT I
670 FOR I = 1 TO N
680 RA = 96 / (1 + XA(I,1) / Z0)
685  IF XA(I,2) = 0 THEN XA(I,2) = .001
690 RB = Z0 * 96 / XA(I,2)
700 TH =   ATN (RA / RB)
710 TH =   ABS (TH)
720 YB = 192 -   ABS (RB) *  SIN (2 * TH)
730 XB = 140 + RB - RB *  COS (2 * TH)
740 DX = XB - 140
750 DY = 96 - YB
760  IF DY = 0 THEN DY = .001
770 TS =   ATN (DX / DY)
772  IF DY < 0 THEN TS = PI + TS: GOTO 790
780  IF DX < 0 THEN TS = 2 * PI + TS
790 RC =   SQR ((DX * DX) + (DY * DY))
791  PRINT "ANGLE =";TS / (4 * PI);" WAVELENGTHS"
792  PRINT "RC =";RC / 96
796 REM ***********************************
797 REM   THIS IS THE LINELENGTH CORRECTION
800 LA = 300 / XA(I,0)
810 CA = (LE / LA) * 4 * 3.14159
820 TS = TS - CA
821 REM   THIS IS THE ATTENUATION CORRECTION
```

Program 13.5 (*continued*)

physical length by 0.69 since waves propagate in a teflon-filled cable at 69-percent of the velocity of light. For polyethelene cable, use 0.65, and for foam, between 0.85 and 0.99, depending upon the density. For foam and spiral cables, it is best to consult the oppropriate catalog or to measure the velocity or electrical length using a short-circuit or open-circuit termination.

The next request is for the one-way voltage attenuation factor. If you wish to ignore the losses, the factor is 1. (This topic will be discussed later.) The next inputs are for frequency F, resistance R, and reactance X, and in that order.

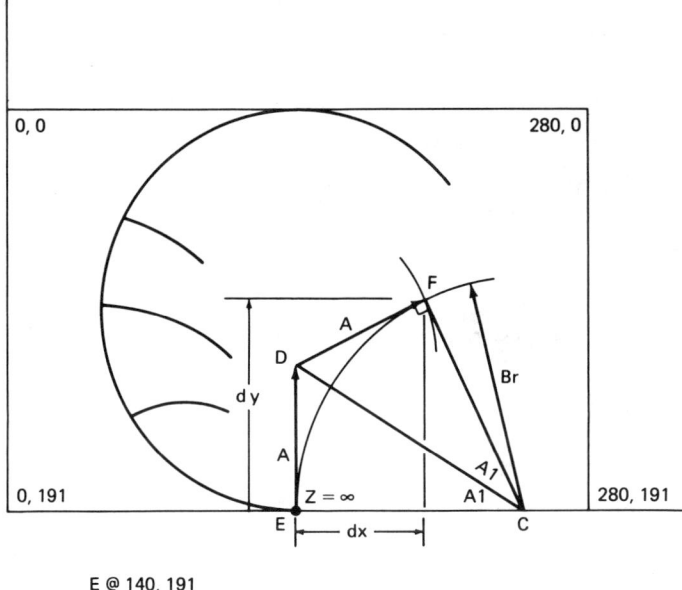

Fig. 13.8 Smith Chart plot by computer.

When the Nth set of F, R, and X has been input, the printer will "wake up" and print the frequency and the corrected VSWR position in wavelengths from the generator for each of the points. The Apple will then blank the screen and plot a box to show the corrected impedance for the first entered point.

At this point, the Apple will stop and wait for an input—any number—and a RETURN. It doesn't actually do anything with the input since this is simply a mechanism to stop the plot routine and give the user a chance to see where the point fell. If it presented all of the points at once, it would be difficult to identify them, although it would be possible to do so from the printed list. (The author has often made a rough sketch to identify the points with a smooth curve after the the plot has been printed.) After the last point has been presented, one more input will send the Apple to line 200, which initiates the plotting of a rudimentary Smith Chart. This will contain only the $R = Z_0$ and $X = Z_0$ curves, the real axis, the periphery, and a dotted 2:1 VSWR circle.

```
821  REM   THIS IS THE ATTENUATION CORRECTION
824  RC =  ABS (RC * AN)
830  YC = 95 - (RC *  COS (TS))
840  XC = 140 + (RC *  SIN (TS))
850  US = (1 + (RC / 96)) / (1 - (RC / 96))
855  P = YB + 3:Q = YB - 3
860  XB(I,0) = US
870  XB(I,1) = XC:XB(I,2) = YC
871  TU =  ABS (TS)
873    IF TS < 0 THEN TU = (2 * 3.14159) - TU
874  XB(I,3) = TU / (4 * 3.14159)
880     NEXT I
890     PR# 1
900     FOR I = 1 TO N
910     PRINT "FREQ=";XA(I,0)
920     PRINT "CORRECTED VSWR=",XB(I,0)
925     PRINT "AT WAVELENGTHS FROM GEN.=",XB(I,3)
930     NEXT I
940     PR# 0
1000     HGR2 : HCOLOR= 3
1010     POKE  - 12524,0
1020     POKE  - 12525,64
1030     POKE  - 12529,255
1035     FOR I = 1 TO N
1040  M = XB(I,1) - 2:U = XB(I,1) + 2
1050  P = XB(I,2) - 2:Q = XB(I,2) + 2
1060     HPLOT M,Q TO M,P TO U,P TO U,Q TO M,Q
1062     INPUT W
1070     NEXT I
1080     GOTO 200
```

Program 13.5 (*continued*)

Figure 13.9 shows the result for a fixed 10-ohm termination on a −2.5-meter line at various frequencies. This is a good example of how a line of this length can spread out the data. The minus sign attached to the line length deserves a word of comment. Since the program was written for transforming measured data at the end of a cable to data of the antenna feedpoint, the positive direction in line, length is toward the load. In this case, however, we wanted the impedance at a point closer to the generator, and the sign of the electrical length is therefore negative.

An important point can be made with this figure. The clockwise circulation on the Smith Chart is the natural direction for circulation with increasing frequency. If your data circulates counterclockwise over any substantial bandwidth, go back and check it because something is terribly wrong. A CCW circulation in a transmission line implies a truly negative line length, that is, the phase shift decreases with increasing frequency. In a lumped element circuit, it implies negative inductors and capacitors whose reactance and susceptance (respectively) decrease with increasing frequency.

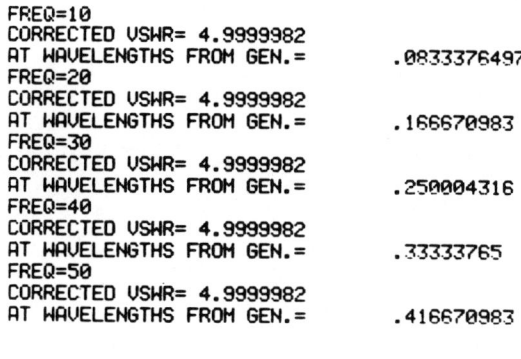
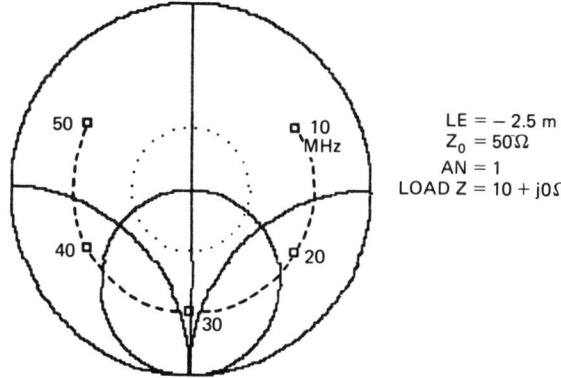

Fig. 13.9 Curves resulting from a fixed 10-Ω termination on a −2.5-meter line at various frequencies.

For an admittance rather than an impedance plot, one may simply replace Z_0 with Y_0 in line 623, 680, and 690. The prompts at 642 and 644 should be changed to read admittance, susceptance, and mhos. Of course, one can convert impedance to admittance mathematically and use either chart. In general, parallel matching elements are easier to insert in a coaxial line, particularly line stubs. These are usually used at the higher frequencies. At lower frequencies, lumped elements are preferred because of their size. For example, at the top of the broadcast band, 1,500-kHz, the wavelength is 200-meters, and the 0.152-wavelength stub Fig. 13.7 would have to be 30.4-meters, or 99.7-feet, long for an air line. A coil would be much more compact and convenient. When matching with lumped elements, a switch is frequently made between Z and Y and back for convenience, using whichever is more illustrative.

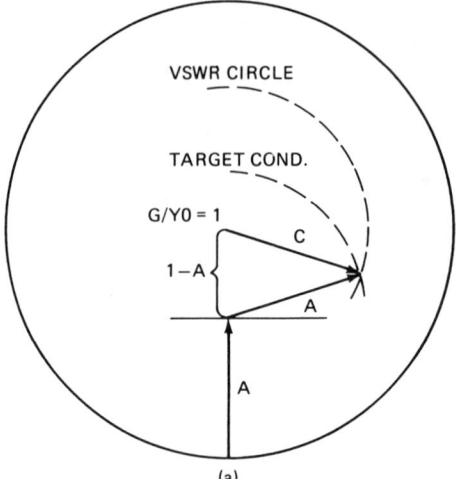

(a)

FOR AN OBLIQUE TRIANGLE WITH 3 SIDES KNOWN

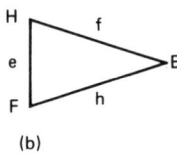

(b)

Fig. 13.10 Geometry for finding the matchpoint.

Problem 2 at the beginning of the chapter is encountered so frequently in antenna work that it deserves special treatment. Although this problem can be solved graphically on the Smith Chart, the enhanced accuracy of a direct computer solution is sometimes worthwhile. The geometry of Fig. 13.10 is employed in this solution. It may be seen that this is actually a Smith Chart solution carried out as if there a Smith Chart within the computer.

The right place to put a matching element to bring the conductance to a desired level is at one of the two intersections of the VSWR circle and the target conductance circle. This is shown in Fig. 13.10(a). Using the formulas and definitions of Fig. 13.2, we can derive a triangle with three sides known and whose formulae are strict geometric equivalences, as shown in Fig. 13.10(b).

The runout following Program 13.6 is a repeat of the problem in Fig. 13.7 and shows the enhanced precision obtainable. A precision VSWR can be obtained by running the point through the Smith Chart plot

```
135  HOME
140  PI = 3.14159265
150  PRINT " INPUT LINE Y0 IN MILLIMHOS"
160  INPUT Y0
170  PRINT " INPUT TARGET CONDUCTANCE IN MILLIMHOS"
180  INPUT G
190  PRINT " INPUT LOAD VSWR"
200  INPUT VS
202  IF G / Y0 > VSWR OR Y0 / G > VSWR THEN  PRINT "THE LOAD CANNOT BE MAT
     CHED THIS WAY"
210  REM ============================================
220  REM    CALCULATE THE SIDES
230  F =   ABS (2 / (1 + VS) - 1)
235  REM   F=C
240  REM   A+(1-A)=1
250  H = 1 / (1 + (G / Y0))
260  E = 1 - H
270  S = (1 + F) / 2
272  REM ////////////////////////////////////////////
274  REM   CALCULATE THE ANGLES
280  MM =  SQR ((S - F) * (S - E) / (S * (S - H)))
290  HA = 2 *   ATN (MM)
300  LL = (PI - HA) / (4 * PI)
310  MN =  SQR ((S - E) * (S - H) / (S * (S - F)))
320  FA = 2 *  ATN (MN)
330  FB = (PI - FA) / 2
340  MO =  SQR ((S - F) * (S - H) / (S * (S - E)))
350  EA = 2 *  ATN (MO)
360  REM &&&&&&&&&&&&&&&&&&&&&&&&&&&&&&&&&&&&&&&&&&&
370  REM   CALCULATE THE MATCHING ELEMENT
380  FB = (PI - FA) / 2
390  B = H *   TAN (FB)
400  BY = Y0 / B
410  LS =  ATN (Y0 / BY)
420  LE = LS / (2 * PI)
595  PRINT ""
600  PRINT " THE DESIRED CROSSINGS ARE"
610  PRINT LL;" AND ";(.5 - LL)
620  PRINT " WAVELENGTHS FROM THE GENERATOR"
630  PRINT "MEASURED FROM THE MINIMUM"
635  PRINT "SUSCEPTANCE POINT"
640  PRINT "AND THE REQUIRED SUSCEPTANCES"
650  PRINT "ARE MINUS AND PLUS ";BY
660  PRINT "MILLIMHOS RESPECTIVELY."
670  PRINT ""
680  PRINT "THIS COULD BE OBTAINED FROM"
690  PRINT " A SHORT-CIRCUITED STUB OF LENGTH"
700  PRINT LE;" AND ";(.5 - LE);" WAVELENGTHS RESPECTIVELY"
2000 END
```

Program 13.6

```
FOR Y0 AND G=20 MILLIMHOS
AND FOR VSWR=2.026
THE DESIRED CROSSINGS ARE
.152527366 AND .347472634
WAVELENGTHS FROM THE GENERATOR
MEASURED FROM THE MINIMUM
SUSCEPTANCE POINT
AND THE REQUIRED SUSCEPTANCES
ARE MINUS AND PLUS 14.4164271
MILLIMHOS RESPECTIVELY.

THIS COULD BE OBTAINED FROM
A SHORT-CIRCUITED STUB OF LENGTH
.150597588 AND .349402412 WAVELENGTHS RESPECTIVELY
```

Program 13.6 (*continued*)

routine for a line of zero length. Solutions for both the line length that requires a shunt inductance and the line length that requires a shunt capacitance are obtained, as well as the magnitude of the required susceptance and the lengths of shorted stub required to realize either solution. In general, the shorter line length and the shorter stub will have the broader bandwidth.

FINDING LINE LENGTH AND ATTENUATION

It was noted earlier that electrical line length and attenuation may be measured. The following technique is frequently used to make these measurements.

In discussing 13.9, we noted that line length of itself tends to string out data points with frequency. In a typical cable measurement, the far end of the cable is short-circuited and the impedance or admittance measured as a function of frequency on the near end of the line. Figure 13.11(a) shows the result obtained when data is run through Program 13.5 with an assumed LE of -5-meters. The longer the line is, the more closely spaced in frequency the points should be in order to avoid an ambiguity of a half wave between points. In Fig. 13.11(b), we see the points plotted versus frequency. If the line length were unknown, it could be solved using the formulas

$$(n + D_1)\lambda_1 = (n + D_2)\lambda_2 \tag{13.3}$$

$$n_1\lambda_1 - n_2\lambda_2 = D_2\lambda_2 - D_1\lambda_1 \tag{13.4}$$

after making a guess at n (which must be a multiple of a half integer, that is, 0.5, 1, 1.5...).

On a strictly mathematical basis, Eq. 13.4 may be used. However, since

THE SMITH CHART   207

```
FREQ=296
CORRECTED VSWR= 18.9999782
AT WAVELENGTHS FROM GEN.=        4.93333726
FREQ=298
CORRECTED VSWR= 18.9999782
AT WAVELENGTHS FROM GEN.=        4.96667059
FREQ=300
CORRECTED VSWR= 18.9999782
AT WAVELENGTHS FROM GEN.=        5.00000392
FREQ=302
CORRECTED VSWR= 18.9999782
AT WAVELENGTHS FROM GEN.=        5.03333726
FREQ=304
CORRECTED VSWR= 18.9999782
AT WAVELENGTHS FROM GEN.=        5.06667059
```

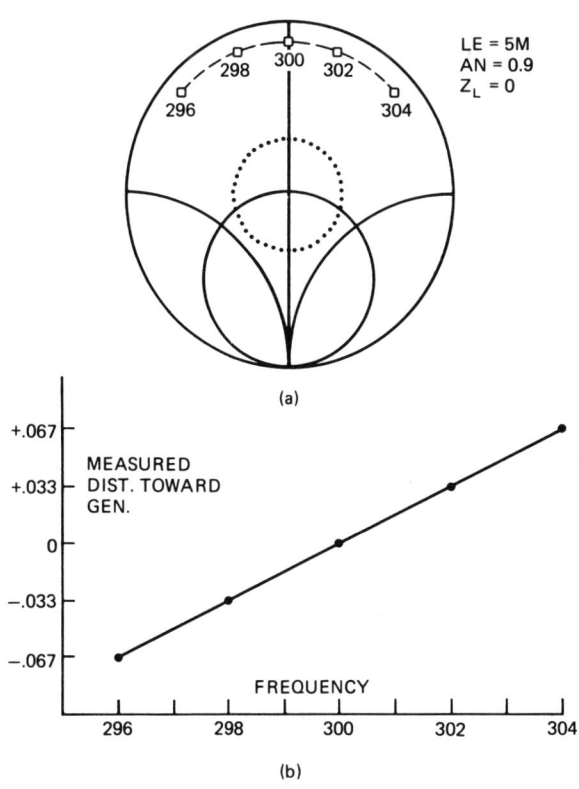

Fig. 13.11 Finding line length and attenuation.

one is obtaining the difference of two measured numbers that are nearly alike and dividing by two numbers that are nearly alike, the results for real measurements become "iffy." On the other hand, a guess at n in Eq. 13.3 will quickly produce a reasonable answer to fit all of the points. It is always advisable to use more than two points.

With regard to attenuation, the short circuit itself will have a voltage reflection coefficient of 1. The signal reaching the short, however, will be lower than the launched signal because of the one-way attenuation of the cable, and the input will see a backward wave attenuated by two trips through the cable, thus yielding a finite VSWR. The two-way voltage attenuation can be calculated from the VSWR with the formula

$$\text{MAGNITUDE VR} = (\text{VSWR} - 1)/(\text{VSWR} + 1) \tag{13.5}$$

The one-way cable loss is equal to the square root of VR Since $VR = 0.9$ in Fig. 13.11, the one-way voltage attenuation (AN) is 0.9486, or -0.458-dB.

In subsequent chapters, we shall see the Smith Chart used for graphical presentations and solutions of lumped circuits as well as distributed circuits.

14
Elementary Radiators

Up to this point we have simply treated antenna elements as Huygens wavelets. We presumed that they radiated without any question of how or why. In this section we shall be taking a brief look at the how and why for a few elementary radiators. In keeping with other materials in this text, the emphasis will be on ease of understanding rather than mathematical rigor. It can be argued, for example, that an antenna can be made to radiate because the speed of light is finite. Without recourse to relativistic arguments, it is fairly obvious that radiation is a necessary consequence of the finite velocity of light.

To begin with, let us suppose that we have a pair of objects that can be charged with opposite polarities, for example, a pair of Ping-Pong balls. These objects will attract one another according to Coulomb's Law, formulated as follows:

$$F = (Q1*Q2)/(4*PI*e*R*R) \qquad (14.1)$$

in which:

- F = force, in newtons
- Q = charge, in coulombs
- e = dielectric constant of the medium
  $[(1/(36*PI))E - 9$ farads/meter in free space]
- R = space between the charges, in meters

If the charges are of opposite algebraic sign, the force acts to pull the objects together. If they are of the same sign, it acts to push them part. Dust and dandruff stick to a comb because the electrons rubbed off the comb charge the particles, and thus the comb and particles have an opposite charge.

If the two particles are held apart, stationary in (otherwise) empty space, the dipole field shown in Fig. 14.1 takes shape. The lines connecting the two charged particles may be considered to be the electric lines of force that carry the coulomb force. The lines forming the closed loci surrounding the Ping-Pong balls are the constant voltage contours.

The figure is finite, but its field would become infinite if the balls stood still long enough.

New, if we forced the Ping-Pong balls apart along the axis connecting them, we would be doing some work since the motion would be opposed by the coulomb force and, for any distance of motion, force times distance equals work.

Note that as we force the Ping-Pong balls apart, the figure would have to expand. The force lines at the "waist" have to move outward to preserve the proportions of the diagram. The farther the force line moves away from the center, the faster it will have to move, until a line sufficiently far out will have to expand at the speed of light. This will be true no matter how slowly we move the Ping-Pong balls. A point will always be encountered where the lines must travel at the speed of light; the lines farther out than this will expand uniformly, however, so that the spacing between lines no longer increases with distance.

One of the sequelae of this phenomenon that the outer lines will be a little straighter than the static picture would reveal and will therefore require a little more force than the coulomb law attraction would require for a static case.

Suppose that we stop pushing and, after a moment's rest, let the Ping-Pong balls begin pulling themselves back together again. Although we can extract work from this action, the force involved will be a trifle less than the static coulomb force because somewhere out along the equator the lines cannot contract faster than the speed of light and thus cannot pull as directly as they would in the static case. Once we have returned to the original spacing, we will be short some work as a result of the extra effort in the force-out and the effort we didn't recoup in the spring-back. That missing work will have been radiated as an electromagnetic wave that is now speeding merrily away from us at the speed of light.

One point can be made regarding the creation of the system. When we originally charged the Ping-Pong balls, we took some electrons from one and transferred them to the other. Although we said that the "static" picture in Fig. 14.1 would obtain if we waited "long enough," it would actually take an infinite amount of time for the lines to expand to infinity at the speed of light. Consequently, in the real world, even the act of charging the balls will radiate some energy. The radiated energy will appear as an adjunct to the work required to do the charging.

The motion of the charge (which is measured in coulombs or ampere-seconds) constitutes a flow of current. The negatively charged sphere $(-)$ carries electrons away from the center of the system, and the positively charged sphere $(+)$ carries a *deficiency* of electrons away (which corresponds to a flow of electrons toward the center). (The unfortunate choice

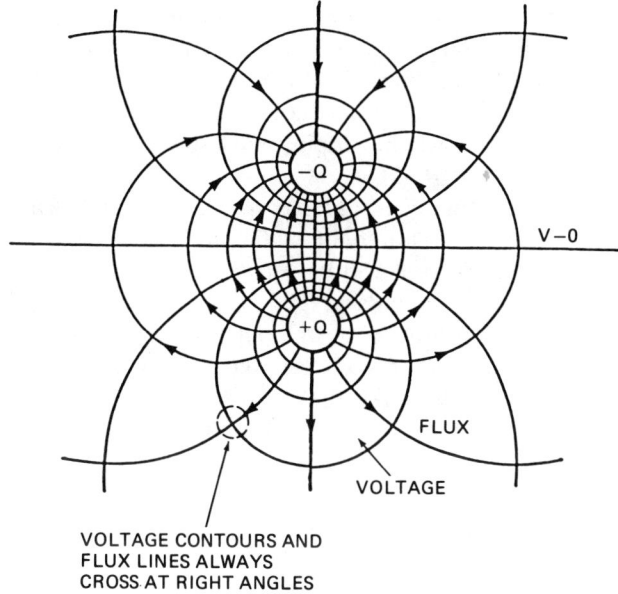

Fig. 14.1 The dipole field.

of indicating an excess of electrons by a minus sign was made by Benjamin Franklin.) Thus the motions are additive. In Fig. 14.1, if the negatively charged sphere is going upward, the net electron flow is upward.

From Ampere's law, we know that the flow of current creates a magnetic field that encircles the direction of current flow. For a current of I amperes flowing on a short wire of length S meters, Ampere's law can be written as follows:

$$H = (I*S*SIN(TH))/(4*PI*P*P) \qquad (14.2)$$

where:

H = magnetic field, in amperes per meter
P = distance to the point, in meters
TH = angle measured from the axis of the current to the point

The geometry is such that angle TH describes a cone about the axis of the current. In Fig. 14.1, an angle TH of 90-degrees would correspond to the plane marked "$V = 0$". The magnetic flux lines are circles about the

212 EXPLORING ANTENNAS AND TRANSMISSION LINES

axis of the current and are therefore crosswise to the electric field lines. Of course, the same arguments can be applied to the magnetic lines as were made for the electric lines. Both fields lose energy at any time they are changing. Once the change has come to an end, the field becomes static in the volume about the origin. This static boundary propagates ouward at the velocity of light.

The electrical and magnetic effects in the changing field are inseparable. If the charges are held stationary, the magnetic field will disappear because the current will be zero and only the electrostatic field remains within the static boundary. The escaping "puff" of energy takes the form of a growing smoke ring or doughnut. The magnetic lines fall within the body of the doughnut circling the system axis, and the electric lines are wound about them like the wires on a toroidal inductor.

For the propagating wave, the situation is very similar to the conditions that led up to Eqs. 11.26 and 11.27. The velocity of propagation in meters per second is represented either by

$$V = SQR(L*C) \tag{14.3}$$

or

$$V = SQR(u*e) \tag{14.4}$$

and the characteristic impedance by

$$Z0 = 1/(SQR(L/C)) \tag{14.5}$$

or

$$Z0 = 1/(SQR(u/e)) \tag{14.6}$$

The value of u is substituted for inductance per unit length and is measured in the same units, namely, henrys per meter. For free space, this can be evaluated by *dc* or low-frequency measurements of inductors or of the attraction between current-carrying wires. The value is 4*PI*1E − 7 henrys per meter. The value of *e* can be obtained by measuring the capacitance of an air-dielectric capacitor. In 1856, Webber and Kohlrausch showed that these values predicted the velocity of light in free space to be 3E8-meters per second.

Substituting these values in Eq. 14.6 gives a value of 337-ohms per square for the characteristic impedance of free space. A frequently asked question is "per square *what*?" The answer is per square *anything*!

Consider a plane wave and an imaginary plane passing through it. On the plane, let us inscribe a rectangular window with one side parallel to the electric lines. Since the resistance of the window will be directly proportional to its height and inversely proportional to its width, all squares will have the same resistance regardless of size. The concept of a "collecting aperture" will be investigated shortly. It refers to the amount of power that will pass through an "aperture" in space.

It is possible to measure directly the electric field strength in a passing plane wave in volts/meter. The power passing through a square-meter aperture in watts per square meter is given by

$$P = (V*V)/377 \tag{14.7}$$

Of course, since the plane wave is subject to inverse-square-law spreading, the power density will not be the same at all ranges from the source. As the wave travels greater distances, the power becomes more spread out and less dense. The strength of plane waves is specified in either volts per meter (usually, microvolts/meter); or watts per square meter, with about equal preference for both. The volts-per-meter notation is generally more useful in low- to medium-frequency cases where wire or whip antennas are used. These can be assigned an electrical dimension related to their physical dimensions that will tell directly what voltage will be induced by a field of V volts-meter. This dimension is usually referred to as *electrical height* even when the antenna is physically horizontal.

The watts/square meter notation is used whenever the antenna is an area type of device such as a parabolic dish, a horn, or a lens. For such antennas, it is possible to specify an area called the *effective collecting aperture*, usually measured in square meters. The effective collecting aperture is usually related to the physical area.

One concept remains to be gleaned from our Ping-Pong ball analogy. It may be seen that there is a certain amount of energy stored in the more or less static field about an antenna. This field contains reactive energy that is neither dissipated nor radiated away. It is confined to the immediate vicinity of the antenna and simply recycles with changes in the drive. In the more rigorous treatment to follow, this "near field" will be taken into consideration along with the radiating components.

THE ELEMENTARY DIPOLE

Now let us consider a small, simple antenna consisting of a short length of straight wire excited with a current. To begin with, we will assume that the antenna is short with respect to wavelength and excited with a uniform current. We shall not provide the derivations of Maxwell's Equations here. Those wishing to probe deeper into the matter are directed to the sources in the accompanying footnote.[1]

We shall use the notation from the Kraus reference in the footnote for the arrangement shown in Fig. 14.2. The Antenna at the center is a short straight dipole with a current expressed by

$$I = I_0 * \text{EXP}(JWT) \tag{14.8}$$

where:

$j = \sqrt{-1}$
$W = 2\pi \times$ frequency
$t =$ time, in seconds

As in the transmission line equations, the time delay in the forward wave is expressed by

$$(T - R/C) \tag{14.9}$$

where:

R = distance to the point, in meters
C = velocity of light

Equations 14.10 through 14.13, which follow, represent the solution for the dipole derived from Maxwell's equations. The dimensions of E are in volts/meter and of H in amperes/meter, where as R, L, and the wavelength are in meters. Note that we have had to partially abandon BASIC notation in the interest of making the equations compact enough for reasonable presentation.

$$E_R = \frac{I_0 L \cos\theta e^{j\omega(T-\frac{R}{C})}}{2\pi\varepsilon_o} \left(\frac{1}{CR^2} + \frac{1}{j\omega R^3} \right) \tag{14.10}$$

---

[1] John D. Kraus, *Antennas* (New York: McGraw Hill, 1950).
Samuel Silver, ed., *Microwave Antenna Theory and Design* (republished by Dover Publications, New York).
S. A. Shelkunoff and H. T. Friis, Antennas, Theory and Practice (New York: John Wiley and Sons, 1952).

$$E_\theta = \frac{I_o L \sin\theta e^{j\omega(T-\frac{R}{C})}}{4\pi\varepsilon_o} \left(\frac{j\omega}{C^2 R} + \frac{1}{CR^2} + \frac{1}{j\omega R^3} \right) \quad (14.11)$$

$$H_\phi = \frac{I_o L \sin\theta e^{j\omega(T-\frac{R}{C})}}{4\pi} \left(\frac{j\omega}{CR} + \frac{1}{R^2} \right) \quad (14.12)$$

$$E_\phi = H_R = H_\theta = 0 \quad (14.13)$$

where:

e = exp ( )
ε_o = dielectric constant of space

Equation 14.13 defines the polarization, showing that there is no circumferential voltage component and no theta or radial component to H. Similarly, the zeros of Eq. 14.13 are to be expected. For example, considerations of symmetry tell us that there is no mechanism for developing a voltage gradient about the circumference of the dipole, nor is there a mechanism for developing a magnetic gradient in the radial or the theta directions.

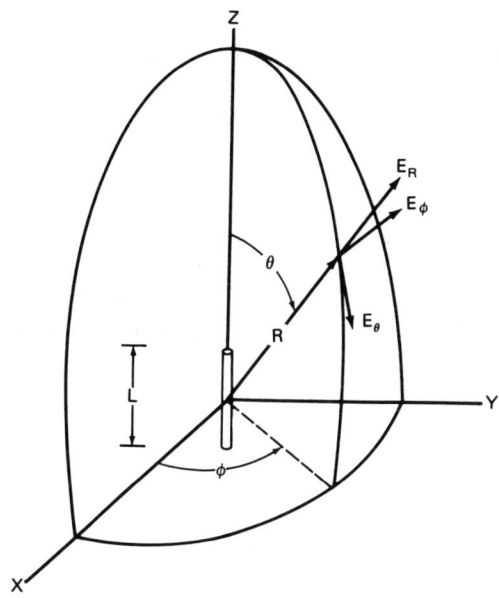

Fig. 14.2 The coordinate system.

216 EXPLORING ANTENNAS AND TRANSMISSION LINES

If we examine the expressions in parentheses, we observe that some of the terms diminish as the square or cube of range R. These terms obviously are significant only in the immediate vicinity of the antenna since they diminish so rapidly with range. These are the "near field" or "induction field" terms mentioned previously. They represent the "quasi-static" part of the field, which, in effect, does not radiate. These near-field terms are a completely different animal from the Fresnel Zone mentioned in connection with Fig. 3.5. In that case, we were concerned with the range at which the phase error across an aperture to a distant point was less than some limit. Here the range is much smaller, on the order of a fraction of a wavelength, and the distinction is between a nonradiating and a radiating field.

If we drop the terms with the reciprocal squares and cubes of R we obtain the *far-field* case, as follows:

$$E_\theta = \frac{j\omega I_o L \sin\theta e^{j\omega(T-\frac{R}{C})}}{4\pi\varepsilon_o C^2 R} \qquad (14.14)$$

$$H_\phi = \frac{j\omega I_o L \sin\theta e^{j\omega(T-\frac{R}{C})}}{4\pi C R} \qquad (14.15)$$

Taking the ratio, we obtain

$$\frac{E\theta}{H\phi} = \frac{1}{\varepsilon_o C} = \sqrt{\frac{\mu_o}{\varepsilon_o}} = 377\text{-ohms} \qquad (14.16)$$

It is noteworthy that the radial voltage term, E_R, has completely disappeared. From Eq. 14.10, we see that this term varies as the cosine of theta and is therefore maximum off the ends of the dipole. Since from experience we know that a dipole does not radiate much along its axis, these characteristics accord with experience. In the far field, little or nothing exists along the axis of the dipole, whereas considerable coupling may exist in the near field.

An interesting special case, usually called the *quasi-static case*, occurs at low frequencies. If we let the frequency approach zero, and assuming that $R \gg L$, we obtain

$$E_R = \frac{Q_o L \cos\theta}{2\pi\varepsilon_o R^3} \qquad (14.17)$$

$$E_\theta = \frac{Q_o L \sin\theta}{4\pi\varepsilon R^3} \qquad (14.18)$$

ELEMENTARY RADIATORS

$$H_\phi = \frac{I_o L \sin\theta}{4\pi R^2} \quad (14.19)$$

Note that

$$Q_o = \int I dt \quad (14.20)$$

Equations 14.17 and 14.18 are identically the expressions for a pair of electrostatic charges $+Q$ and $-Q$ separated by a distance L. Equation 14.19 is identical to Ampere's Law, given earlier as Eq. 14.2 thus corroborating our initial Ping-Pong ball analogy.

In the far field, the dipole may be considered as a true point source of energy since the phase is invariant for constant R.

Another point will prove of interest in subsequent problems. Both E and H are directly proportional to the frequency and to L for this electrically small antenna. The latter fact has a significant bearing on what is required to obtain reasonable amounts of radiation from electrically small antennas. The fact that the characteristic impedance of free space can be verified by the manipulation of Eq. 14.16 is also interesting.

As we have seen previously, standing waves can exist in free space as can the antenna patterns previously calculated. The expressions show only the outgoing wave since there would be no reflected wave in free space.

RADIATION RESISTANCE

In electrical circuit terms, the antenna must have some mechanism that will account for the energy lost to radiation. This is the *radiation resistance* of the antenna. If one imagines the antenna being enclosed by a large sphere, Eqs. 14.14 and 14.15 can be used to calculate the electric and magnetic field strength on any unit area of the sphere. The power incident upon any unit area is termed the *Poynting vector*. The integral of all of the differential powers represents the total power radiated by the antenna. Note that the product of E (volts/meter) and H (amperes/meter) is watts/square meter.

Since the radiation is excited by a current in the radiator, the power dissipation can be suitably described as a resistance in series with this current. Although this resistance can be described for a number of types of antennas, it is physically measurable in only a few simple types. Mathematically, this process is described as follows:

$$I^2 R_R = \int_0^{2\pi} \int_0^{\pi} P(\theta\phi) R^2 \sin\theta \, d\theta \, d\phi \quad (14.21)$$

218 EXPLORING ANTENNAS AND TRANSMISSION LINES

where:

R_R = radiation resistance
$P(\theta\phi) = E_\theta H_\phi$ at the angle

Now the average value of the Poynting vector is

$$P(\theta\phi) = \frac{1}{2}\text{REAL}[E_\theta H\phi] \qquad (14.22)$$

But from Eq. 14.16,

$$E_\theta = H_\phi \sqrt{\frac{\mu_o}{\varepsilon_o}} \qquad (14.23)$$

Thus, the transmitted power is

$$P_T = \frac{1}{2}\sqrt{\frac{\mu_o}{\varepsilon_o}} \int_0^{2\pi}\int_0^{\pi} |H_\phi|^2 R^2 \sin\theta\, d\theta\, d\phi \qquad (14.24)$$

and for the short dipole,

$$H_\phi = \frac{\omega I_o L \sin\theta}{4\pi C R} \qquad (14.25)$$

Substituting yields

$$P_T = \frac{1}{32}\sqrt{\frac{\mu_o}{\varepsilon_o}} \frac{\omega^2 I_o^2 L^2}{\pi^2 C^2} \int_0^{2\pi}\int_0^{\pi} \sin^3\theta\, d\theta\, d\phi$$

Now

$$\omega^2 = 4\pi^2 F^2 = \frac{4\pi^2 C^2}{\lambda^2} = \beta^2 C^2 \qquad (14.27)$$

$$\beta = 2\pi/\lambda \qquad (14.28)$$

Integrating Eq. 12.26 and substituting gives

$$P_T = \sqrt{\frac{\mu_o}{\varepsilon_o}} \frac{\beta^2 I_o^2 L^2}{12\pi} = \left(\frac{I_o}{\sqrt{2}}\right)^2 R_R \qquad (14.29)$$

$$P_T = \sqrt{\frac{\mu_o}{\varepsilon_o}} \frac{\beta^2 L^2}{6\pi} = \left(\frac{120\pi}{6\pi}\right)\left(\frac{4\pi^2}{\lambda^2}\right)L^2$$

Thus

$$R_R = 80\pi^2 \left(\frac{L}{\lambda}\right)^2 \tag{14.31}$$

It is well to remember some of the restrictions placed upon the original development of the equations. It was assumed that the dipole was very short with respect to wavelength and that the current on the dipole was uniform. With these restrictions, we see that if the length is 0.1-wavelength, the radiation resistance will be 7.7-ohms. At the length of 0.01-wavelength, the radiation resistance will be 0.079-ohm. In other words, if the antenna is short, it will take very large currents to radiate any significant power. This factor is very significant in aircraft and land mobile antennas for the HF band and for all antennas for the LF and VLF bands. For example, at 2-MHz, the wavelength is 150-meters. A 3-meter "buggy whip" is about the tallest antenna that can be safely operated on a jeep without tying the tip down. At this height one can pass under most bridges and branches without contact. At 2-MHz, the antenna is only 2-percent of a wavelength!

We have assumed the current to be uniform. On a very short antenna, the current distribution is very nearly triangular if the antenna is not heavily end-loaded with capacitance. Although the radiation patterns scarcely change from those calculated in Eqs. 14.14 and 14.15, the average current is nearly halved, and since the current is squared in the R_R calculation, Eq. 14.31 has its leading constant reduced to 20. The error in computing R_R is relatively small up to a dipole span of about 0.2-wavelength. Past this point the sinusoidal form of the standing wave on the antenna must be accounted for.

The radiation resistance can be calculated for dipole, monopole-ground-plane, and loop antennas, but the physical significance of the term for horn or other types of antennas is questionable.

15
Wire Antennas

The previous chapter dealt with the radiation resistance of small, elementary radiators. In this chapter we shall consider some of the properties of antennas of a more conventional size. We shall consider some properties beyond the radiation resistance.

The analysis of antennas by the integral-equation method is a time-honored mathematical tour de force that is beyond the scope of this text. The reader inclined to pursue the matter in more detail is directed to earlier references to Kraus's *Antennas* in Chap. 9 or Shelkunoff and Friis' *Antennas: Theory and Practice* in Chap. 13. The original work analyzed in these references was done by Abraham in 1898 and Hallen in 1938.

A reasonable concept of the behavior of a dipole antenna can be obtained by considering the antenna to be a transmission line terminated in an impedance of some sort. This concept will be somewhat more valid if the transmission line is a conical one. A conical transmission line bears the same relationship to spherical waves that a parallel transmission line does to plane waves. Its characteristic impedance is given by

$$Z0 = 120*LOG(COT(HC/2)) \tag{15.1}$$

where:

LOG = natural logarithm, base *e*
HC = the half-cone angle

The geometry is illustrated by Fig. 15.1, which uses the coordinate system of Chap. 14. A few points are of interest. If angle *HC* is 9-degrees, then Z_0 = 300-ohms. For a very "fat" cone in which *HC* = 46.8-degrees, Z_0 = 100-ohms. If this antenna were cut in half and the lower cone replaced with an infinitely extensive and infinitely conductive ground plane, the antenna would have a characteristic impedance of 50-ohm. In fact, a real antenna built with such a cone angle would have a very broad impedance bandwidth in a 50-ohm system. Unfortunately, this would be a clumsy design, and the pattern would tend to deteriorate whenever the length of the cone becomes larger than a half wavelength.

The 9-degree conical design is very broadband. It is frequently used as

Fig. 15.1 The conical dipole.

a source antenna for driving a paraboloid or a corner reflector in the UHF TV band. Usually the elements are made out of a pair of triangular pieces of sheet metal in the form of a "bow tie." The bow is creased slightly in the centerline along the long axis for mechanical stiffness. When used with a 300-ohm transmission line, the impedance match can be good over a very broad band, for instance, from 470 to 960-MHz.

Program 15.1 is an empirical model that does a reasonable job of simulating the measurable performance of whip antennas in shipboard or land mobile applications. Below 0.1-wavelength, the resistance is smaller than it would be in most installations because the ground plane is seldom good enough for such low resistive terms to be achieved. On a steel-decked naval ship, however, careful measurements will yield resistances as low as about 2.5-ohms.

The model data of this program produces a spiral on a Smith Chart because of the attenuation term in line 1110. Although the model could have been constructed with a short length of line and a high impedance termination, a better fit for experimental data would be obtained by adding a quarter wavelength of line to the electrical length being simulated and by using an attenuation constant with a short-circuit load. It is difficult to make the attenuation fit the data to be simulated with a line length of only a tiny fraction of a wavelength.

This model is useful for producing realistic simulated antenna data for network evaluation. The writer first devised it to generate data for an antenna simulator to be used for antenna coupler testing.

```
1060  PR# 1
1100  TV = .26
1110  RG = .66 + .34 * EXP ( - 1 * (5 * (TV - .25)) ^ 2)
1120  IF (TV - .25) = .5 THEN TV = .74999
1130  GOSUB 1999
1520  PRINT " FOR ANTENNA HEIGHT= ";(TV - .25);" WAVELENGTHS"
1530  PRINT "THE   RESISTANCE IS ";R;" AND THE REACTANCE IS ";X
1540  PRINT ""
1542  IF TV < .3 THEN TV = TV + .01: GOTO 1110
1550  TV = TV + .05
1560  IF TV < 1.03 GOTO 1110
1570  STOP
1990  REM  ****************************************
1992  REM       CONVERSION FROM REFLECTION
1993  REM       COEFFICIENT AND LINELENGTH
1994  REM       TO R+JX
1995  REM  $$$$$$$$$$$$$$$$$$$$$$$$$$$$$$$$$$$$$$$$
1999  PI = 3.14159
2000  T4 = TV * 4 * PI
2001  Z0 = 200
2010  H2 = RG * COS (T4)
2015  IF (1 + H2) = 0 THEN H2 = - .99999
2020  T5 = (2 * PI) - T4
2030  X2 = - RG * SIN (T5)
2040  T3 = ATN (X2 / (1 + H2))
2050  AB = (1 + H2) / (2 * COS (T3))
2052  RR = AB / COS (T3)
2060  R = Z0 * ((1 / RR) - 1)
2062  IF (TV - .25) < .1001 THEN R = 40 * PI * PI * (TV - .25) ^ 2: GOTO 2
      070
2064  IF (TV - .25) < .1999 THEN R = (R + (40 * PI * PI * (TV - .25) ^ 2))
      / 2
2070  RX = AB / SIN (T3)
2080  X = Z0 / RX
2100  RETURN

]RUN
 FOR ANTENNA HEIGHT= .01 WAVELENGTHS
THE   RESISTANCE IS .039478351 AND THE REACTANCE IS -3178.70135

 FOR ANTENNA HEIGHT= .02 WAVELENGTHS
THE   RESISTANCE IS .157913404 AND THE REACTANCE IS -1582.8605

 FOR ANTENNA HEIGHT= .03 WAVELENGTHS
THE   RESISTANCE IS .355305159 AND THE REACTANCE IS -1048.00208

 FOR ANTENNA HEIGHT= .04 WAVELENGTHS
THE   RESISTANCE IS .631653616 AND THE REACTANCE IS -778.380378

 FOR ANTENNA HEIGHT= .05 WAVELENGTHS
THE   RESISTANCE IS .986958775 AND THE REACTANCE IS -614.839156

 FOR ANTENNA HEIGHT= .1 WAVELENGTHS
          TANCE IS 3.9478351 AND THE REACTANCE IS -274.064424
```

Program 15.1

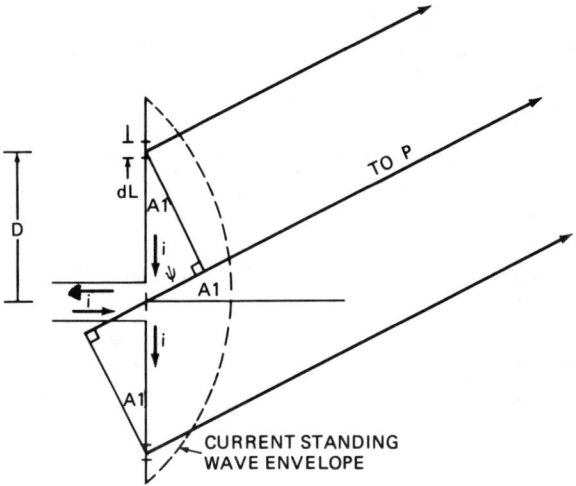

Fig. 15.2 Geometry of the dipole antenna.

A physical model can be constructed with a length of cable and a carborundum noninductive resistor for a termination. Unless the line is made attenuating, the points for the higher order resonances will not usually be too well simulated.

THE DIPOLE ANTENNA

Figure 15.2 shows the geometry of the dipole antenna. We will consider it to be constructed of two straight, equal lengths of wire fed in the center by a balanced transmission line. The meaning of the balance will be discussed later. It may be seen from the small arrows at the center that the out-of-phase transmission line currents turn the corner and appear to be in-phase when viewed from a distant point. Since the dipole is physically symmetrical, we would expect it to be electrically symmetrical as well; consequently, we may use the simplification of contra rotating vectors as in Eqs. 15.2 and 15.3.

$$\psi = \frac{2\pi D}{\lambda} \sin(A1) \qquad (15.2)$$

$$c/V_P = 2\cos\psi H\phi \qquad (15.3)$$

Each element of the dipole may be considered to be a small dipole in

224 EXPLORING ANTENNAS AND TRANSMISSION LINES

itself, as discussed in Chap. 14. In Eq. 15.4, the coordinate system has been changed (see Eq. 14.15) to make it match the coordinates used for previous arrays.

$$H_\theta = \frac{j\omega I_o dL\cos(A1)}{4\pi CR} \qquad (15.4)$$

The current along the radiator will obviously not be uniform. If we were to model the antenna as a transmission line, we might expect to have a forward-going wave that propagates out toward the open-circuited end of the line and a backward wave that reflects off this open circuit toward the feedpoint in the center, indicated by Eq. 15.5

$$I_o = \frac{E_F}{Z_o}\varepsilon^{-\frac{j\omega D}{C}} - \frac{\Gamma E_F}{Z_o}\varepsilon^{j\omega D} \qquad (15.5)$$

As we saw in Chap. 12, an absence of dissipation would imply a set of standing waves on the antenna with the current minima going to zero and a cosine distribution. A perfect standing wave with a unity reflection coefficient would imply a perfect cosine standing wave and the attainment of infinite impedance at the voltage loops and zero impedance at the voltage nulls. The classical analysis of the impedance of the dipole antenna is sometimes carried out with these restraints, but doing so implies that there is no radiation whatsoever from the antenna. This is obviously not the case.

The alternative model proposed here treats the dipole as a transmission line with a characteristic impedance of 600-ohms. Although the termination is considered to be an infinite impedance, the line is presumed to be lossy, and the loss is presumed to be stated as EXP(−0.08) per radian. (Loss stated in exponents of the natural log base is usually termed loss in nepers as opposed to loss in decibels.) The characteristic impedance and the loss were selected to yield an impedance of 75-ohms for a dipole span of a half wavelength and approximately 2500-ohms for a span of a full wavelength. These values are approximately what one would measure on a practical half-wave wire dipole antenna and a full-wave dipole antenna, respectively.

The input impedance is a function of the length-to-diameter ratio of the dipole. A "fat" dipole with a length-to-diameter ratio of 10 to 20 will demonstrate a lower impedance at the high-impedance points and a higher impedance at the low-impedance points. It will act with a higher attenuation and a lower VSWR. Conversely, the larger ratios correspond to a higher VSWR on the transmission line and will give higher impedances at the high-impedance points and somewhat lower impedances at the low-impedance points. The values cited here correspond approximately to those that would be obtained with a length-to-diameter ratio of 100 or more. The first resonance is much less sensitive to the length-to-diameter ratio when L is close to a quarter wavelength than the high-impedance resonance is when L is close to a half wavelength.

For zero attenuation, the standing wave pattern is a series of sine functions, of course, and the phase is either zero or PI because of the contrarotating vectors of equal amplitude. With a finite attenuation, however, the distribution is no longer a perfect sinusoid, and the phase angle walks somewhat. As the attenuation gets higher, this effect becomes more pronounced. Analyses of the pattern of a dipole antenna often assume that the current distribution is sinusoidal and that no phase shift occurs in the standing wave. This, of course, is tantamount to assuming that the antenna doesn't radiate. In this treatment, we shall assume the presence of a finite attenuation induced by radiation. The current distributions will be computed accordingly.

Lines 230 through 302 of Program 15.2 represents a solution of the wave equations (e.g., Eq. 12.2 solved by mean's of Euler' Equation). This is a straightforward manipulation considering the fact that BASIC will not accept the EXP(jWT) function directly. The use of Euler's Equation is probably the easiest way out. For the calculation of $A3$ in lines 215 and 225, the backward wave is presumed to have come from a point in space $2*L$ from the feedpoint.

The output in Fig. 15.3 shows the result of one such calculation. The 20-meter value of L represents half a wavelength at 7.5-MHz, and the input current is appropriately low. At the point at which $D1 = 10$-meters, we see that the current is maximum and the phase angle 90-degrees. At the tip of the radiator where $D1 = 20$-meters, the current is essentially zero, as befits an open circuit.

The simulation is not truly complete since the resonances of a real antenna usually occur when the length is slightly less than the quarter-wave spacings. On a "fat" antenna, the shortening can exceed 10-percent, where as on a typical slender wire antenna, it will be only a few percent. The standard 35-foot (10.67-meter) whip used by the U.S. Navy experiences a first or quarter-wave resonance just below 6-MHz when its electrical length is 0.213-wavelength, for a forshortening of 3.7-percent. This whip tapers from a diameter of about 6-inches at the base to about 4-inches at the tip for a length-to-diameter ratio of about 60. A thin wire will show forshortening of perhaps 2-percent.

```
130  HOME
135  PRINT "INPUT L IN METERS"
136  INPUT L
140  C = 3E8
145  PI = 3.14159265
150  PRINT " INPUT D1 IN METERS"
155  INPUT D1
160  PRINT "INPUT F"
170  INPUT F
180  PR# 1
190  A4 =  - .08
192  PRINT "FOR ATTENUATION CONSTANT= ";A4
194  PRINT ""
200  REM  ***********************************************
210  REM       CALCULATE CURRENT
215  D2 = (2 * L) - D1
220  A2 = D1 * F * 2E6 * PI / C
225  A3 = D2 * F * 2E6 * PI / C
230  RO =   EXP (A4 * A2) *  COS (A2)
240  JO =   EXP (A4 * A2) *  SIN (A2)
250  RI =   EXP (A4 * A3) *  COS (A3)
260  JI =   EXP (A4 * A3) *  SIN (A3)
270  IR = RO - RI
280  IJ = JO - JI
290  IN =  SQR ((IR * IR) + (IJ * IJ))
295   IF IR = 0 THEN TH = PI / 2: GOTO 310
300  TH =  ATN (IJ / IR)
302   IF IR < 0 AND IJ > 0 THEN TH = PI + TH
305  PRINT "FOR L= ";L;" AND D1= ";D1;" AT F= ";F;" MHZ"
310  PRINT "CURRENT = ";IN;" AT ANGLE ";(TH * 180 / PI);" DEGREES"
315  PRINT ""
320  D1 = D1 + 1
325   IF D1 = L THEN D1 = L - .0001
330   IF D1 <  = L GOTO 215
2990  PR# 0
3000  STOP
```

Program 15.2

We can calculate the voltage wave and determine the impedance at any location by dividing E by I. The technique used in Program 15.1 is easier to simulate physically since the power is dissipated in a large resistor. It would be possible to construct a coaxial line with an iron center conductor to simulate this model, but the power handling capabilities of such a line would be limited. In order to obtain high impedance with an outer conductor of reasonable size, the center conductor would have to be a fine wire. In this case, any significant input power will make it heat up and tend to burn out easily, as well as change the attenuation.

RADIATION PATTERNS

Once the current distribution and phasing have been computed, we can proceed to calculate the radiation pattern. The pattern is obviously the summation of (more properly the integral of) the contributions of all of the dL elements in Fig. 15.2, with due consideration for the phase shifts resulting from the angle in space, as shown in Eqs. 15.2 and 15.3. Since we are considering a pair of symmetrical contrarotating vectors from dL and $-dL$, we need not calculate anything but the real part of the phase shift due to space. Of course, the phase of the exciting current must be preserved. We lump into a constant all of the terms in Eq. 15.4 except I_o, which is our calculated current distribution (including the phase), and $\cos(A1)$, which is inherent in the radiation of dL. It is seldom worthwhile to calculate the phase in the far field, and the pattern is usually normalized to unity at the peak of radiation. The antenna is treated as a linear array, each element being represented by a dL.

Program 15.3 begins with a computation of the current. This computation is similar to that of Program 15.2 except that the real part of the current is stored in $X(I,0)$ and the imaginary part in $X(I,1)$. Since we intend to add the results in the summation (integration across the length of L), it is more convenient to preserve the currents in complex form. If the space phase of Eq. 15.3 had not canceled its imaginary component as a result of symmetry, the polar notation would have been handier.

Lines 500 to 630 amount to a calculation of the summation in direction $A5$ (this corresponds to $A1$ in fig. 15.2, but $A1$ has already been used in the current calculation). Line 645 sorts out the strongest voltage for EM to be used later to normalize the pattern.

The branch to the high-resolution graphics routine at line 450 is prompted by a desire to see something on the screen while the Apple is munching on the numbers. If the unit were allowed simply to fall through this setup from line 410 and the print statement at 615 were deleted, the program would run noticeably faster, but, for an L large in terms of wavelength, the screen would completely blank and remain so for a few seconds until the computation is completed.

The final routine from lines 710 to 830 places the axes on the figure and marks off the angle scale in steps equal to calculation increments AD as defined in line 510. The dotted circle marks the -3-dB contour on the chart.

For the longer L examples, line 150 was altered to make the sampling dL a twentieth of a wavelength. Lines 500, 520, and 640 were altered to make $AD = PI/144$ and to calculate 72 steps in $A5$ because the features of the lobes become finer as the length is increased.

```
FOR ATTENUATION CONSTANT= -.08

FOR L= 20 AND D1= 0 AT F= 7.5 MHZ
CURRENT = .395077438 AT ANGLE 6.41697281E-07 DEGREES

FOR L= 20 AND D1= 1 AT F= 7.5 MHZ
CURRENT = .446983309 AT ANGLE 34.0554433 DEGREES

FOR L= 20 AND D1= 2 AT F= 7.5 MHZ
CURRENT = .597483732 AT ANGLE 55.6075423 DEGREES

FOR L= 20 AND D1= 3 AT F= 7.5 MHZ
CURRENT = .781558478 AT ANGLE 67.5591106 DEGREES

FOR L= 20 AND D1= 4 AT F= 7.5 MHZ
CURRENT = .967019048 AT ANGLE 74.7267011 DEGREES

FOR L= 20 AND D1= 5 AT F= 7.5 MHZ
CURRENT = 1.13878935 AT ANGLE 79.4470899 DEGREES

FOR L= 20 AND D1= 6 AT F= 7.5 MHZ
CURRENT = 1.28816758 AT ANGLE 82.7893944 DEGREES

FOR L= 20 AND D1= 7 AT F= 7.5 MHZ
CURRENT = 1.40929983 AT ANGLE 85.2835017 DEGREES

FOR L= 20 AND D1= 8 AT F= 7.5 MHZ
CURRENT = 1.49802251 AT ANGLE 87.2159742 DEGREES

FOR L= 20 AND D1= 9 AT F= 7.5 MHZ
CURRENT = 1.55145281 AT ANGLE 88.7537208 DEGREES

FOR L= 20 AND D1= 10 AT F= 7.5 MHZ
CURRENT = 1.56783354 AT ANGLE 90 DEGREES

FOR L= 20 AND D1= 11 AT F= 7.5 MHZ
CURRENT = 1.54646652 AT ANGLE 91.0218689 DEGREES

FOR L= 20 AND D1= 12 AT F= 7.5 MHZ
CURRENT = 1.48767181 AT ANGLE 91.8645992 DEGREES

FOR L= 20 AND D1= 13 AT F= 7.5 MHZ
CURRENT = 1.3927475 AT ANGLE 92.5597006 DEGREES

FOR L= 20 AND D1= 14 AT F= 7.5 MHZ
CURRENT = 1.26391835 AT ANGLE 93.1296131 DEGREES

FOR L= 20 AND D1= 15 AT F= 7.5 MHZ
CURRENT = 1.1042691 AT ANGLE 93.5905621 DEGREES

FOR L= 20 AND D1= 16 AT F= 7.5 MHZ
CURRENT = .917660725 AT ANGLE 93.954354 DEGREES

FOR L= 20 AND D1= 17 AT F= 7.5 MHZ
CURRENT = .708630052 AT ANGLE 94.2295318 DEGREES

FOR L= 20 AND D1= 18 AT F= 7.5 MHZ
CURRENT = .482274393 AT ANGLE 94.4221295 DEGREES

FOR L= 20 AND D1= 19 AT F= 7.5 MHZ
CURRENT = .244123243 AT ANGLE 94.53616 DEGREES

FOR L= 20 AND D1= 19.9999 AT F= 7.5 MHZ
CURRENT = 2.45125324E-05 AT ANGLE 94.5735833 DEGREES
```

Fig. 15.3 Output of Program 15.2.

```
130  HOME
135  PRINT "INPUT L IN METERS"
136  INPUT L
140  C = 3E8
145  PI = 3.14159265
150  PRINT "ENTER FREQUENCY IN MHZ"
155  INPUT F
160  DL = C / (F * 1E6 * 50)
165  N = INT (L / DL)
170  PRINT N,DL
175  DIM X(N,3)
180  PR# 1
190  A4 = - .08
191  PRINT "FOR FREQUENCY= ";F;" MHZ AND L= ";L;" METERS"
192  PRINT "STEP SIZE IS ";DL;" METERS AND ATTENUATION CONSTANT IS ";A4
194  PRINT ""
195  PR# 0
200  REM  **********************************************
210  REM        CALCULATE CURRENT
212  FOR I = 1 TO N
215  D2 = (2 * L) - D1
220  A2 = D1 * F * 2E6 * PI / C
222  X(I,2) = A2
225  A3 = D2 * F * 2E6 * PI / C
230  R0 = EXP (A4 * A2) * COS (A2)
240  J0 = EXP (A4 * A2) * SIN (A2)
250  RI = EXP (A4 * A3) * COS (A3)
260  JI = EXP (A4 * A3) * SIN (A3)
270  IR = R0 - RI
280  IJ = J0 - JI
290  X(I,0) = IR
300  X(I,1) = IJ
310  D1 = D1 + DL
370  NEXT I
380  DA = X(2,2) - X(1,2)
390  FOR I = 1 TO N
400  X(I,2) = X(I,2) + DA / 2
410  NEXT I
420  REM  ::::::::::::::::::::::::::::::::::::::::::::::
430  REM        CALCULATE PATTERN
440  REM  eeeeeeeeeeeeeeeeeeeeeeeeeeeeeeeeeeeeeeeeeeeeee
450  HGR2
460  HCOLOR= 3
```

Program 15.3

Only a quadrant of the pattern is plotted. The pattern posesses mirror symmetry about the axis plotted horizontally, which represents $A5 = 0$ or the equator of the pattern. The vertical axis represents the axis of the wire, and the pattern is a figure of revolution about the wire. This should be relatively obvious since there is no mechanism for shaping the pattern in the plane of the equator. If the pattern is integrated with program 3.1,

```
470  POKE   - 12524,0
480  POKE   - 12525,64
490  POKE   - 12529,255
500  DIM Y(36,0)
510  AD = PI / 72
520  FOR J = 0 TO 36
530  FOR I = 1 TO N
540  PS = X(I,2) *  SIN (A5)
550  EP =   COS (A5) *  COS (PS)
560  ER = ER + X(I,0) * EP
570  EJ = EJ + X(I,1) * EP
580  NEXT I
590  Y(J,0) =  SQR ((ER * ER) + (EJ * EJ))
595  ER = 0:EJ = 0
600  IF  ABS (Y(J,0)) > EM THEN EM =  ABS (Y(J,0))
615  PRINT "AT ANGLE ";A5 * 180 / PI;"DEGREES , Y IS ";Y(J,0)
617  A5 = A5 + AD
630  NEXT J
635  A5 = 0
640  FOR J = 0 TO 36
645  IF Y < 0 THEN Y = 0: IF X < 0 THEN X = 0
650  Y = 190 - 190 *  ABS (Y(J,0) / EM) *  SIN (A5)
660  X = 190 *  ABS (Y(J,0) / EM) *  COS (A5)
670  IF A5 = 0 THEN Y0 = Y:X0 = X: GOTO 690
675  IF Y < 0 THEN Y = 0: IF X < 0 THEN X = 0
677  IF X < 0 THEN X = 0: IF Y < 0 THEN Y = 0
680  HPLOT X0,Y0 TO X,Y
685  X0 = X:Y0 = Y
690  A5 = A5 + AD
700  NEXT J
710  HPLOT 0,0 TO 0,190
720  HPLOT 0,190 TO 190,190
730  A5 = 0
740  Y = 190 - 190 *  SIN (A5)
750  X = 190 *  COS (A5)
760  Y1 = 190 - 185 *  SIN (A5)
770  X1 = 185 *  COS (A5)
780  HPLOT X,Y TO X1,Y1
790  Y2 = 190 - 134 *  SIN (A5)
800  X2 = 134 *  COS (A5)
810  HPLOT X2,Y2
820  A5 = A5 + AD
830  IF A5 <  = PI / 2 GOTO 740
3000 STOP
```

Program 15.3 (*continued*)

take note that *A* in that routine is measured from the axis whereas *A*5 is measured from the equator.

Determining the actual field strength at some point in space is usually accomplished by determining the gain of the antenna by pattern integration and then applying the Friis Transmission Formula.

REVIEWING THE PATTERNS

Having discussed the program, let us run it a few times to see what it tells us about the radiation patterns of some dipole antennas.

In Fig. 15.4(a), we see the pattern generated by a dipole that is only a twentieth of a wavelength across. The half-power beam width is 90-degrees (± 45-degrees from the equator). We see that the pattern does not depart appreciably from the cosine term of Eq. 15.4; in other words, the pattern is nearly identical with that of the elementary and infinitesimal dipole. All very small dipoles have the same pattern. As the length grows to a span of a half wave or a quarter wave on each side of the centerline, the beam width shrinks to 77-degrees. When L is equal to a half wavelength (again, a full wavelength span), we see that the pattern continues to narrow—to 47-degrees.

In Fig. 15.5(a), we see the very narrow pattern formed when is five-eighths of a wavelength. The beam width has shrunk to 32-degrees. This is the narrowest lobe that will be generated on the equator for any length. It has a particular significance for broadcasting that will be discussed shortly. A five-eighths-wavelength high vertical antenna will place the strongest signal on the horizon that is achievable with a single tower.

In Figs. 15.5(b) and (c), we see that as L increases to three-quarters of a wavelength and then to a full wavelength, the maximum on the equator decreases rapidly, becoming zero at a full wavelength. In the latter case, the pattern has become a pair of cones about the antenna axis.

Figure 15.6 shows that the cone becomes more and more endfire as the length increases. At selected positions, the pattern becomes a cone with minor lobes.

Figure 15.7 shows the progression as the wire grows longer. At 10-wavelengths, the minor lobes have settled to ripples, and the cone angle ceases to shrink much after 5-wavelengths. Note the change in scale compared to the earlier patterns. The inward swing ceases because the signal has become more and more attenuated and the backward wave dwindles into insignificance. At 10-wavelengths and an attenuation EXP(-0.08) per radian, the wave has dwindled by 43.7-dB, and the effect of the backward wave has nearly vanished. To all intents and purposes, it ceases to be significant beyond 5-wavelengths. The antenna is now a traveling-wave rather than a standing-wave antenna.

THE TRAVELING-WAVE ANTENNA

For the shorter lengths where a significant part of the signal reflects from their open end, antennas carry a standing wave. If the wire becomes long

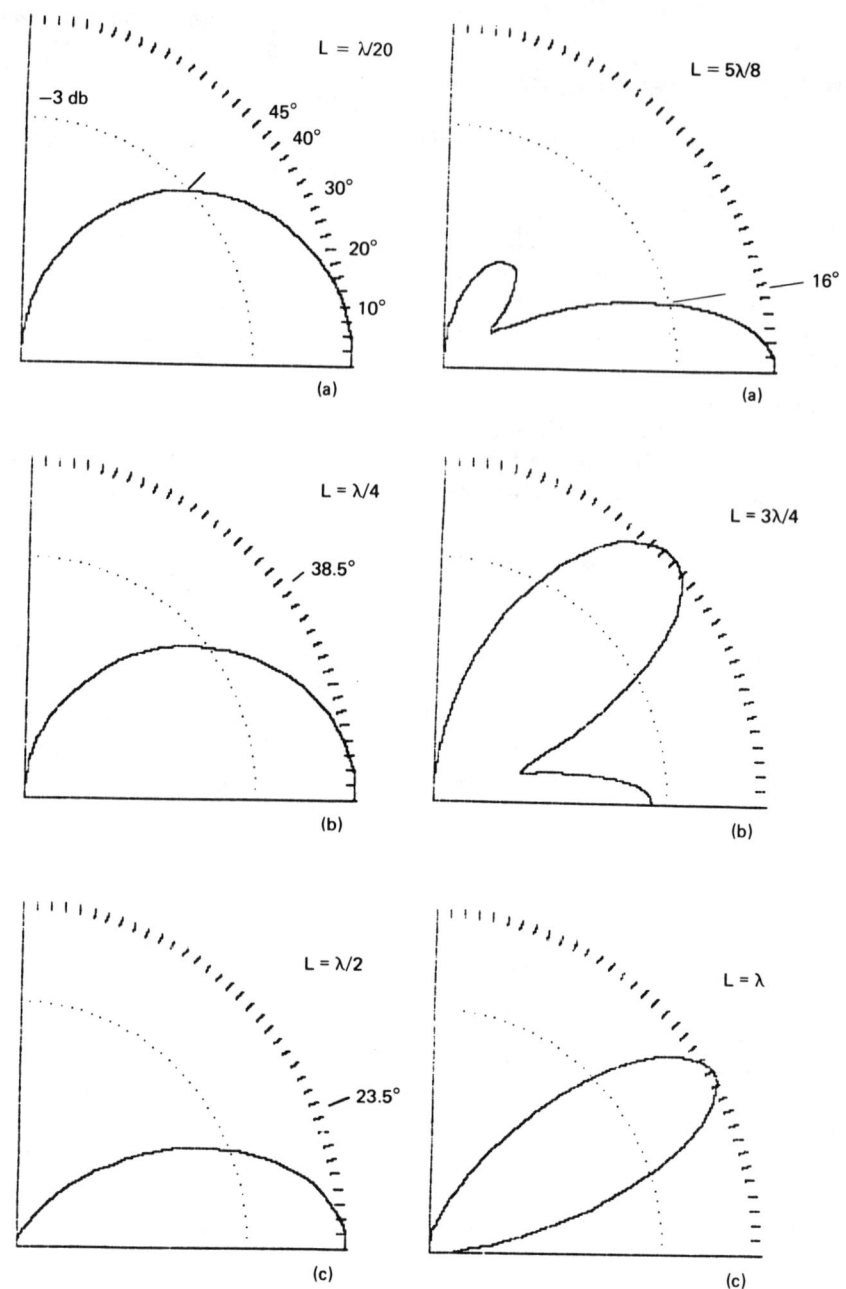

Fig. 15.4 Dipole pattern when (a) $L = \lambda/20$, (b) $L = \lambda/4$, and (c) $L = \lambda/2$. Freq. = 7.5 MHz; Step size = 0.8 m; Atten. const. = -0.08.

Fig. 15.5 Dipole pattern when (a) $L = 5\lambda/8$, (b) $L = 3\lambda/4$, and (c) $L = \lambda$. Freq. = 7.5 MHz; Step size = 0.8 m; Atten. const. = -0.08.

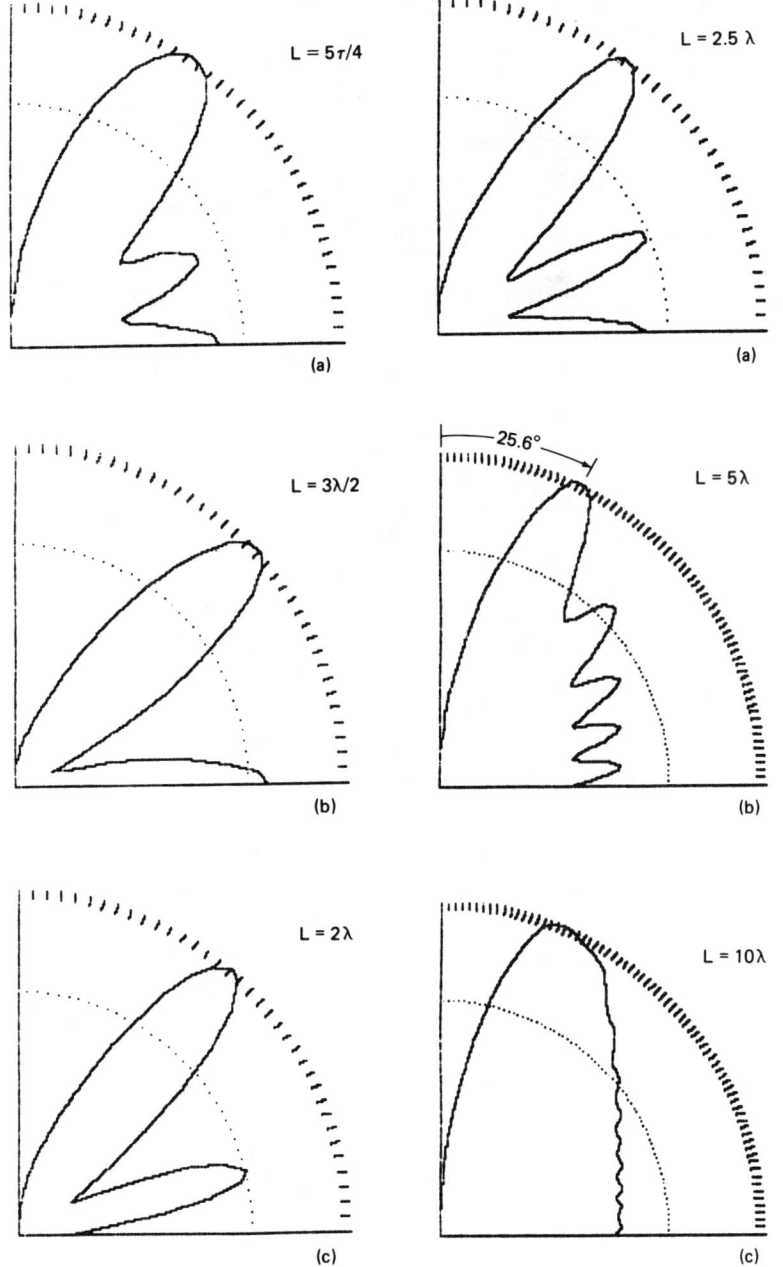

Fig. 15.6 Dipole pattern when (a) $L = 5\lambda/4$, (b) $L = 3\lambda/2$, and (c) $L = 2\lambda$. Freq. = 7.5 MHz; Step size = 2 m; Atten. const. = -0.08.

Fig. 15.7 Dipole pattern when (a) $L = 2.5\lambda$, (b) $L = 5\lambda$, and (c) $L = 10\lambda$. Freq. = 7.5 MHz; Step size = 2 m; Atten. const. = -0.08.

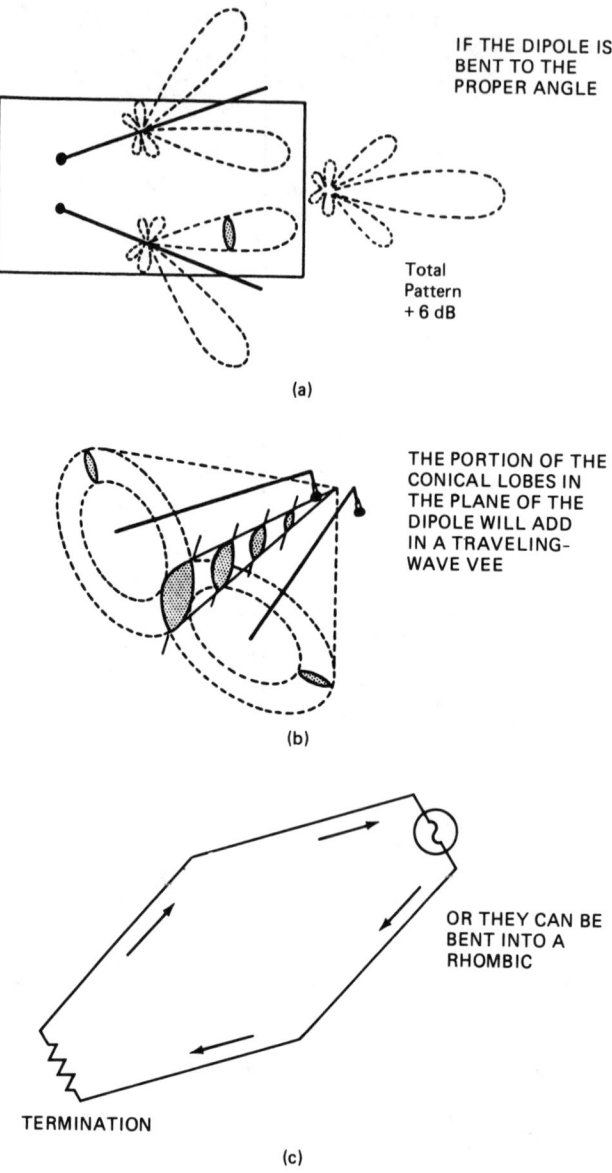

Fig. 15.8 The traveling-wave antenna.

enough, however, the backward wave becomes negligable and the antenna turns into a traveling wave device since only the outward wave is of any significance. A similar effect can be obtained by terminating the antenna. As shown in Fig. 15.8, if the dipole is bent in the middle to form an acute angle, it can be arranged for the two conical lobes to intersect and add, thereby raising the gain in the intersecting section by 6-dB since the voltages will also add [see Fig. 15.8(c)].

The antennas thus formed are called "VEE" antennas because of their shape. The VEE can be either terminated or unterminated, depending upon the radiation needed to minimize the backward wave.

Two VEE antennas can be connected together to form a *rhombic*, as shown in Fig. 15.8(c). The rhombic must usually be terminated even when the sides are as long as those of a VEE because the far end will capture some of the energy that had been radiated by the feed end. The rhombic was long a favorite for long-haul HF networks because of its gain and directivity and because it is very broadband. A terminated VEE or a rhombic can usually be operated over a two-to-one frequency band without adjustment.

The general characteristic of traveling-wave antennas is that they tend to be broadband. Since the backward wave is largely attenuated, the impedance stays relatively constant with changing frequency. Most broadband antennas such as the log-periodic, the spiral, etc., are generically traveling-wave antennas. At low frequencies, the energy travels relatively farther out on the antenna, and, as frequency increases, the major portion of the energy is radiated off closer to the feed.

16
Mutual Impedance

The self-impedances discussed in the previous chapter are the terminal impedances presented by the antenna to a generator when the antenna is in free space. They are also the source impedances of the antennas in the receiving condition. However, the latter description relies on the argument of reciprocity developed by Rayleigh[1] and Helmholtz and expanded by Carson.[2] Many of the natural phenomena around us exhibit reciprocal behavior. For instance, the lever, gear trains, and ac transformers all tend to be reciprocal. This theorem is generally stated: "If an emf is applied to the terminals of antenna A and the current measured at the terminals of another antenna B, then an identical current will be obtained at the terminals of antenna A if the same emf is applied to the terminals of antenna B." It is assumed that the medium is linear, passive, and isotropic.

If we consider the equivalent circuit (Fig. 16.1), we note that any four-terminal network may be simulated with a T network insofar as amplitude and phase of input voltage and output current are concerned. We may write the mesh equations as follows:

$$V_A = I_A(Z_1 + Z_T) - I_B Z_T \tag{16.1}$$

$$0 = -I_A Z_T + I_B(Z_2 + Z_T) \tag{16.2}$$

Solving Eq. 16.2 gives

$$I_B = \frac{I_A Z_T}{Z_2 + Z_T} \tag{16.3}$$

Substituting Eq. 16.3 in Eq. 16.1 gives

$$V_A = I_A(Z_1 + Z_T) - \frac{I_A Z_T^2}{Z_2 + Z_T} \tag{16.4}$$

whence

$$V_A(Z_2 + Z_T) = I_A(Z_1 + Z_T)(Z_2 + Z_T) - I_A Z_T^2 \tag{16.5}$$

Simplifying

$$V_A(Z_2 + Z_T) = I_A(Z_1Z_2 + Z_1Z_T + Z_2Z_T + Z_T^2) - I_AZ_T^2 \quad (16.6)$$

Thus

$$I_A = \frac{V_A(Z_2 + Z_T)}{Z_1Z_2 + Z_1Z_T + Z_2Z_T} \quad (16.7)$$

Substituting Eq. 16.7 into Eq. 16.3 yields

$$I_B = \frac{V_AZ_T}{Z_1Z_2 + Z_1Z_T + Z_2Z_T} \quad (16.8)$$

Now note that the equation is symmetrical, since if we place the generator at B and substitute Z_2 for Z_1 everywhere, we obtain

$$I_A = \frac{V_BZ_T}{Z_1Z_2 + Z_1Z_T + Z_2Z_T} \quad (16.9)$$

Thus if $V_A = V_B$ then $I_A = I_B$.

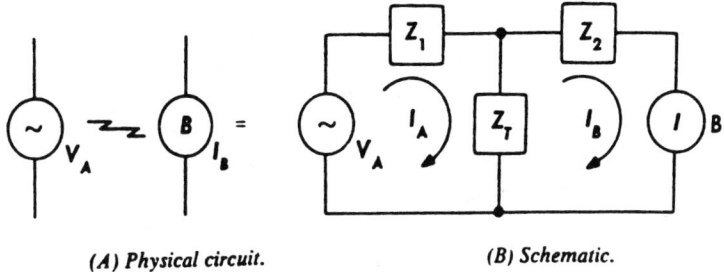

(A) Physical circuit. (B) Schematic.

Fig. 16.1 Two-antenna equivalents.

---

[1] Lord Rayleigh, *The Theory of Sound*, vols. 1 and 2 (New York: Dover Publications. Inc.)
[2] J. R. Carson. "Reciprocal Theorems in Radio Communication," *Proc. IRE.*, No. 17, June, 1929, pp. 952–956.

238 EXPLORING ANTENNAS AND TRANSMISSION LINES

In the circuit, the generator and ammeter impedances are considered zero. If they are merely identical, however, they may simply be lumped into Z_1 and Z_2.

The foregoing illustrates and proves the theorem of reciprocity without any restrictions on the values of source and load impedances (Z_1 and Z_2) or transfer impedance Z_T. This transfer impedance is *not* the same as the mutual impedance used in transformer theory. Mutual impedance of two coupled circuits is defined as the negative of the emf (V_{21}) induced in circuit 2 divided by the current in circuit 1 while circuit 2 is open-circuited. Thus, in a transformer analogue (Fig. 16.2), the mutual impedance is

$$Z_{21} = -\frac{V_{21}}{I_1} \qquad (16.10)$$

which is equivalent to the two-antenna case shown in Fig. 16.3.

Now the equations for the two-mesh mutually coupled circuit are

$$V_1 = I_1 Z_{11} + I_2 Z_{22} \qquad (16.11)$$

$$0 = I_1 Z_{21} + I_2 Z_{22} \qquad (16.12)$$

Arguing from reciprocity, we may write equivalent equations for Equations 16.10 and 16.11, thereby demonstrating symmetry, and thus

$$Z_{12} = Z_{21} = Z_M \qquad (16.13)$$

Therefore, the self-impedances (Z_{11} and Z_{22}) of each antenna and the circuit impedances of each antenna are involved, and the antenna currents of each influence the other. The term $I_2 Z_{21}$ is equal to $-V_{21}$ from Eq. 16.10. The actual computation of mutual impedance involves the integration of the induced components across both antennas and is a function of orientation and spacing. This solves out into cosine integral and sine integral terms and will not be explored here. Tables of mutual coupling for antennas in the configurations of Fig. 16.4 are to be found in most antenna handbooks for half-wave radiators. Fig. 16.4 shows the standard types of antenna geometries.

Another case frequently of interest is the radiator above ground, which is commonly shown as in Fig. 16.5. Tables of mutual coupling are also available for this type.

Computations of mutual coupling are frequently required since

MUTUAL IMPEDANCE 239

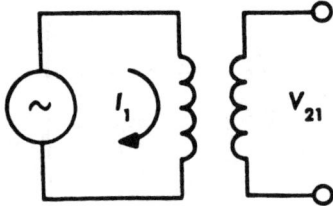

Fig. 16.2 Definition of mutual impedance.

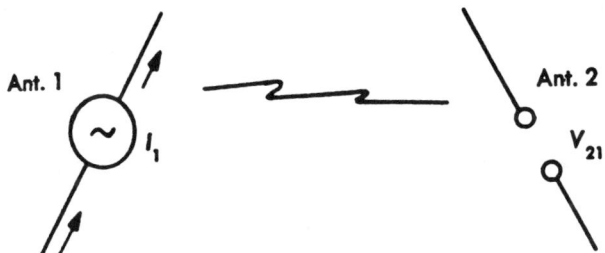

Fig. 16.3 Equivalent definition for antennas.

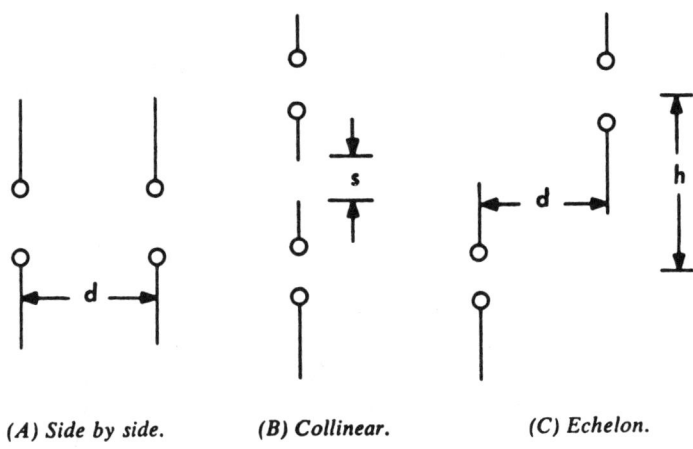

(A) Side by side.   (B) Collinear.   (C) Echelon.

Fig. 16.4 Standard antenna geometries used in mutual-coupling tables.

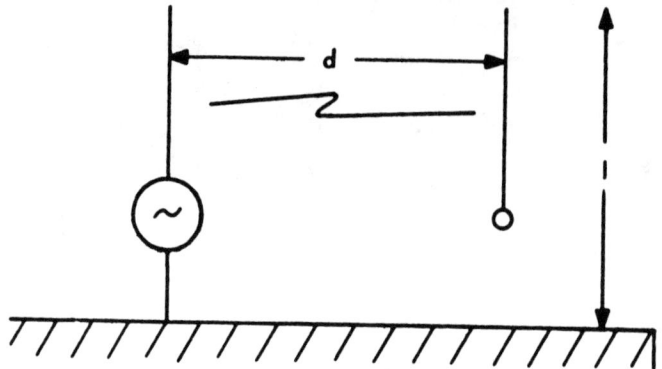

Fig. 16.5 Radiator above ground.

antennas are very often used in arrays. Unfortunately, such calculations have a tendency to suffer in accuracy since they depend on a knowledge of both self-impedance and current in the individual radiators. Very often the radiators will not be of resonant length (as in parasitic arrays like the Yagi) and will not be operating with a sinusoidal current distribution; as a result, tabulated data may not be available. However, mutual coupling is measurable, and empirical techniques are generally used in array antennas. Where more than two radiators are involved, Eqs. 16.10 and 16.11 become

$$V_1 = I_1 Z_{11} + I_2 Z_{12} + \cdots + I_N Z_1 N \tag{16.14}$$

$$V_2 = I_1 Z_2 + I_2 Z_{22} + \cdots + I_N Z_2 N$$

$$\cdots \cdots$$

$$V_N = I_1 Z_1 N + I_2 Z N_2 + \cdots + I_N Z_{NN}$$

The matrix of Eq. 16.14 bears witness to the complexity of mutual-impedance computation in arrays of relatively modest size, particularly those in which beam steering is accomplished by adjusting element-current phase. Such problems are usually handled by considering only the mutual coupling to closely adjacent elements for a typical central and an edge or corner element. As a rule, they are solved by computer programs and will constitute the largest portion of a steerable array development program.

In general, mutual coupling effects are noticeable only up to distances where the $1/r^2$ and $1/r^3$ terms described in Chaps. 14 and 15 become

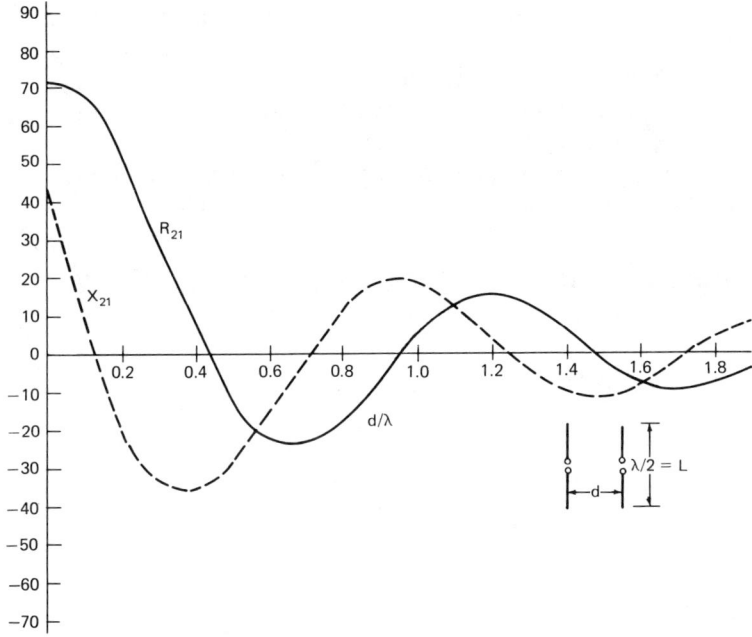

Fig. 16.6 Mutual coupling between thin λ/2 parallel dipoles.

neglibile. For spacings less than 3λ/4, however, they are frequently dominant and must be taken into account if proper operation of an array is to be attained.

It would not be practical to treat the many cases of mutual coupling here. The curve of Fig. 16.6, however, is an example of the results for parallel half-wave dipoles employing zero-diameter wire. The relevant equations are as follows:

$$R_{21} = 30\{2\text{Ci}(\beta d) - \text{Ci}[\beta(\sqrt{d^2 + L^2} + L)] - \text{Ci}[\beta(\sqrt{d^2 + L^2} - L)]\} \quad (16.15)$$

$$X_{21} = 30\{2\text{Si}(\beta d) - \text{Si}[\beta(\sqrt{d^2 + L^2} + L)] - \text{Si}[\beta(\sqrt{d^2 + L^2} - L)]\} \quad (16.16)$$

where Ci = the cosine integral; Si = the sine integral; and $\beta = 2\pi/\lambda$.

242 EXPLORING ANTENNAS AND TRANSMISSION LINES

It may be seen that the coupling is very tight at close spacing and not insignificant even at spacings in excess of a wavelength. The coupling for the end-on case is considerably smaller and decays more rapidly.

Mutual coupling formulas frequently employ cosine and sine integrals such as those seen in Eqs. 16.15 and 16.16. Some of the standard sine, cosine, and exponential integral relationships used to compute the values of these functions are shown in the following section.

SINE, COSINE, AND EXPONENTIAL INTEGRAL RELATIONS

$$\text{Ci}(x) = \int_{\infty}^{z} \frac{\cos u}{u} du = \text{cosine integral} \qquad (16.17)$$

$$\text{Ci}(x) = \ln\gamma x - \frac{x^2}{2!2} + \frac{x^4}{4!4} - \frac{x^6}{6!6} + \cdots \qquad (16.18)$$

where

$$\gamma = e^C = 1.781$$
$$\ln\gamma = C = 0.577 = \text{Euler's constant}$$

$$\text{Ci}(x) = 0.577 + \ln x - \frac{x^2}{2!2} + \frac{x^4}{4!4} - \frac{x^6}{6!6} + \cdots \qquad (16.19)$$

When $x < 0.2$,

$$\text{Ci}(x) \simeq 0.577 + \ln x$$

When $x \gg 1$,

$$\text{Ci}(x) \simeq \frac{\sin x}{x}$$

$$\text{Cin}(x) = \int_0^z \frac{1 - \cos u}{u} du \qquad (16.20)$$

$$\text{Cin}(x) = \ln\gamma x - \text{Ci}(x) \qquad (16.21)$$

$$\text{Cin}(x) = \frac{x^2}{2!2} - \frac{x^4}{4!4} + \frac{x^6}{6!6} - \cdots \qquad (16.22)$$

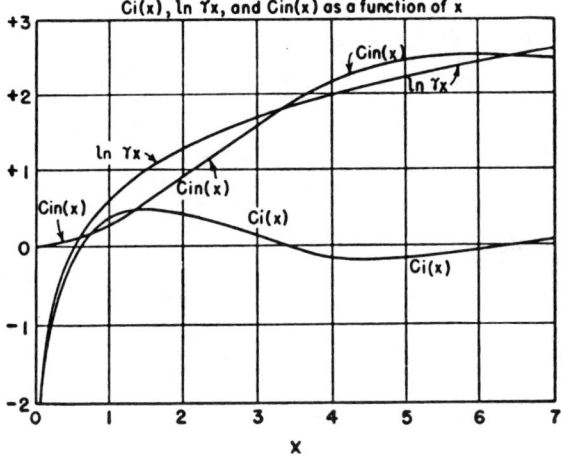

Fig. 16.7 Comparison of curves for Ci(x) and Cin(x).

$$\text{Ci}(x) = \ln\gamma x - \text{Cin}(x) \tag{16.23}$$

Curves for Ci(x), lnγx, and Cin(x) are compared in Fig. 16.7.

$$\text{Si}(x) = \int_0^z \frac{\sin u}{u} du = \text{sine integral} \tag{16.24}$$

$$\text{Si}(x) = x - \frac{x^3}{3!3} + \frac{x^5}{5!5} - \frac{x^7}{7!7} + \cdots \tag{16.25}$$

When $x < 0.5$,

$$\text{Si}(x) \simeq x$$

When $x \gg 1$,

$$\text{Si}(x) \simeq \frac{\pi}{2} - \frac{\cos x}{x}$$

$$\text{Ein}(z) = \int_0^z \frac{1 - e^{-u}}{u} du = \text{exponential integral} \tag{16.26}$$

where $z = x + jy$

```
140 D = - 1
150  PRINT " INPUT ANGLE IN RADIANS"
160  INPUT X
165 CI = .577 + LOG (X)
170 M = 2
175 N = M
180 A = 1 / N
190 B = N
200 A = A * X / B
210 N = N - 1
220  IF N > 0 GOTO 190
230 CI = CI + (D * A)
240  PRINT "FOR M= ";M;" CI= ";CI;" AND A= ";A
250 M = M + 2
255 D = - 1 * D
260  IF A > 1E - 6 GOTO 175
2000 STOP
```

```
]RUN
 INPUT ANGLE IN RADIANS
?15
FOR M= 2 CI= -52.9649498 AND A= 56.25
FOR M= 4 CI= 474.3788 AND A= 527.34375
FOR M= 6 CI= -2162.33995 AND A= 2636.71875
FOR M= 8 CI= 5783.1295 AND A= 7945.46845
FOR M= 10 CI= -10107.8094 AND A= 15890.9389
FOR M= 12 CI= 12464.547 AND A= 22572.3564
FOR M= 14 CI= -11454.355 AND A= 23918.902
FOR M= 16 CI= 8166.61931 AND A= 19620.9743
FOR M= 18 CI= -4657.54689 AND A= 12824.1662
FOR M= 20 CI= 2176.38379 AND A= 6833.93068
FOR M= 22 CI= -849.264383 AND A= 3025.64817
FOR M= 24 CI= 281.242746 AND A= 1130.50713
FOR M= 26 CI= -79.9843839 AND A= 361.22713
FOR M= 28 CI= 19.8445424 AND A= 99.8289263
FOR M= 30 CI= -4.25209501 AND A= 24.0966374
FOR M= 32 CI= .871780438 AND A= 5.12387545
FOR M= 34 CI= -.0952926269 AND A= .967073065
FOR M= 36 CI= .067805013 AND A= .16309764
FOR M= 38 CI= .0430784487 AND A= .0247265644
FOR M= 40 CI= .0464664635 AND A= 3.38801483E-03
FOR M= 42 CI= .0460448589 AND A= 4.21604633E-04
FOR M= 44 CI= .0460927178 AND A= 4.7858973E-05
FOR M= 46 CI= .046087742 AND A= 4.97588566E-06
FOR M= 48 CI= .0460882175 AND A= 4.75587476E-07
```

Program 16.1

$$\text{Ein}(jy) = \int_0^{jy} \frac{1 - e^{-u}}{u} du \tag{16.27}$$

$$\text{Ein}(jy) = \text{Cin}(y) + j\text{Si}(y) \tag{16.28}$$

$$\text{Ei}(\pm jy) = \text{Ci}(y) \pm j\text{Si}(y) \tag{16.29}$$

Because of the nature of these functions, it would be worthwhile to investigate how they are evaluated by a computer. We have elected to demonstrate the solution to Eq. 16.19 since it represents a series progression and therefore lends itself to computer evaluation. Nevertheless, there is a snare here for the unwary! You will note that the numerator contains a power of x whereas the denominator contains a factorial multiplied by the largest term of the factorial. If x is of any significant size, the number in the numerator gets awfully large awfully fast. Similarly, the number in the denominator gets very large very fast. For modest values of x, the initial terms of the alternating series are large. As a result, the series tends to converge very slowly, and it is necessary to evaluate quite a few terms.

Program 16.1 shows the approach taken to minimize some of these difficulties. The real kernel of the program is to be found in lines 190 through 220. Rather than letting the numerator and denominator grow uncontrollably, the sequence is to multiply by x in the numerator and then to divide by the next smaller term of the factorial in the denominator until the denominator diminishes to unity. The extra division by the base of the factorial is initially performed in line 180. After the term has been calculated, it is accumulated in alternating sequence into Ci in line 230. The alternation is performed by lines 140 and 255. The program runs until the increment being added to Ci becomes less than 1E − 6.

The alternating sign is supplied in line 230, where the value of term A is multiplied by D. The value of the latter is toggled between +1 and −1 in line 255. Note, in line 165, that the Apple reads LOG as the natural logarithm.

This form of program illustrates the fact that some computations have to be approached by a special algorithm in order to be solvable with reasonable hardware.

The run at the conclusion of the program shows how slowly the series converges. Note that it was not until the thirty-eighth term that the error moved into the third decimal place. As a matter of fact, when x is larger than about 4, the error grows for the first sequence of terms. Note that the error in this example peaks at the twelfth term.

This program can be slightly altered to run the sine integral and Cin as well. It actually runs at a reasonable speed until x gets larger; it then slows geometrically. The program begins to run out of gas despite the precautions at $x = 25$ or so because of round-off errors. Beyond this point, the accuracy falls below 1 percent.

17
Electrically Small Antennas

In the course of practical antenna work, particularly in the communication field, one is frequently called upon to work with electrically small antennas. By "electrically small," we mean antennas smaller than the first resonance, or one-quarter wavelength. An electrically small antenna can actually be very large in physical terms. For example, a VLF station such as OHU in Hawaii operating at 14-kHz broadcasts a wavelength of 21,429-meters (70,286-feet), but the 1200-foot high antenna is only 0.017-wavelength long. Despite the fact that it is taller than the Empire State Building, the antenna is electrically small!

Consider another case. The *CB* band at 27-MHz has a wavelength on the order of 11-meters, or 36-feet. Any *CB* antenna shorter than 9-feet is thus electrically small. Any practical antenna to be mounted on a walky-talky has to be electrically small.

The only radio required by international treaty for passenger and cargo ships operates in the 415- to 510-kHz band. At 500-kHz, the wavelength is 600-meters, or 1968-feet. Any antenna shorter than 492-feet is therefore electrically small. For a maritime radio operating in this band, it is impossible to install an electrically standard size radiator because anything taller than 192-feet above the waterline cannot pass through the Panama Canal.

Suffice it to say that a great many situations exist for which electrically small antennas are absolutely necessary. In fact, electrically small antennas are actually much more common than full size antennas. Consider car radios alone. The top of the standard broadcast band is 1605-kHz, with a wavelength of 186-meters or 613-feet. The typical 1-meter whip used with most car radios is thus only 0.0054-wavelength long.

What are the characteristics of an electrically small antenna? In a definitive work, Harold Wheeler described their properties on the basis of the space about the antenna volume. Suppose that we have an antenna that fills a certain volume in space. The properties of the space *outside* of that volume determine the minimum Q that can be attained by a lossless antenna *within* the volume.[1]

Consider a small whip antenna. The whip itself will exhibit some capacitance, which is measurable down to dc. If the current of the whip

---

[1]Harold Wheeler, "Fundamental Limitations of Small Antennas," *Proceedings of the IRE*, Dec. 1947, p. 1479.

were uniform from base to top, it would present a resistive impedance of

$$RR = 80*PI*PI*(L/LA)\wedge 2 \qquad (17.1)$$

where:

L = radiator length
LA = wavelength

Note that this equation is the same as Eq. 14.31.

A whip will draw a nearly uniform current only if it is very heavily top-loaded. A typical shipboard medium-frequency (490–510-kHz) antenna will consist of a top loader made up of horizontal wires and a "downlead," as shown in Fig. 17.1. In effect, all of the radiation comes from the down-lead since the currents on the top loader essentially cancel in the far field.

For an unloaded whip or mast, the current distribution is not uniform but triangular. The formula for radiation resistance thus becomes

$$RR = 40*PI*PI*(L/LA)\wedge 2 \qquad (17.2)$$

To give an idea of how small this resistance can become for an electrically small antenna, consider the case of a 15-foot (4.57-meter) whip at 2-MHz. In this case, the height is 4.57/150, or 0.030-wavelength. The radiation resistance works out to be 0.355-ohms.

For a simple, straight whip or mast antenna, the capacitance tends to run about 2pF/foot + 5pF. Thus, a 15-foot whip antenna might measure something like 35-pF. At 2-MHz, this works out to $-j2273$-ohms. Thus, the impedance predicted for the antenna in ohms is

$$Z = 0.355 - j2273$$

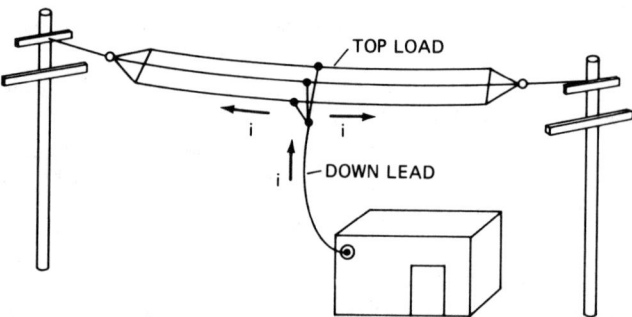

Fig. 17.1 The top-loaded antenna.

This value represents a Q of over 6000! In actual practice, however, one will never measure such a high Q. To begin with, the base of the antenna will always be shunted by some stray capacitance, and the loss resistance in the ground plane will elevate the terminal resistance of the antenna somewhat. A standard U.S. Navy value for the impedance, in ohms, of a 15-foot whip antenna at 2-MHz is

$$Z = 6.2 - j1125$$

It is noteworthy that the standard value for the 25-foot whip is given as

$$Z = 3.2 - j600$$

Offhand, one would expect the resistance term for the 25-foot antenna to be higher by the ratio $(25/15)^2$ or 2.78. The fact that the taller antenna has less resistance lends credence to the idea that much of the resistance term occurs in the ground, probably as losses.

In order to dissipate any significant amount of power in an electrically short antenna, it is necessary to correct the power factor at the very least. Consider the $(6.2 - j1125)$-ohm value. In order to radiate 100-watts, will be necessary to drive the current as follows:

$$P = I*I*R$$
$$100 = I*I*6.2$$
$$I = 4.016 \text{ ARMS}$$

The voltage at the antenna base is then approximately

$$V = 4.016*1,125$$
$$= 4518 \text{ VRMS} = 6,390 \text{ VP}$$

Now if we were to supply an inductor in series with the antenna terminals with an inductive reactance of $+j1125$-ohms the net input impedance of the system would be 6.2-ohms plus whatever loss resistance would be donated by the inductor. Suppose that the inductor has a Q of 200. Then the loss resistance is 1125/200, or 5.625-ohms, so that the total circuit input impedance becomes 6.2 + 5.6, or 11.8-ohms. The required input voltage is 4.016*11.8, or 45-VRMS. The actual input power is 190-watts, of which 100 is delivered to the antenna and 90 is dissipated in the inductor.

It is characteristic of electrically small antennas that very large voltages and currents are required to dissipate, and perhaps radiate, modest

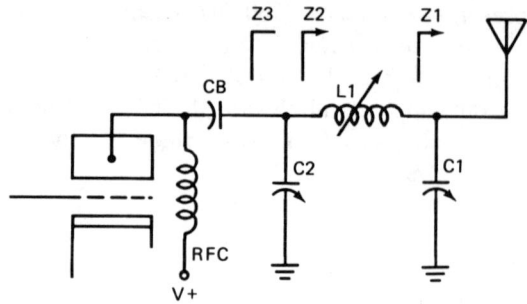

Fig. 17.2 The PI network.

amounts of power because their power factor is so poor. Typically, such antennas must be supplied with an antenna coupler to be able to operate with reasonable efficiency. In older equipment, the final amplifier of the transmitter was usually equipped with a PI network to permit tuning of the antenna to match the plate impedance of the tube. A simplified amplifier circuit is shown in Fig. 17.2. Capacitor CB is the blocking capacitor that keeps the very high plate voltage of the tube off the antenna. The tube dc plate current is fed through the RF choke labeled RFC. Components $C1$, $C2$, and $L1$ form the "PI," so called because of the resemblance of the shape to the Greek letter pi.

The PI network is not uniquely determinate since there are an infinite number of values that can transform one impedance into another. Let us try a simple example in which the antenna load is selected for the first resonance with an impedance of $37 + j0$-ohms. $C1$ could have a variety of values, but we arbitrarily choose to give it a 45-degree phase angle with the antenna. Since the capacitor goes in parallel, we must add the complex susceptances. Next, to obtain the series equivalent circuit, we assume that $ZA = 37\text{-}\Omega = 0.027$-mho and arbitrarily set $C1$ so that $B1 = +j0.027$-mho. Then,

$$Z1 = 18.52 - j18.52 \tag{17.3}$$

The susceptance is obtained from

$$Z = \frac{1}{Y} = \frac{1}{G + jB} \tag{17.4}$$

Since the imaginary part must be removed from the denominator, we multiply both sides of the fraction by the conjugate, as shown in Eq. 17.5, and then factor the numerator into real and imaginary parts to obtain Eq. 17.6, as follows:

$$Z = \frac{(G - jB)}{(G + jB)(G - jB)} = \frac{G}{G^2 + B^2} - j\frac{B}{G^2 + B^2} \quad (17.5)$$

$$= R - jX \quad (17.6)$$

The plate impedance of a transmitting tube runs in the neighborhood of some hundreds of ohms to a few thousand ohms. We here assume that we wish to attain 1000-ohms. We add a series inductor of the right size and find that the real part of the parallel equivalent is 0.001-mho, or the reciprocal of 1,000-ohms. Now by adding $X_{L1} = +j153.3$, we obtain

$$Z_2 = 18.51 + j134.8\text{-ohms} \quad (17.7)$$

$$Y_2 = 0.001 - j0.00728\text{-mho} \quad (17.8)$$

It remains then to set C2 to cancel the residual inductive reactance and thus achieve our match. Setting C2 for $B2 = +j0.00728$, we obtain

$$Y_3 = 0.001 + j0\text{-mho} \quad (17.9)$$

$$Z_3 = 1,000\text{-ohms} \quad (17.10)$$

The source of the "magic number" for the inductor value will be discussed shortly.

It is fairly easy to see that this technique with networks requires frequent transitions between the parallel equivalent and the series equivalent circuits. The routine of Program 17.1 permits such rapid transformations. Since BASIC does not have a rectangular/polar/rectangular operation, the program is written to follow the steps of equations 17.2 through 17.4. On a calculator, one can simply do a rectangular-to-polar transition, take the reciprocal of the magnitude, and do a polar-to-rectangular transition.

Now, the "magic number" routine can be found in Eqs. 17.11 through 17.16. The procedure is relatively straightforward and can be easily performed on a calculator or with a basic program. Since X2 may be either positive or negative, Eq. 17.12 has two answers except for the special case when $X1 = 0$. The routine, which will work just as well for a

```
145  HOME
150  PRINT "ENTER 1 FOR IMPEDANCE TO ADMITTANCE"
160  PRINT "ENTER 2 FOR ADMITTANCE TO IMPEDANCE"
170  INPUT A
180  IF A = 1 GOTO 300
190  IF A = 2 GOTO 500
200  GOTO 145
280  REM ::::::::::::::::::::::::::::::::::::::::::::
290  REM       THE IMPEDANCE ROUTINE
300  PRINT "ENTER RESISTANCE"
310  INPUT R
320  PRINT "ENTER REACTANCE"
330  INPUT X
340  G = R / ((R * R) + (X * X))
350  B =  - X / ((R * R) + (X * X))
360  IF B < 0 GOTO 390
370  PRINT "THE EQUIVALENT ADMITTANCE IS ";G;" +J ";B;" MHOS."
380  GOTO 700
390  PRINT "THE EQUIVALENT ADMITTANCE IS  ";G;" -J "; ABS (B);" MHOS"
400  GOTO 700
480  REM  ::::::::::::::::::::::::::::::::::::::::::::
490  REM       THE ADMITTANCE ROUTINE
500  PRINT "ENTER CONDUCTANCE IN MHOS"
510  INPUT G
520  PRINT "ENTER SUSCEPTANCE IN MHOS"
530  INPUT B
540  R = G / ((G * G) + (B * B))
550  X =  - B / ((G * G) + (B * B))
560  IF X < 0 GOTO 590
570  PRINT "THE EQUIVALENT IMPEDANCE IS ";R;" +J ";X;" OHMS"
580  GOTO 700
590  PRINT "THE EQUIVALENT IMPEDANCE IS ";R;" -J "; ABS (X);" OHMS"
600  GOTO 700
700  PRINT " ENTER ANY NUMBER TO GO BACK TO START"
710  INPUT A
720  GOTO 145
2000 STOP
```

Program 17.1

target resistance if one simply swaps R for G and B for x, is as follows.

Suppose we have $R + j(X1)$-ohms and wish to add enough reactance so that the real part of the equivalent admittance is G. Then,

$$Y = \frac{1}{R + j(X1) + j(X2)} = \frac{1}{R + j[(X1) + (X2)]} \qquad (17.11)$$

$$Y = \frac{R}{R^2 + [(X1) + (X2)]^2} - j\frac{X}{R^2 + [(X1) + (X2)]^2} \qquad (17.12)$$

$$G = \frac{R}{R^2 + [(X1) + (X2)]^2} \qquad (17.13)$$

$$B = -j\frac{X}{R^2 + [(X1) + (X2)]^2} \qquad (17.14)$$

Thus,

$$GR^2 + G[(X1) + (X2)]^2 = R \qquad (17.15)$$

and

$$[(X1) + (X2)]^2 = \frac{R}{G} - R^2 \qquad (17.16)$$

As an example, let $G = 0.001$-mho and $R = 18.52$-ohms. Then,

$$\begin{aligned}[(X_1) + (X_2)]^2 &= \frac{18.52}{0.001} - (18.52)^2 \\ &= 18{,}520 - 343 \\ &= 18{,}177 \\ (X_1) + (X_2) &= 134.8\end{aligned}$$

The basic technique used here is the one most commonly employed for impedance matching. One simply adds a reactance or susceptance until the opposite equivalent circuit contains the correct real part. A last element is then added to cancel the residual reactance.

In a PI network, the impedance transformation is related to the ratio between the two capacitors. In the present example, the transformation is from a small antenna impedance to a large plate impedance, and the value of $C1$ is much larger than the value of $C2$. The PI has the advantage of strongly supressing harmonics since it has a low-pass filter format. For the best filtering action, the network is adjusted so that the phase angles of the input and output are equal.

If the PI is used to match a low resistance in series with a large capacitive reactance, one usually sets $C1$ to the minimum possible value. In this condition, the network becomes essentially an L net consisting only of $L1$ and $C2$.

Consider, for example, the antenna impedance of $6.2 - j1125$. Suppose that we would like to transform that value to $50 + j0$-ohms and that $C1$ can be set to a negligibly small value of susceptance. Using Eq. 17.14, we can solve for the required reactance. This turns out to be 16.48-ohms. Since we would like to cancel the residual with $C2$, we make the inductive reactance 16.48-ohms larger than the antenna capacitive reactance, or $+j1142$-ohms, which is 90.8-µH at 2-MHz. The residual susceptance works out to $-j0.053$-mhos, or 4,230-pF at 2-MHz.

The above calculation did not take into account any loss. If we assume that the coil has a loss resistance of about 5-ohms that is to be added to the 6.2-ohms from the antenna, the new required reactance will be 20.85-ohms for an inductive reactance of $+j1146$-ohms, or 91.2-µH. The residual susceptance is $-j0.037$-mhos, which can be canceled with a 2962-pF capacitor. We see that the inductance has changed only slightly but the capacitance has changed dramatically. In networks of this type, the losses usually cannot be neglected. As a matter of fact, we see that the calculation would really have to be iterative since the change in inductance would change the loss resistance, and so forth.

An alternative arrangement would be to use the inductor in shunt with the antenna and a series capacitor to cancel the residual. After an iteration, assuming a Q of 200 for the inductor, we find that the shunt inductor would be 58.6-µH and the series capacitor, 37.2-pF. These values are significantly different from the series inductor/shunt capacitor scheme.

For fixed installations and some shipboard installations, top loading is frequently used. It not only doubles the radiation resistance but also reduces the reactance, thereby making matching easier and more efficient. A typical top-loaded shipboard MF antenna (415 to 510-kHz) will have a loader capacitance of 250-pF.

For packset and automotive applications, top loading is physically unsatisfactory because the antenna may catch and snag in trees and brush. Therefore inductive loading is frequently employed (Fig. 17.3). If the inductor is at the antenna base, the current distribution does not change from the triangular case; therefore, the radiation resistance is not raised. However, the external loading inductor usually has a higher Q than can be achieved within the equipment case. Thus, a higher efficiency may result. In addition, circulating currents in the radiator base capacitance are greatly reduced.

For center loading, however, the current distribution is squared off as shown in Fig. 17.3(b), and a higher radiation resistance results. Since the reactance is also lowered, this system gains in efficiency and bandwidth over the base-loaded type at the cost of some physical awkwardness. In both cases, the length of the inductor is simply added to the total height. If the inductor diameter/wavelength ratio is very tiny, the horizontally polarized (loop-mode) radiation is negligible and the principal radiation from the inductor lumps in with the vertically polarized dipole mode.

The loading of the types previously described may be carried out in varying degrees ranging from small amounts to resonant loading in which the imaginary portion of the impedance goes to zero. In the latter case, if any appreciable bandwidth is to be covered, the loading must be made

ELECTRICALLY SMALL ANTENNAS 255

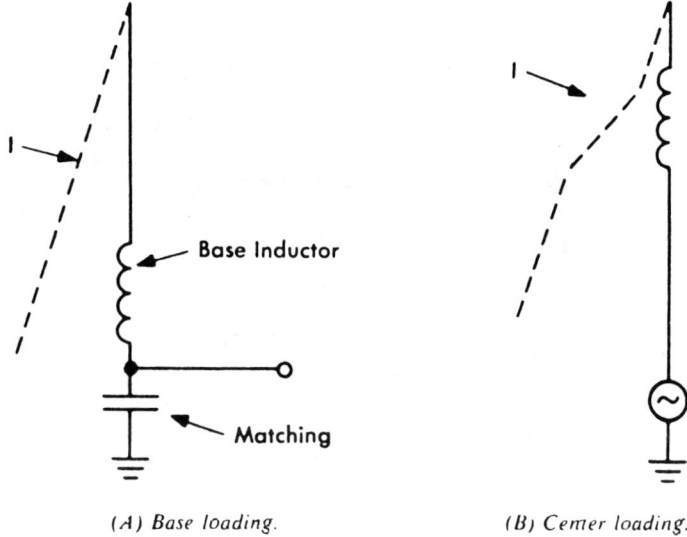

(A) Base loading.   (B) Center loading.

Fig. 17.3 Base- and center-loaded whips.

variable. Antennas tunable over the entire high-frequency band and more have been constructed in this fashion. In general, the resistive term $(R_R + R_{loss})$ is not directly suitable for transmission-line feeding, and therefore appropriate transformations are required. These are commonly incorporated into the antenna structure itself.

An example is the line-loaded monopole or hula hoop or D.D.R.R. radiator (Fig. 17.4). In this antenna, an unbalanced, conductor-above-ground transmission line is used to load a short vertical radiator of height h ($h \cong 0.01\lambda$, commonly). The principal radiation is vertically polarized[2] because of the large resonant currents on the vertical section h.[3] With the transmission line resonant, a generator may be tapped at a distance d from the radiator h to find a suitable resistive impedance.

A second type is the resonant helix, sold commercially as the *Helacyl II* by ITT. In this antenna, a base-loaded monopole is reduced in height until only a vestigial radiator remains, and the base inductor is greatly enlarged until it accounts for the majority of the radiation. The circuits of Fig. 17.5 show two embodiments of this antenna along with equivalent circuits. Tuning is accomplished by varying inductance to resonance and then tapping off an appropriate resistive impedance. Because of the

---
[2] J. M. Boyer, "The Hula Hoop Antenna" *Electronics*, vol. II, 1963.
[3] R. W. Burton and R. W. P. King, "Theoretical Considerations and Experimental Results for the Hula Hoop Antenna," *Microwave Journal*, Nov., 1963.

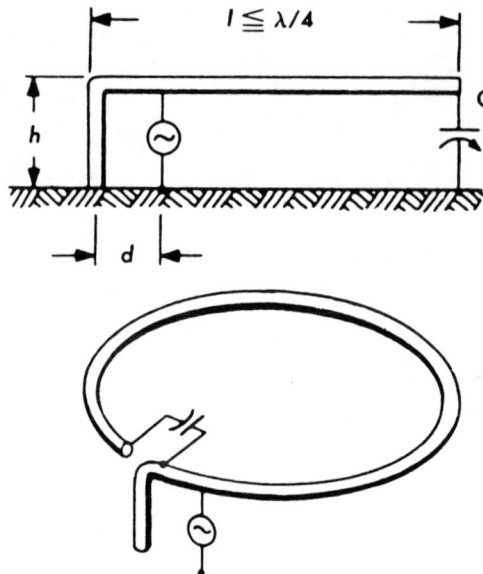

Fig. 17.4 Hula hoop antenna.

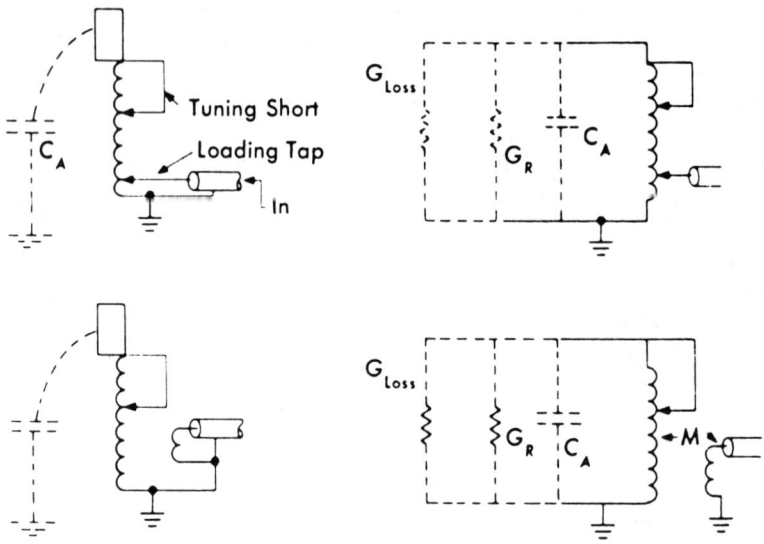

Fig. 17.5 Resonant helix antennas.

shunting effect of the input line, these adjustments interact. Conductance G_{loss} is made up of skin, core, contact, and short-circuit losses. The term G_R is the radiation conductance. The units are very similar except that input is by means of mutual coupling which may be either variable for full matching or fixed (in which case a compromise resistive term is obtained).[4]

All three of these antennas are very high Q and relatively inefficient devices. A discussion of these factors will be found in the section on short-antenna limitations.

It is particularly noteworthy that each of these loaded antennas radiates in the monopole mode in real half-space and is thus vertically polarized. The equivalent lump capacitance and the inductor are small enough (at the lowest frequencies) to behave as lumped elements. Let us consider such an antenna with height $h = \lambda/45$ and inductor turns $N = 75$. Then, the one-turn area $A = 7.43 \times 10^{-6}\lambda^2$.

For the monopole, from Eq. 14.31 (current is uniform due to lumped capacitance effect),

$$R_{R_M} = \tfrac{1}{2}(80\pi^2)\left(\frac{L}{\lambda}\right)^2 \tag{17.17}$$

But $L = 2h$, and therefore

$$R_{2_M} = 160\pi^2\left(\frac{h}{\lambda}\right)^2 = 1585 \times \frac{1}{2025}$$
$$= 0.782\text{-ohm}$$

For the loop, from Eq. 17.28,

$$R_{R_L} = 31{,}200(NA/\lambda^2)^2 \tag{17.18}$$
$$= 31{,}200(75 \times 7.43 \times 10^{-6})^2$$
$$= 17.3 \times 10^{-4}\text{-ohm}$$

---

[4] See H. P. Westman, ed., *Reference Data for Radio Engineers*, 5th ed., pp. 25–11 to 25–13. Also see A. G. Kandoian and W. Sichak, "Wide Frequency Range Tuned Helical Antennas and Circuits," *Electrical Communication*, vol. 30, Dec., 1953, pp. 294–299; W. Sichak and J. J. Nail, *U.S. Patent No. 2,781,514*; and A.G. Kandoian, *U.S. Patent Nos. 2,866,197* and *2,875,433*.

Since the currents are identical, the ratio between the two components is

$$\frac{R_{R_M}}{R_{RL}} = \frac{7820}{17.3} = 451$$

A few more conclusions may be drawn from this sample. Let us assume that an ωL of 4000-ohms represents a practical limit of loading inductor size. Then, for an inductor maximum true Q of 400, skin and other losses would total 4000/400, or 10-ohms, and efficiency $\eta = R_R/R_{sum} = 0.782/10 = 0.078$.

The bandwidth between 2:1 VSWR points is

$$a = Q\frac{\Delta f}{f_R} \quad \text{and} \quad \pm\frac{\Delta f}{f_R} = \frac{0.35}{400} = 8.75 \times 10^{-4} \tag{17.19}$$

In the broadcast band at $f_R = 10^6$-Hz, the bandwidth would be only ±87.5-Hz. Thus, the antenna Q is too high to permit voice transmission. De-Q'ing the antenna to ±5-kHz between 2:1 VSWR points would reduce the Q to

$$Q = \frac{0.35 \times 10^6}{5 \times 10^3} = 70 \tag{17.20}$$

in turn reducing the efficiency, since

$$R_{loss} = \frac{\omega L}{Q} = \frac{4000}{70} = 57.2\text{-ohms} \tag{17.21}$$

and therefore,

$$\eta = \frac{0.782}{57.2} = 0.0137 \tag{17.22}$$

thus illustrating a typical bandwidth/efficiency tradeoff in an electrically very small antenna. Physically, it would be 21.9-feet tall and approximately 2-feet in diameter at 1-MHz. Naturally, in high-power installations, provision must be made for adequately disposing of the heat generated in the loss resistance.

A fairly common electrically small antenna is the loop. A coil wound around a ferrite rod serves as the antenna for nearly all portable transistor radios. Since the ferrite rod is not suitable for transmitting, the loop must be constructed with very large high-conductivity conductors to

ELECTRICALLY SMALL ANTENNAS 259

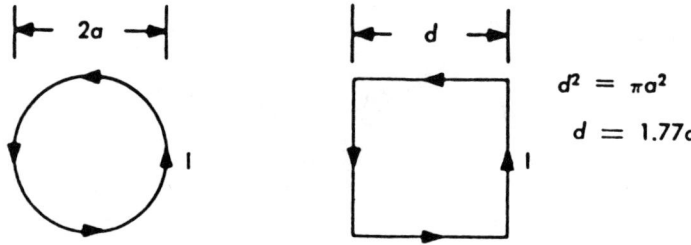

Fig. 17.6 Equivalent circular and square loops.

be used for transmitting because of the very large current that must be circulated to radiate any significant power.

The properties of the loop antenna for very small loops may be deduced by a variety of methods, as follows:

1. By assuming the loop to be a magnetic dipole (conducting fictitious magnetic currents).
2. By using the known field of an electric current element to find the voltage induced in the loop. (See Schelkunoff and Friis, *Antennas: Theory and Practice*, p. 319.)
3. By using the known fields of four linear current elements found in Chap. 14 in Eqs. 14.17 and 14.18 to form a square loop of equivalent area. (See J. D. Kraus, *Antennas*, pp. 155–157.)

The last method is perhaps the simplest since it follows the development of this book. Figure 17.6 shows a circular loop and its equivalent square loop. Now, for $a \leq \lambda$, we may use the infinitesimal dipole relations, and I may be assumed constant around the loop. Note also that, in the yz plane, the radiation from elements 1 and 3 cancels at all angles θ (Fig. 17.7). The pattern of dipoles 2 and 4 may then be written as follows:

$$E\phi = -E_{\phi_0} e^{j\psi/2} + E_{\phi_0} e^{-j\psi/2} \tag{17.23}$$

In this equation

$$\psi = \frac{2\pi d}{\lambda} \sin\theta \tag{17.24}$$

(see Fig. 17.8). It can be shown that

$$E_\phi = -2j E_{\phi_0} \sin[(2\pi d/\lambda)\sin\theta] \tag{17.25}$$

The far-field loop radiation is

$$E_\phi = \frac{120\pi^2 I_0 \sin\theta}{r} \frac{A}{\lambda^2} e^{j\omega(t - r/c)} \tag{17.26}$$

where

$A = d^2$
I_0 = peak current in loop

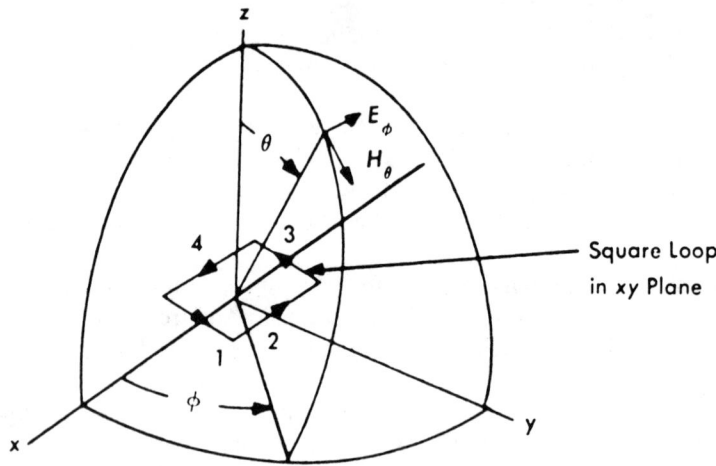

Fig. 17.7 Loop geometry.

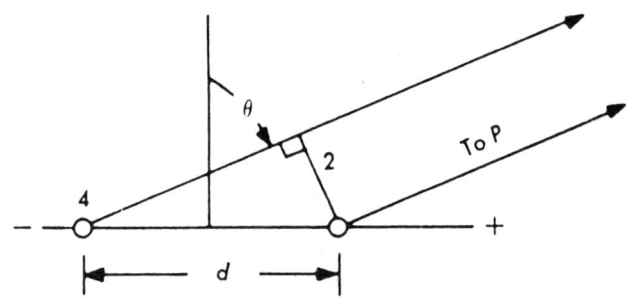

Fig. 17.8 Dipoles 2 and 4.

And, since $H_\theta = E_\phi/120\pi$,

$$H_\theta = \frac{\pi I_0 \sin\theta}{r} \frac{A}{\lambda^2} e^{j\omega(t - r/c)} \qquad (17.27)$$

Thus the formulas for the loop and dipole (along the z axis) differ only in that the loop expressions are greater by the factor $j2\pi$ and the polarizations are crossed. The loop thus appears to behave like a fictitious magnetic dipole along the axis of the loop. The time quadrature relation is used to good advantage in high-frequency direction finders where the dumbell pattern of a loop is added to the circular pattern of a dipole to yield a cardioid with a single null for direction finding without ambiguity.

The radiation resistance of a small loop of n turns may be shown to be

$$R_R = 31{,}200\left(\frac{nA}{\lambda^2}\right)\text{-ohms} \qquad (17.28)$$

Thus the radiation resistance of a loop ($n = 1$) of $d = \lambda/10$ (with $A = 0.01\lambda^2$) is 3.1-ohms.

The relation in Eq. 17.28 is accurate to a few percent up to this point. Note that, for $d = \lambda/10$, the circumference is 0.4λ. Thus the current can no longer be considered to be constant in amplitude and phase.

A few antenna types such as the "cloverleaf" (see Fig. 17.9), in which a series of smaller loops are connected to form an effective large loop, are practically used with medium to large diameters. The principal advantages are horizontal polarization with an omnidirectional pattern (for a horizontal loop). The cloverleaf is widely used in FM broadcasting.

A 0.1-wavelength represents about the top limit of loop size. A more typical value might be a meter on a side. At 2-MHz, this represents 1/150 wavelength and $4.4E - 5$-wavelengths. For $n = 4$, $R_R = 9.86E - 4$-ohms. Because of the very tiny radiation resistance, the loop must carry very large circulating currents. For a reasonable efficiency, moreover, the conductors must be very large and low in resistance. Fortunately, the loop can be matched with a vacuum capacitor that has a very high Q, and, by constructing the loop of heavy-gauge tubing with careful silver plating, an effective radiator can be constructed.

Figure 17.10 shows such an antenna. This loop is designed to cover the range from 2 to 16-MHz. It radiates a horizontally polarized signal and is used for high-incidence-angle skywave propagation. The coupling link is balanced with respect to earth, and the antenna is ground-independent since it does not require any earth connection or counterpoise.

262 EXPLORING ANTENNAS AND TRANSMISSION LINES

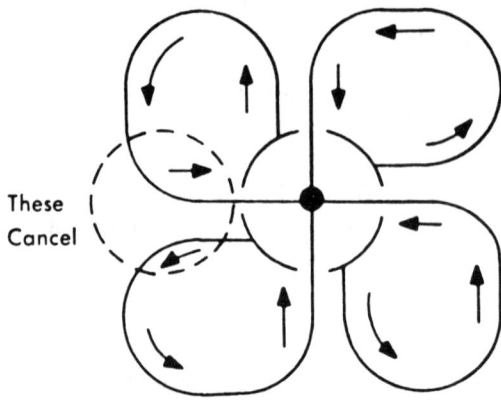

Fig. 17.9 Cloverleaf antenna.

Fig. 17.10 A high-power transmitting loop antenna.

ELECTRICALLY SMALL ANTENNAS   263

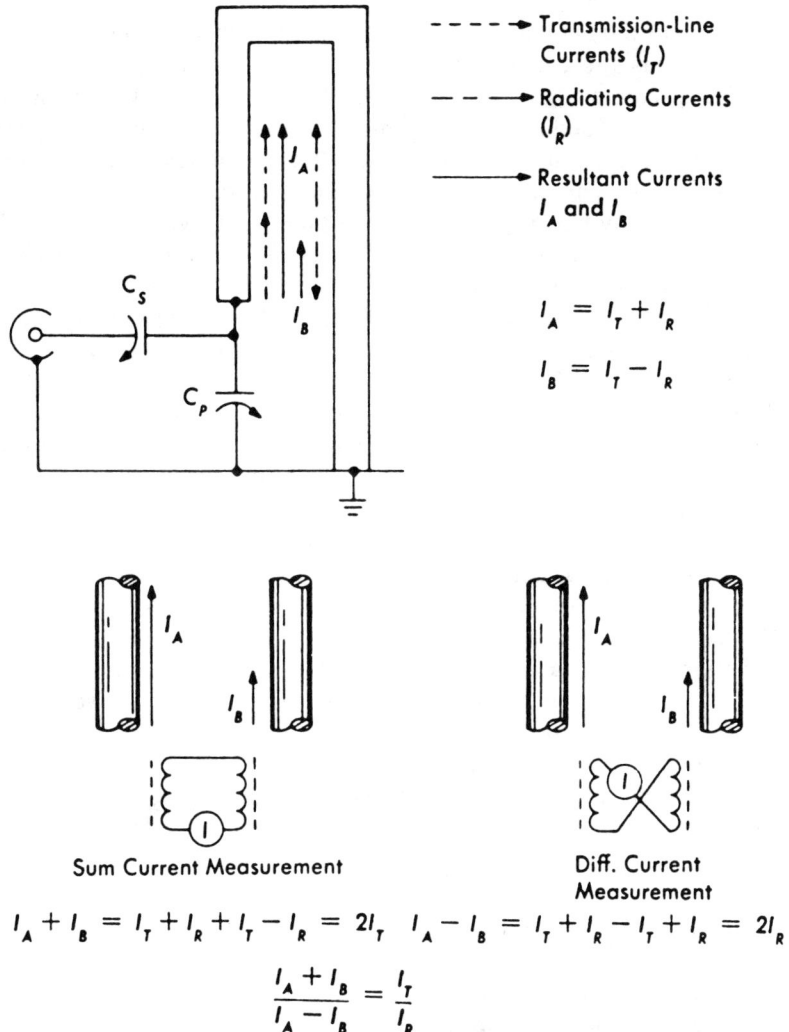

Fig. 17.11 Principles of hairpin monopole antenna.

THE HAIRPIN MONOPOLE

Still another structure has been developed by General Dynamics. In practical high-power installations, this antenna has a variety of practical advantages in that the design is readily controlled to provide optimum Q to be in keeping with system bandwidth requirements and that no sliding contacts or rubbing joints are required.

Operation of the antenna may be best explained by referring to Fig. 17, which illustrates the current excitation of the radiator. Fundamentally, the radiating element is a short-circuited length of a two-wire balanced transmission line. The unit is excited in an unbalanced fashion by grounding one leg and exciting the other. One therefore finds that the currents on the radiator legs are unequal. This inequality is due to the existence of two modes: a truly balanced transmission-line mode, which is essentially nonradiating, and an unbalanced (in-phase) mode, which radiates in a monopole fashion. Experimentally, sum and difference currents on the radiators are measurable with a pair of shielded current-sampling loops.

At low frequencies, the radiator appears as a small inductor, and the capacitor C_P is used for power-factor restoration while capacitor C_S is used for loading control. The antenna may easily be made to look like any resistance between several thousand and a few ohms by proper balance adjustment of C_P and C_S, and a 50-ohm impedance is relatively simple to obtain.

The tuning curves of Fig. 17.12 illustrate the values of C_P and C_S required to obtain 50 ± 1-ohms on the 10-kW antenna. As shown, the antenna tunes the 2- to 30-MHz range in two bands: 2 to 8-MHz and 8 to 30-MHz.

In the higher band, a short-circuiting switch serves to reduce the length of the transmission line for the transmission-line currents; however, this short does not affect the radiating currents. Therefore, the full 14-foot radiator length is still available on the high band.

The tuning curves shown are relatively stable in varying environments.

Figure 17.13 shows a photograph of the 10-KW hairpin designed for shipboard use.

IMPEDANCE CALCULATION

The calculation of the terminal impedance of an electrically small antenna may proceed by either theoretical or practical empirical means. Calculation of radiation resistance by the techniques shown is quite satisfactory for both loop and whip (or dipole) types, and standard handbook inductor formulas yield reasonably accurate results for loop reactance. The reactive terms for whip or dipole antennas may be calculated by

1. Calculated R_R and empirical Q curves
2. Determining radiator Z_k and assuming an open-circuit Z_M for slender antennas, or Z_M equal to the flat-plate end capacitance.
3. Using measured or calculated antenna static capacitance, assuming lumped performance.

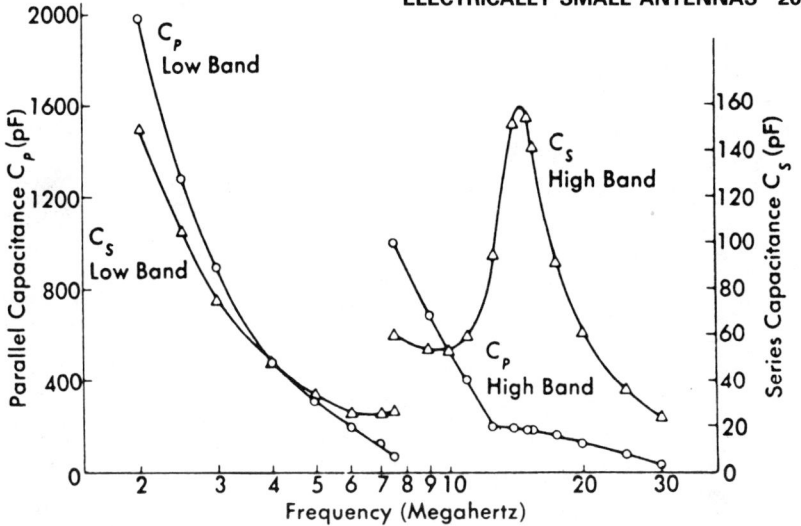

Fig. 17.12 Tuning curves for 10-kW hairpin monopole.

The latter may be computed with a formula established by G.W.O. Howe:

$$C_A = \frac{2\pi\varepsilon_0 h}{\ln(h/a) - 1} \text{ farads} \quad (17.29)$$

where

- h = antenna height, in meters
- a = antenna radius, in meters
- ε_0 = absolute dielectric constant of free space, or approximately $1/36\pi \times 10^{-9}$ farad per meter

All three methods work reasonably well in free space or on a large, flat, unobstructed ground plane. In practical vehicular, ship, or packset installations, the capacitance may be only half as great and the radiation plus loss resistance may be as much as 250 times too small.[5] The excessive resistance term results on whips when a lossy ground return is present. Dipoles and monopoles over solid-conductor ground planes seldom show a resistive term greater than $5R_R$ unless a large loading inductor is used. The additional measurable resistance may be associated with large circulating currents in this case.

---

[5] J. Kuecken, "Packset Radio Antenna Measurements," *1966 IEEE International Conv. Record*, Pt. 5, pp. 261–271.

266 EXPLORING ANTENNAS AND TRANSMISSION LINES

Fig. 17.13 10-kW hairpin antenna designed for shipboard use.

Obviously, with antennas that behave as a small capacitor with a small radiation resistance in series, the presence of any appreciable stray capacitance will grossly affect the performance. The current in C_R of Fig. 17.14 makes no contribution to system radiation but contributes losses in the inductor loss resistance. If, in addition, C_R is itself lossy, the inefficiency is compounded.

FUNDAMENTAL LIMITATIONS

The preceding discussion has touched on dipole and loop types (or electric and magnetic dipoles) and presented examples of loading effects on the electric dipole. This problem has been considered generically by Wheeler[6] on a theoretical basis. He considers volumetric antennas (Fig.

---

[6]H. A. Wheeler, "Fundamental Limitations of Small Antennas," *Proc. IRE*, Dec. 1947, pp. 1479–1484.

17.15) with a volume Ah and arrives at the conclusion that the radiation power factor (which is the reciprocal of the Q value used here) is nearly identical for each type, differences being due to fringing. This value is given for the electric dipole as

$$Q_e = \frac{3}{4}\left(\frac{\lambda^3}{\pi^2 K_a A h}\right) \tag{17.30}$$

where K_a is the ratio of the effective capture area to the actual area ($1.27h/r$ for $h \gg r$ or 1.0 for $h \ll r$).

The capacitance is

$$C_A = \varepsilon \frac{K_a A}{h} \tag{17.31}$$

where $\varepsilon = 1/36\pi \times 10^{-9}$ farad per meter.

For the magnetic dipole, the value of Q is given as

$$Q_m = \frac{3}{4}\left(\frac{\lambda^3}{\pi^2 K_b A h}\right) \tag{17.32}$$

where $K_b = 1 + 0.9r/h$ for $h > r$. Note that the magnetic dipole radiates in the loop mode, i.e., it is cross polarized to the electric dipole. Therefore, the electric dipole corresponds to the loaded helical antennas previously described. For the example of Eq. 17.17,

$$Ah = 7.43 \times 10^{-6}\lambda^2 \times \frac{\lambda}{45} = 16.5 \times 10^{-8}\lambda^3$$

$$K_a \cong 1.27 \frac{(\lambda/45)}{\lambda \times 1.54 \times 10^{-3}} = 18.35$$

Then,

$$\begin{aligned} Q_e &= \frac{3}{4}\left(\frac{\lambda^3}{\pi^2 K_a A h}\right) \\ &= \frac{3}{4}\left(\frac{1}{\pi^2 \times 18.35 \times 16.5 \times 10^{-8}}\right) \\ &= 25 \times 10^3 \end{aligned} \tag{17.33}$$

Wheeler notes that, in half-space, the radiation-damped power factor is doubled. Therefore, $Q_{e_{hs}} = 12.5 \times 10^3$. This value is somewhat at

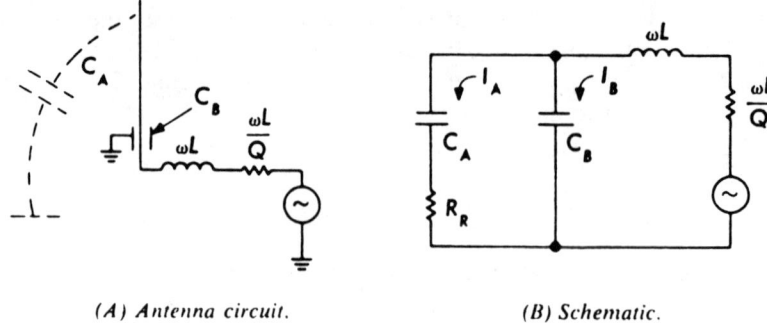

(A) Antenna circuit.   *(B) Schematic.*

Fig. 17.14 Presence of stray capacitance.

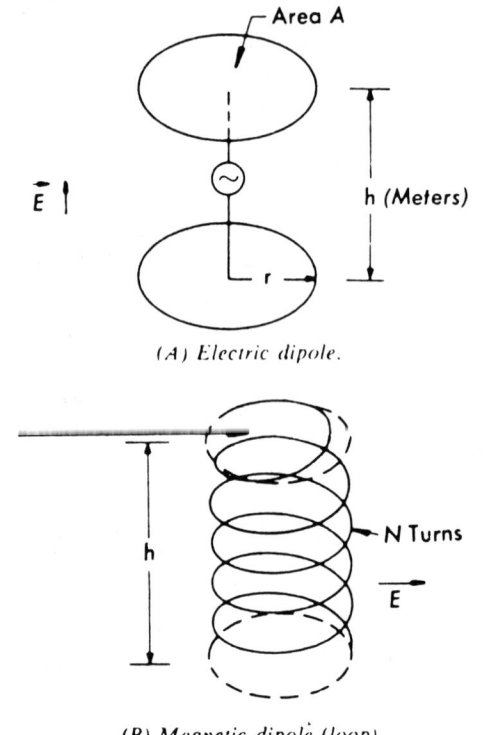

(A) Electric dipole.

(B) Magnetic dipole (loop).

Fig. 17.15 Wheeler's volumetric antennas.

variance with the $Q_{e_{hs}}$ obtained semiempirically as 5.12×10^3 in Eq. 17.33. In view of the large value of K_a and the empirical derivation of the other estimate, such discord is not unexpected and would yield an error in efficiency computation of 3.9-dB on a practical antenna with losses of 11 to 19-dB.

This chapter has shown the origin of the large losses and bandwidth efficiency relationship in electrically small antennas. The drawbacks of electrically small antennas are so great that their use is seldom justified except at low frequencies where sheer physical size becomes excessive, or in aircraft, automotive, or shipboard applications at high frequencies where small physical size is worth the premium in efficiency and bandwidth. Regarding the latter, it is noteworthy that the automotive broadcast whip probably has the largest popularity of any generic type and that probably the majority of all antennas are electrically small.

18
Mobile and Vehicular Antennas

The mobile antenna operates under certain special restraints that do not apply to fixed antennas. We shall touch briefly upon the requirements for these antennas and discuss the steps that are taken to make them more reliable and effective.

LAND MOBILE ANTENNAS

Anyone choosing an antenna to be mounted upon a jeep, tank, truck, or automobile must first and foremost face the fact that a great many places under which these vehicles must pass do not provide more than 13.5-feet of clearance. Anyone who has driven a vehicle through an Oak-Beam Test is aware of its violence. The Oak-Beam Test, which is imposed by the U.S. Army Signal Corps upon mobile antennas, requires one to make 20 passes at 25-mph beneath a 4 × 4 oak beam suspended 8-feet above the ground. A 10-foot antenna mounted on a jeep will be struck just about in its middle. High-speed photographs will reveal that the tip continues to move forward for a split second and actually tries to wrap itself around the beam before it starts to unwind. When it does so, however, it exceeds the speed of sound and cracks like a bullwhip. It will usually whip all the way down to the pavement behind the vehicle before it lashes forward across the roof, also with enormous gusto. Amazing though it may seem, a good fiberglass whip will survive 20 such blows, although it will be pretty frazzled at the end. A speed of 25-mph does not seem very fast, but it is fast enough to make a collision between the antenna and a fixed object very violent.

It is also well to remember that oak beams and trees are not the only things that a whip can collide with. Particularly in Europe, the catenary on electric trains is less than 15 feet above the roadway. In this case, one not only has the collision to think about but the possibility that the vehicle may be connected to a 2200-volt line! It is generally considered that a 15-foot whip is the tallest that can be used on a moving vehicle, and for ordinary road travel, it is usually tied down. As a general rule, a whip that does not rise more than 12-feet above the roadway will not strike many things although it will not pass in and out of most garages and buildings. A good safety precaution is to have the antenna insulated to withstand a high-pot test of 15-kV to protect against contact with power lines, etc.

The base spring also deserves some attention. To protect the antenna, it should be as limber as possible but at the same time stiff enough to hold the antenna up against 120-mph winds, that is, at 60-mph into a 60-mph headwind. It should also be stiff enough to prevent the whip from bending over and flailing horizontally in a fast turn and from lashing forward in a fast stop. For large whips, the author has used springs with different stiffnesses in different directions so that the whip would knock down backward fairly easily but be very stiff forward and to the sides.

In general, an automotive antenna needs to be omnidirectional. At frequencies below about 20-MHz, this objective does not pose much of a problem since the vehicle is too small to support resonances. Above this frequency, however, the vehicle will easily support resonances, and an effort must be made to keep the current on the vehicle to a minimum.

It is possible to keep the current down in a number of ways. If only a narrow band is to be accommodated, the current can be minimized by making the whip a half wave long rather than a quarter wave. The quarter-wave antenna is handy because its impedance is usually around 37-ohms, and it can be driven from a 50-ohm coax with only a modest mismatch and no matching network at all. When a slender half-wave antenna is end-fed, it will have an impedance on the order of 1 to 2-Kohms. It obviously requires some matching network to drive from a 50-ohm transmitter output. This approach is usable only at frequencies above 32-MHz where the halfwave is less than 15-feet. The antenna can also be base-loaded to forshorten it slightly so that an end-fed radiator can be used down to about 20-MHz if it has been equipped with end-loading.

The requirement for a base-matching network and the desire to obtain broader bandwidths has led to continuing research and development. During the early 1940s, for example, the choke dipole (Fig. 18.1) was developed to permit center feeding. The choke section serves to stop the currents flowing on the outside of the feed coax, thereby acting as a balun.

The attainable limits of impedance on such a device are not adequate, however, to produce a truly adequate choking action. For instance, the open-circuit impedance of the choke is given by the lossless case,

$$Z_{choke} = jZ_{oc} \tan\beta l \qquad (18.1)$$

but from handbook formulas,

$$Z_{oc} = 139 \, log_{10} (D_2/d_2) \qquad (18.2)$$

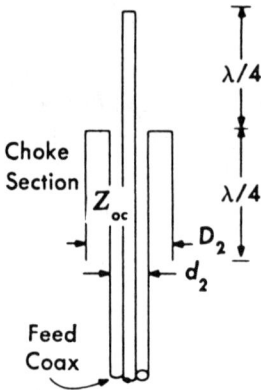

Fig. 18.1 Coaxial choke dipole.

If, for mechanical reasons, we assume that $D_2/d_2 < 5$, then the maximum value becomes

$$Z_{oc}(\max) = 139 \log_{10} 5 \qquad (18.3)$$
$$= 96.5 \text{ ohms}$$

Now if we assume that the end-fed impedance of the relatively "fat" dipole is on the order of 500-ohms and that a modestly effective choke must present an impedance of at least 10 times this impedance, we may solve for the limiting values of βl thus:

$$5000 = |Z_{oc} \tan \beta l| \qquad (18.4)$$
$$= |96.5 \tan \beta l|$$

Since $\beta l = 90° \pm 1.108°$, the frequency bandwidth is only ± 1.02 percent even in the lossless case! Above and below this, the choke will not adequately stop the current. In a practical antenna, finite losses in even a very good choke would probably prevent attainment of this degree of isolation. For instance, if losses in the choke section limit the attainable VSWR to less than 50, the maximum choke impedance is 50×96.5, or $4825 + j0$-ohms. With more modest D_2/d_2 ratios, the efficacy of this approach is diminished, and, in fact, Rowe[1] shows a multielement skirt array excited by leakage past a series of low-impedance chokes (Fig. 18.2).

---

[1] R. G. Rowe, "Collinear Coaxial Array for 152 Megacycles," *Tele-Tech*, Jan., 1949, pp. 34–35.

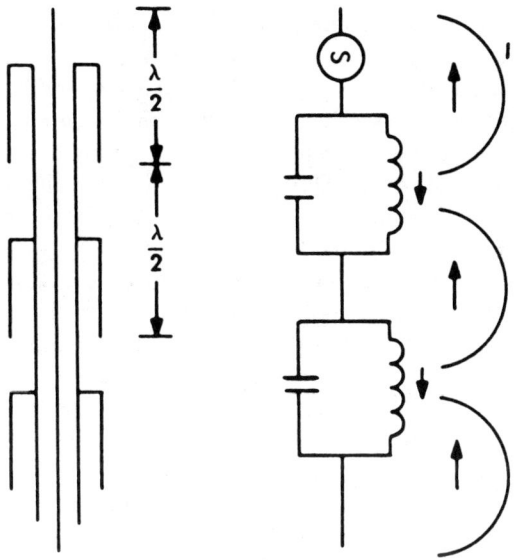

Fig. 18.2 Collinear coaxial array.

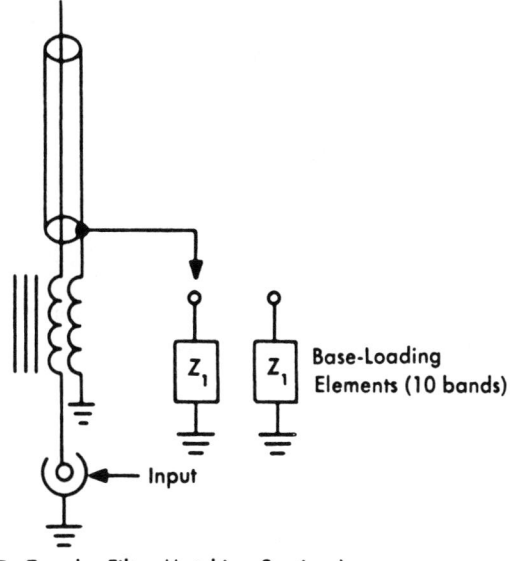

(To T and π Filter Matching Sections)

Fig. 18.3 AT-912 antenna.

274 EXPLORING ANTENNAS AND TRANSMISSION LINES

By the 1950's, we find Brueckmann[2] discussing the optimum height of a center-fed ground-independent vehicular antenna to cover the 30- to 76-MHz band. This antenna (Fig. 18.3) uses a series of base-loading elements switched across one leg of a bifilar choke winding to force the current distribution into a quasi-half-wave ground-independent condition for isolation from the vehicle. Thus, a fairly arbitrary physical height was utilized in the ground-independent mode over a 2.5:1 frequency band.

A second-generation antenna of this type (Fig. 18.4) utilizes a choke sleeve with the miniature coaxial line wound into a helical inductor over a ferrite rod. This approach permits attainment of very high characteristic impedance (600 to 900-ohms) and very slow propagation velocity (v_p is approximately equal to $0.03c$) in the choke section. Thus, excellent isolation is available for good ground-independent operation.[3]

A later effort at loading resulted in foreshortening of the radiating element of the AS-1729 by means of a variable-pitch spiral loading of the radiating elements—at some sacrifice in radiation efficiency and VSWR at the band edges (Fig. 18.5). By concentrating the inductive loading in areas of minimum current, the circulating losses resulting from stray capacitance are minimized, as shown by Czerwinski.[4]

An approach to obtaining broadband operation without tuning is exemplified in the whiplike, center-fed ground-independent antenna[5] (Fig. 18.6). In this antenna, the high-Z_0 helical base choke and the lower radiator sleeve choke are combined in series to obtain a high choke impedance over a wide frequency band. Center-fed ground-independent operation is obtained over a 1.3:1 frequency band with, the VSWR less than 2:1, using a length-to-radius ratio of 40. Broader bands would be feasible with "fatter" radiators.

The ground-independent antenna has not yet been exploited to full advantage in aircraft, marine, and other applications where the advantages of freedom from pattern distortion caused by resonances on the vehicle skin would be just as attractive as they are on a motor vehicle.

For the full-sized AS-1729 antenna, the max-to-min ratio measured by swinging the vehicle on a turntable is usually less than 3-dB in contrast to the over 40-dB notches obtained with a base-fed whip on the same vehicle.

---

[2]H. Brueckmann, "Theory and Performance of Vehicular Center-Fed Whip Antenna," *IRE Trans.*, Vehicular Comm. vol. VC-9, Dec., 1960, pp. 10–20. Also see H. Brueckmann, *U.S. Pat. No. 2,913,722* (Nov. 17, 1959).

[3]See W. P. Czerwinski, "Field Intensity Efficiency of Vehicular Antenna AS-1729/VRC," *Test Report 1564*, U.S. Army Electronics Lab., Ft. Monmouth, N.J., Mar., 1964.

[4]W. P. Czerwinski, "On Optimizing The Efficiency and Bandwidth of Inductivity Loaded Antennas," *IEEE Trans.*, Ant and Prop. Vol. AP 13, No. 5, Sept., 1965, pp. 811 and 812.

[5]J. A. Kuecken, *U.S. Patent No. 3,438,042*.

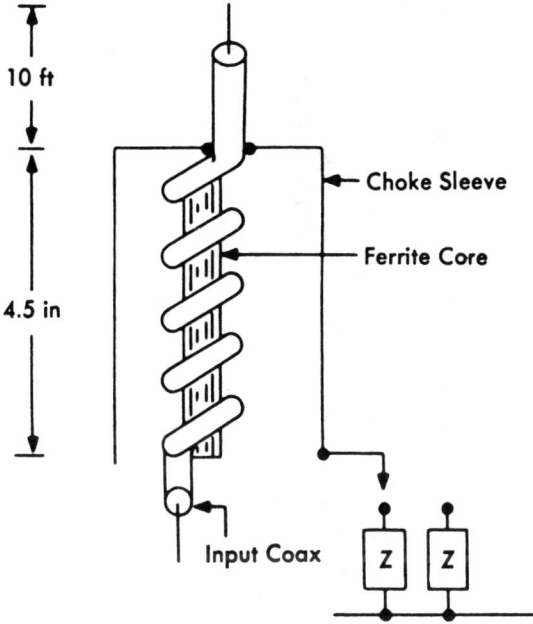

Fig. 18.4 AS-1729 antenna: 30 to 76-MHz in ten switched bands.

Most of the remaining features of motor vehicle antennas are the same as features described for other applications.

THE SHIPBOARD ANTENNA

The limitations on the height of shipboard antennas are much less stringent then those for a land mobile vehicle. To pass through the Panama Canal, anything aboard a vessel may not extend more than 195-feet above the waterline. This restriction is not too onerous, however, except on very large ships. Nearly all commercial or naval ships will have an MF antenna, as discussed in Chap. 17. In addition, the inclusion of one or more HF antennas is common. It is fairly common to equip the HF radios with a whip antenna. Table 18.1 gives the "official" impedance values specified for the three whip antennas commonly used by the U.S. Navy. In general, the 15-foot whip is used only on small vessels such as landing craft, etc. Wherever possible, the 35-foot whip is employed. This is a fairly heroic piece of stuff. The standard 35-foot whip is about 8-inches in diameter at the base and 4.5-inches at the top. It is constructed of fiberglass and would probably make a good mast for a

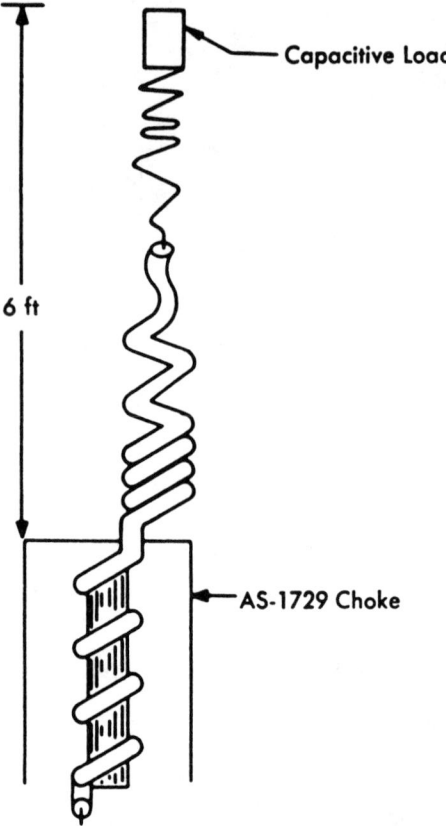

Fig. 18.5 Foreshortened AS-1729.

25-foot sailboat. It is required to be strong enough to withstand shaking when it is horizontal on its mount. As a matter of fact, the antennas installed at the edge of the deck of an aircraft carrier are on hinged mounts and are lowered to be horizontal during aircraft launch and retrieve functions. In this position, the whip has to withstand the accelerations caused by ship roll and pitch, vibration, wind loading, and ice loading.

ANTENNA COUPLERS

From the data on the antennas in Table 18.1, it can be easily seen that some form of antenna matching network will be required to transfer any

MOBILE AND VEHICULAR ANTENNAS   277

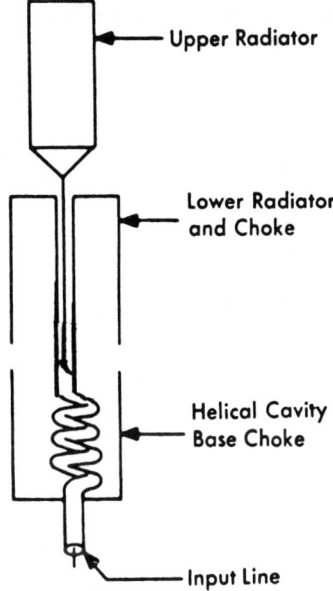

Fig. 18.6 Center-fed ground-independent antenna.

significant power into the whip antennas. Actually, an antenna coupler is required nearly any time a whip antenna is to be used at HF on a movable vehicle. Since the basic properties required of a shipboard antenna coupler generally encompass the properties of couplers for aircraft and land mobile applictions, we shall discuss these couplers here.

To begin with, let us consider some of the basic properties of an antenna coupler. As noted in Chap. 17, the basic matching technique is to add some impedance (susceptance) to the antenna until the conductance (resistance) is at the proper value. A final element is then used to cancel the residual. The illustration of Fig. 18.7 shows two examples of such a process on a Smith Chart.

Ideally, one would like to be able to use the same network to match all possible antenna impedances. All of the matching networks, however, have some limits to their ability to match impedance, even if no restriction is placed upon the size of any of the components. Figures 18.8 and 18.9 show the areas that cannot be matched by various circuits.

Table 18.1 U.S. Navy "Standardized" Antenna Impedances

FREQUENCY, MC	NOMINAL CHARACTERISTIC IMPEDANCE ($R + jX$)		
	15-FT WHIP	25-FT WHIP	35-FT WHIP
2.0	6.2 $-j1125$	3.2 $-j600$	7.5 $-j350$
2.5	6.3 $-j900$	4.2 $-j500$	
3.0	6.5 $-j750$	6.0 $-j400$	12 $-j210$
3.5	6.6 $-j625$	7.0 $-j350$	
4.0	6.9 $-j545$	9.3 $-j280$	18 $-j115$
5.0	7.65 $-j415$	13 $-j190$	28 $-j60$
6.0	8.75 $-j330$	19 $-j120$	40 $+j2$
7.0	10.4 $-j270$	35 $-j75$	80 $+j80$
8.0	12.3 $-j217$	50 $-j20$	190 $+j150$
9.0	15.1 $-j166$	75 $+j0$	580 $+j80$
10.0	19.9 $-j125$	150 $+j40$	300 $-j200$
11.0	24.3 $-j85$	210 $+j25$	120 $-j290$
12.0	31.75 $-j52$	250 $-j5$	50 $-j200$
13.0	42 $-j15$	250 $-j110$	35 $-j130$
14.0	56.5 $+j37$	210 $-j130$	30 $-j80$
15.0	71.5 $+j75$	180 $-j160$	30 $-j40$
16.0	97 $+j123$	160 $-j180$	33 $-j8$
17.0	129 $+j174$	130 $-j190$	45 $+j15$
18.0	174 $+j213$	85 $-j190$	70 $+j40$
19.0	250 $+j244$	55 $-j170$	95 $+j60$
20.0	389 $+j260$	30 $-j170$	120 $+j70$
21.0	506 $+j80$	20 $-j210$	140 $+j50$
22.0	467 $+j0$	19 $-j100$	160 $-j2$
23.0	366 $-j212$	18 $-j65$	160 $-j25$
24.0	277 $-j205$	18 $-j50$	130 $-j40$
25.0	232 $-j188$	20 $-j35$	120 $-j48$
26.0	200 $-j163$	23 $-j9$	110 $-j50$
27.0	175 $-j142$	30 $+j0$	110 $-j50$
28.0	46 $-j122$	35 $+j30$	105 $-j50$
29.0	41 $-j106$	60 $+j60$	100 $-j30$
30.0	39 $-j85$	110 $+j95$	100 $-j20$

MIL-A-23547A(SMIPS):

3.4.4 *Antenna characteristics.* The antenna coupler shall be capable of operation within the limits described in this specification with antennas having the following nominal electrical characteristics. For purposes of comparison, antennas used during testing procedures shall have nominal characteristics identical to their physical counterparts.

3.4.4.1 *15-foot whip antenna.* The antenna coupler shall be capable of operation with a 15-foot whip antenna having the nominal characteristic impedance similar to those values shown (here).

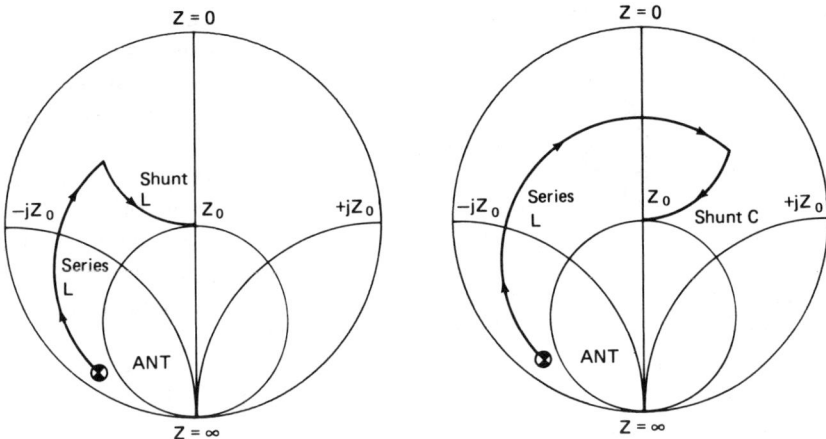

Fig. 18.7 Reactive impedance matching.

Consider, for example, the circuit in Fig. 18.8a. Within the $R = 50$-ohm circle at the bottom of the Smith Chart, a series inductor cannot be used to match to 50-ohms since it would cross the axis at some resistance higher than 50-ohms. Similarly, the upper portion of the forbidden zone lies in the 20-millimho circle. The addition of a series inductor will only drive the point further into the forbidden zone.

In general, two like elements (two capacitors or two inductors) will yield a cusped forbidden zone, whereas two dissimilar elements will yield a "Yin-Yang" shape such as that shown in Figs. 18.8(c) and (d). The circuit in Fig. 18.9(a) is a popular one since it encompasses many of the points that naturally occur on a whip antenna. The circuit is usually modified as shown in Fig. 18.9(b) with a 4:1 impedance transformer. In this condition, it will be seen that all of the points in Table 18.1 lie within the matchable zone.

The operation of this circuit is shown in Fig. 18.9(d). Note that a small resistance in series with a small reactance, as at the first resonance of the whip, will require a very small series capacitive reactance. If the first resonance occurs at too low a frequency, the series C may not be large enough to give the necessary low reactance, therefore resulting in the hatched area in Fig. 18.9(b) that indicates the area forbidden by a finite C.

In the days before the computer and the programmable calculator, the solution of element sizes for an antenna coupler with finite Q in the inductor was rather tedious because of the large number of iterations required. As shown by the broken line in Fig. 18.9(d), the finite Q adds conductance as well as susceptance and thereby changes the values of both elements. Naturally, when L changes, the loss conductance changes as well. Of course, the computer can be programmed to follow an iterative course in which it first solves the problem, substitutes a loss, and then resolves it until the results converge.

Program 18.1 is arranged to solve the antenna coupler program for an "L network" similar to that in Fig. 18.9(b). This network has the additional advantage that the inductor and capacitor are usually smaller than those required by a circuit with a series inductor and shunt capacitor as shown in Fig. 18.8(c). The program has the ability to include a stray capacitance and a series inductor.

It was noted in Chap. 17 that a PI network is not uniquely determined since it can have a large number of values. The L network, on the other hand, *is* uniquely determined, that is, for a given load, there is only one value of inductor and capacitor that can be used to match it. Accordingly, the program makes the user select the value of the series inductor.

After the program has corrected the antenna impedance for the stray capacitance and the series inductor (if any), it goes to line 340 and computes the inductor for a lossless match. It then computes a first-approximation loss resistance in line 360 and goes into the iterative loop from lines 380 to 440, incrementing the loss resistance and solving for the inductor. At line 410, it tests to see whether the inductor Q has arrived at a reasonable level. As soon as that level is reached, it goes to line 500 and prints the results. The routine from lines 610 to 725 computes the voltages, currents, and radiated power for a 1000-watts input. If the series inductor is zero, all of the difference between the radiated power and 1-kW is dissipated in the shunt inductor.

It will be instructive to run a few problems. You will recall that Chap. 17 noted that base capacitance is destructive of efficiency. Figure 18.10 shows a run for the $(6.2 - j1125)$-ohm antenna (a 15-foot whip at 2-MHz) 10, 50, and 100-pF of stray capacitance. You can see that the shunt inductor shrinks, the series capacitor grows, the antenna base voltage falls, and the radiated power falls with increasing base capacitance. It is not difficult to see that the stray capacitance at the base of the antenna should be kept as small as possible.

The largest inductor required for any of the whip conditions is about 40.2-μH. If the antenna coupler is intended to match a 15-foot whip at values no lower than 2-MHz, the 40.2-μH inductor should be adequate.

As noted earlier, the largest capacitor comes at the lowest-frequency, low-resistance resonance. Obviously, the 35-foot whip will go through the first resonance at a lower frequency than a shorter whip. The data of Fig. 18.11 shows the results of extreme low resonance at 6-MHz. In this case,

MOBILE AND VEHICULAR ANTENNAS 281

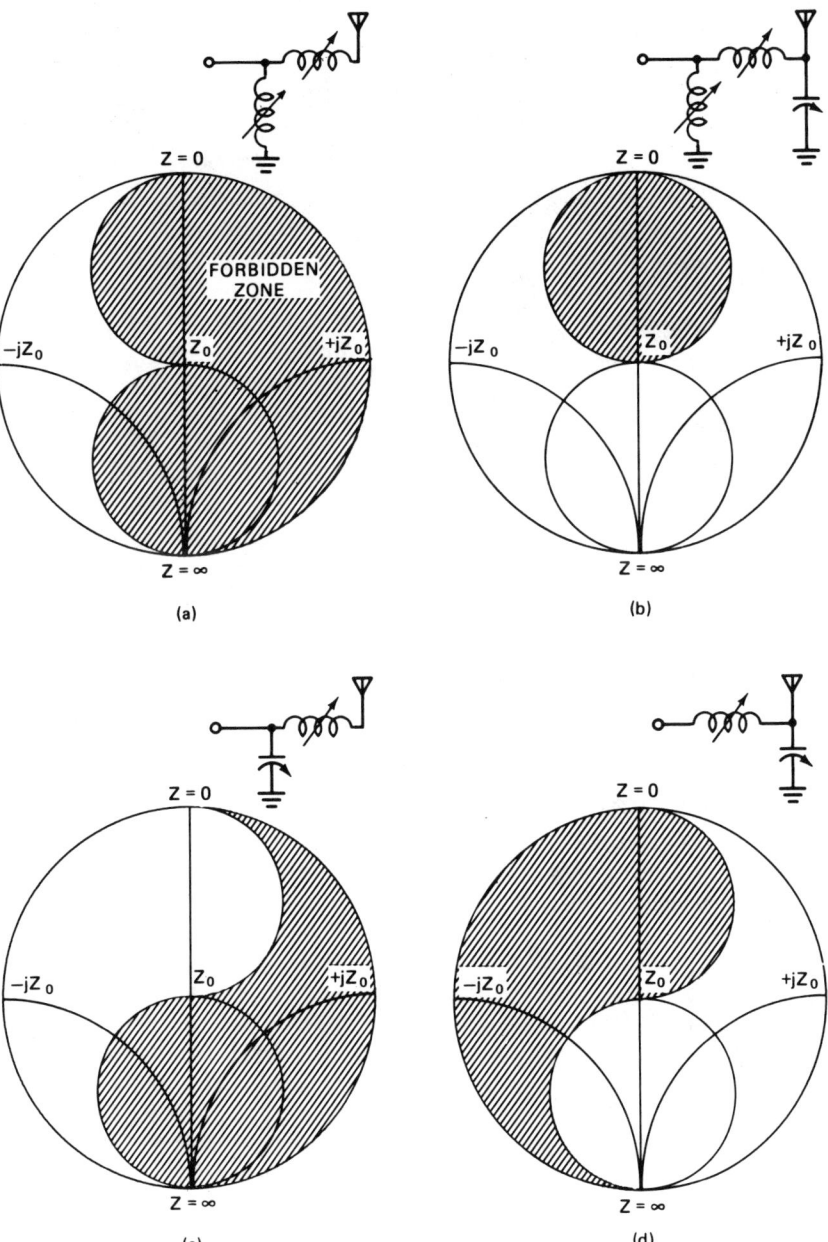

Fig. 18.8 Matching limits for various circuits.

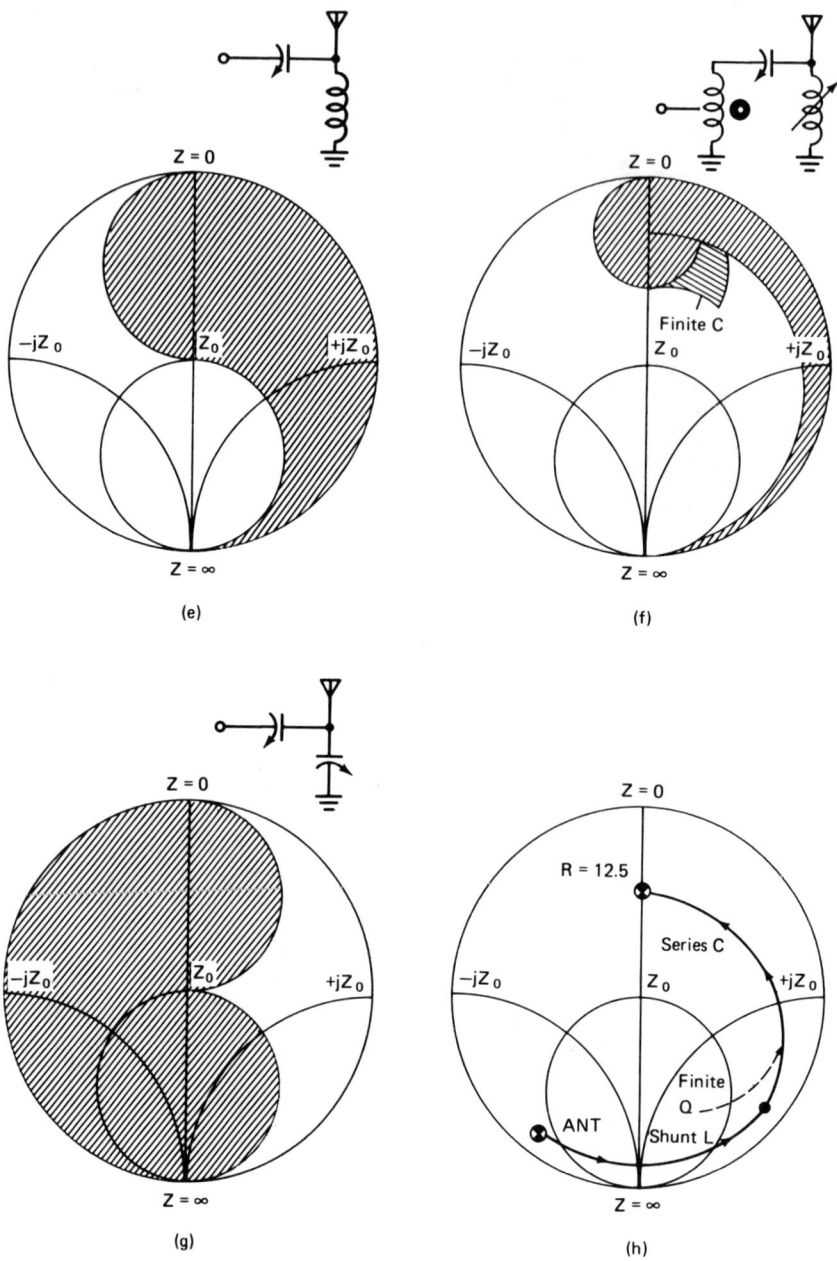

Fig. 18.9 Matching limits for various circuits (cont'd).

```
155  HOME
156  P = 300
160   PRINT "ENTER FREQUENCY IN MHZ"
170   INPUT F
180  W = 2 * 3.14159 * F * 10 ^ 6
190   PRINT "ENTER ANTENNA RESISTIVE TERM"
200   INPUT R
210   PRINT "ENTER REACTIVE TERM"
220   INPUT X
230  D = ((R * R) + (X * X))
240  G = R / D
250  B =  - X / D
260   PRINT "ENTER STRAY CAPACITANCE"
265   PRINT "IN PICOFARADS"
270   INPUT CS
280  BC = W * CS * 10 ^  - 12
300  BN = B + BC
301  DE = ((G * G) + (BN * BN))
302  RE = G / DE:XE =  - BN / DE
303   IF XE < 0 GOTO 307
304   IF RE < 12.5 THEN  PRINT "POINT IS IN THE ": PRINT "FORBIDDEN LOW RES
     ISTANCE ZONE": END
307   GOSUB 1000
310   REM  $$$$$$$$$$$$$$$$$$$$$$$$$$$$$$$$$$$$$$
320   REM    NEXT CALCULATE BL TO MATCH
330   REM  &&&&&&&&&&&&&&&&&&&&&&&&&&&&&&&&&&&&&&
335   REM  BM=BN-BL
340  BM =   SQR ((G / 12.5) - (G * G))
350  BL =  - (BM + BN)
360  GN = G +  ABS (BL / 350): REM  FIRST CUT AT LOSS COMPENSATION FOR QL=3
     00
380  BM =   SQR ((GN / 12.5) - (GN * GN))
390  BL =  - (BM + BN)
400  GL = GN - G
410   IF  ABS (BL / GL) < P GOTO 500
420  GL = 1.01 * GL
430  GN = G + GL
440   GOTO 380
500  XS = BM / ((GN * GN) + (BM * BM))
510  C = 1 / (XS * W)
520  L =  ABS (1 / (W * BL))
530  PR# 1
540   PRINT "AT FREQUENCY=",F,"MHZ"
550   PRINT "ANTENNA RESISTANCE=",R
560   PRINT "ANTENNA REACTANCE=",X
565   PRINT "STRAY CAPACITANCE=",CS,"PICOFARADS
567   PRINT "SERIES INDUCTOR",LS,"MICROHENRIES"
570   PRINT "SHUNT INDUCTOR=",L,"HENRIES"
580   PRINT "SERIES CAPACITOR=",C,"FARADS"
590   PRINT "INDUCTOR Q=",  ABS (BL / GL)
600  PR# 0
```

Program 18.1

```
610  REM  *****************************************
620  REM  CALCULATE VOLTAGES AND CURRENTS
625  REM  FOR  1KW INPUT
630  REM  &&&&&&&&&&&&&&&&&&&&&&&&&&&&&&&&&&&&&&&&&
640  E = SQR (1000 / GN)
650  IA = E / ( SQR ((R * R) + (X * X)))
660  IL = ABS (E * BL)
670  PR = (IA * IA) * R
680  EP = 1.41 * E
685  PR# 1
690  PRINT "SHUNT IND.VOLTAGE=",EP,"VOLTS PEAK"
700  PRINT "ANTENNA CURRENT=",IA,"AMPS RMS"
710  PRINT "COIL CURRENT=",IL,"AMPS RMS"
715  PRINT "RADIATED POWER=",PR,"WATTS"
717  PRINT ""
720  PR# 0
725  PRINT "ENTER ANY NUMBER TO CONTINUE"
800  INPUT A
810  GOTO 155
980  REM  ::::::::::::::::::::::::::::::::::::::::
985  REM          THIS ROUTINE ADDS SERIES INDUCTANCE
990  REM  *****************************************
1000 PRINT "ENTER SERIES L IN MICROHENRIES"
1010 INPUT LS
1020 XL = LS * W * 1E - 6
1022 IF XL <  - XE GOTO 1030
1024 IF RE < 12.5 THEN  PRINT "SERIES INDUCTOR PLACES POINT": PRINT "IN F
     ORBIDDEN ZONE": STOP
1030 RL = XL / 300
1040 DA = (G * G) + (BN * BN)
1050 RC = G / DA
1060 XB =  - BN / DA
1070 RN = RL + RC
1080 XN = XL + XB
1090 DN = (RN * RN) + (XN * XN)
1100 BN =  - XN / DN
1110 GN = RN / DN
1120 RETURN
1922 IF XL <  - XE GOTO 1030
```

Program 18.1 (*continued*)

the inductor is less than 1-μH; for 40 + j2-ohms, however, the series C is 1433-pF, and, for 24 − j4-ohms, it rises to 2159-pF. The requirement for a very large capacitor can be reduced by including a series capacitor between the antenna and the coupler. The bottom run in Fig. 18.11 shows that the inclusion of just 30-ohms of capacitive reactance reduces the requirement for the coupler below 1000-pF. This is a standard size for vacuum capacitors and is considerably cheaper than the 3000-pF size. A switchable series capacitor is very often included in a marine coupler for this reason.

At the 30-MHz end of the band, the inductor may fall to 0.3-μH, and, in some cases, a capacitance as small as 20-pF may be required.

Having examined the workings of the coupler, we may draw some

```
AT FREQUENCY=       2           MHZ
ANTENNA RESISTANCE=             6.2
ANTENNA REACTANCE=              -1125
STRAY CAPACITANCE=              10              PICOFARADS
SERIES INDUCTOR 0               MICROHENRIES
SHUNT INDUCTOR= 4.02877955E-05  HENRIES
SERIES CAPACITOR=               7.64612841E-11  FARADS
INDUCTOR Q=     297.478941
SHUNT IND.VOLTAGE=              13126.361       VOLTS PEAK
ANTENNA CURRENT=8.27496399      AMPS RMS
COIL CURRENT=   18.3883275      AMPS RMS
RADIATED POWER= 424.54518       WATTS

AT FREQUENCY=       2           MHZ
ANTENNA RESISTANCE=             6.2
ANTENNA REACTANCE=              -1125
STRAY CAPACITANCE=              50              PICOFARADS
SERIES INDUCTOR 0               MICROHENRIES
SHUNT INDUCTOR= 3.11357482E-05  HENRIES
SERIES CAPACITOR=               8.26667965E-11  FARADS
INDUCTOR Q=     297.588143
SHUNT IND.VOLTAGE=              12141.1581      VOLTS PEAK
ANTENNA CURRENT=7.65388411      AMPS RMS
COIL CURRENT=   22.0075753      AMPS RMS
RADIATED POWER= 363.20804       WATTS

AT FREQUENCY=       2           MHZ
ANTENNA RESISTANCE=             6.2
ANTENNA REACTANCE=              -1125
STRAY CAPACITANCE=              100             PICOFARADS
SERIES INDUCTOR 0               MICROHENRIES
SHUNT INDUCTOR= 2.43130442E-05  HENRIES
SERIES CAPACITOR=               8.9744767E-11   FARADS
INDUCTOR Q=     297.646408
SHUNT IND.VOLTAGE=              11183.7802      VOLTS PEAK
ANTENNA CURRENT=7.05034539      AMPS RMS
COIL CURRENT=   25.9609556      AMPS RMS
RADIATED POWER= 308.185694      WATTS
```

Fig. 18.10 Stray "*C*" effect.

general conclusions about it. First of all, its sensitivity to stray capacitance at low frequencies makes it necessary for it to be placed immediately adjacent to the antenna rather than connected to it by a coaxial cable. Even if we considered the losses to be tolerable, there would be a voltage problem that would burn up nearly any standard cable.

The result of the coupler's need to be immediately adjacent to the antenna is that it has to be placed in an outdoor box. Whereas most navy radios are designed with an indoor environment in mind, the antenna coupler must run out-of-doors where it will be subject to whatever weather the ship may encounter. The box must be waterproof and immune to snow, wind, rain, and ice. Note that for Army equipment, the whole radio must be weather and abuse proof.

286 EXPLORING ANTENNAS AND TRANSMISSION LINES

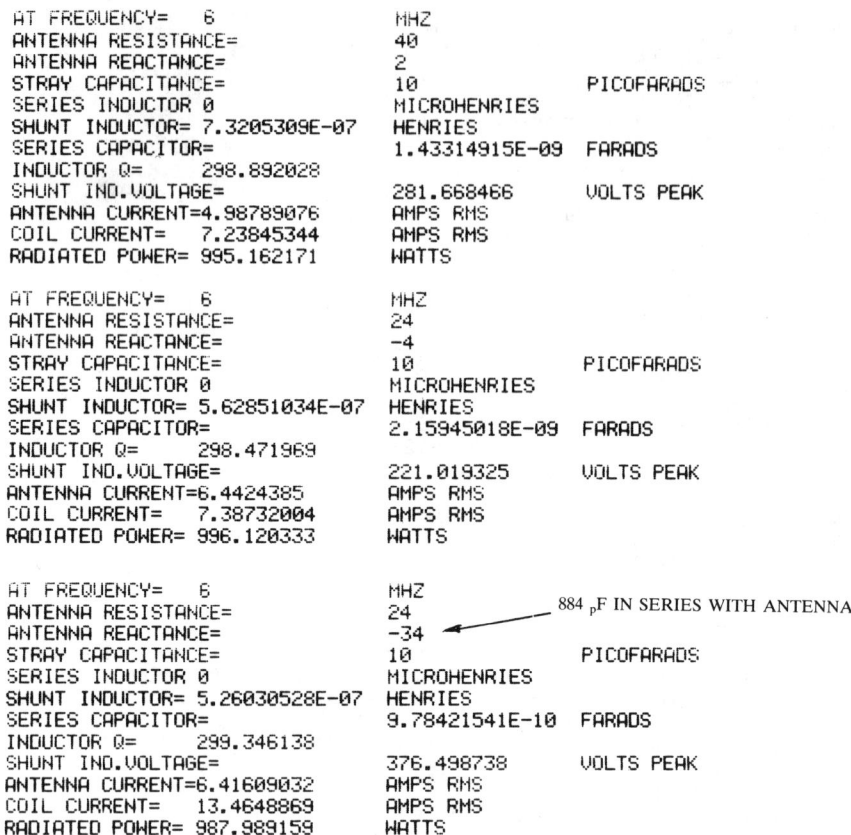

Fig. 18.11 Maximum capacitance.

Another consequence of its outdoor position is that the shipboard antenna coupler must be either remote-controllable or automatic because of its distance from the radio room. The operator is often below decks and cannot run up on deck every time the frequency needs to be changed.

A problem as serious as cold weather is hot weather. With the antenna coupler out-of-doors, baking in a tropical sun, the unit must be prepared to dissipate a large amount of power from RF losses without burning itself up. A kilowatt coupler may have to dissipate as many as 600-watts under certain circumstances, a figure as high as that of a small oven!

In addition to all of the above, the coupler must be prepared to dissipate the static and induced charges upon the antenna from atmospheric electricity. One of the advantages of the shunt inductor coupler is the fact that the antenna is operated at *dc* ground potential. When a series capacitor is used, it must be provided with a *dc* bypass in the form of an RF choke or a large resistor. A whip antenna of 15-feet or more should also always have a lightning ground for safety.

An additional requirement is that for an antenna coupler bypass. Otherwise, in the event that the radioman wishes to listen but not to talk on some new frequency, he will have the problem that the antenna coupler is still tuned to the previous frequency. In this condition, it is liable to be so badly mismatched for the new frequency that he cannot receive much of anything, and he will usually not be able to retune the coupler without transmitting. A coupler bypass that directly connects the whip to the center conductor of the feed coax is used. Although the whip will, as a rule, be badly mismatched at the far end of the long cable, it will usually perform better than it did with the coupler tuned to the wrong frequency.

BROADBAND SHIPBOARD ANTENNAS

In recent years, some naval ships have been fitted with such broadband antennas as log-periodics and bi-cones. These are not likely to become popular on fighting ships since they obstruct the guns and weaponry. A more acceptable design is that of a fan in which three or four wires fan out from a point near the gunnels to one of the yards. Two fans are used, with one fan going up on each side. The apex of one fan is grounded and the apex of the other driven. Such an antenna can have a VSWR below 3:1 over much of the HF band.

Naval RADAR antennas are not much different from their ground-based counterparts except for the fact that they must either operate from a gyro-stabilized base or be capable of electronically cancelling ship motion out of their scan or track operation. The ship is in constant motion at sea, and this motion must be cancelled out of the antenna motion. In some cases, as in a sea-search radar, the elevation beam is simply made broad enough so that stabilization is not required. Such antennas are characterized by a wide horizontal aperture of little height.

AIRBORNE ANTENNAS

Airborne antennas are broadly classified according to the performance of the aircraft. Some slow aircraft and helicopters are equipped with antennas that are not much different from land mobile antennas. In medium-performance aircraft, the antennas are usually streamlined to reduce air resistance. Streamlining can take the form of the nose radome for weather and navigation radar as well as of the blades seen sticking out from the top and sides of airliners. In the extreme, it can take the form of the huge lozenge carried by AWACS and Airborne Early Warning aircraft. In some carrier-based planes, the antenna actually can take up more area than the wings.

In the highest performance aircraft, the antennas are entirely flush with the aircraft surface. Nothing whatever protrudes to increase the drag. The annular slot antenna shown in Fig. 18.12 is an example of the flush antenna with zero drag used on missiles as well as planes. To all intents and purposes, it may be thought of as having evolved from a top-loaded whip in which the top loading has been so increased that almost none of the whip remains. This particular antenna is realtively narrow-band. If its disc were driven simply from the center conductor of the input coax, the bandwidth would be broader. A plut of Q versus slot radius is shown in Fig. 18.13.

Figure 18.14(a) shows the flush slot antenna. It is similar to the one used on F-111 aircraft. It may be seen that it is generically similar to the hairpin antenna described in Chap. 17, although it differs in that it actually excites the entire aircraft skin. A current flows all over the airplane whenever it is in use.

Figure 18.14(b) shows a Boeing 707 style "stinger" antenna. Actually a fairly conventional whip antenna that uses an antenna coupler similar to the one described earlier, it too excites the entire skin of the aircraft. The slot antenna has the advantage of dispensing with the lightning arrestor located at the base of the stinger whip because it is directly grounded to the aircraft skin. The original antenna coupler supplied with the 707 did not *dc*-ground the whip and therefore required the lightning arrestor.

Even if the whip is grounded by means of a shunt coil, it is considered good engineering practice to provide a spark gap that can break down during surges. In surface-based antenna couplers, this is simply a ring gap. On an airborne antenna coupler, however, it is usually necessary to provide a sealed lightning arrestor gap since the breakdown voltage of a spark gap falls with increasing altitude.

The altitude problem must be taken into account with all aircraft antennas because an antenna that performs perfectly at sea level can go into corona at the same power level at 40,000-feet. This problem becomes particularly pernicious in missiles and very high flying aircraft. The breakdown is at its worst at altitudes between 80,000 and 120,000-feet. At still higher altitudes, breakdown is less likely to occur, provided that the affected areas can freely ventilate into space. A slow outgassing has disabled a number of satellite antennas because of corona breakdown.

Of course, a perpetual concern with aircraft equipment is its weight. In

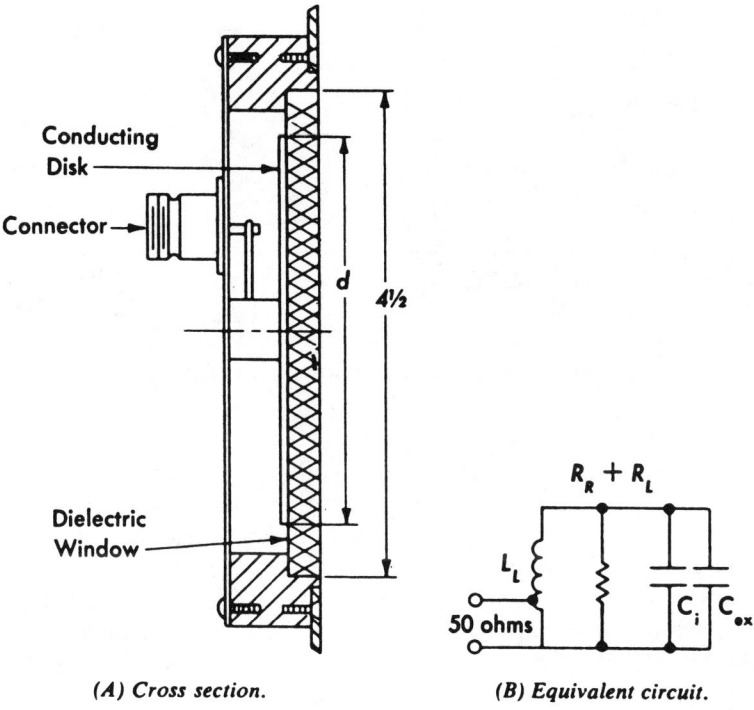

(A) Cross section.   *(B) Equivalent circuit.*

Fig. 18.12 Annular slot antenna.

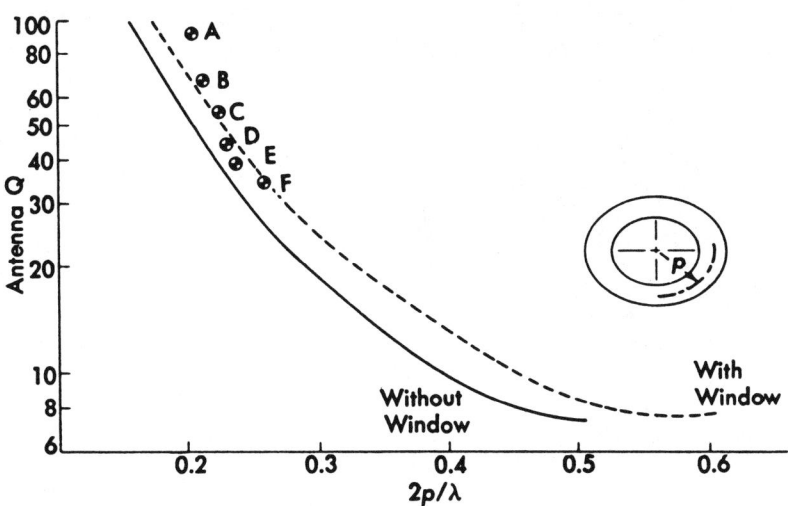

Fig. 18.13 Antenna Q versus slot diameter.

290 EXPLORING ANTENNAS AND TRANSMISSION LINES

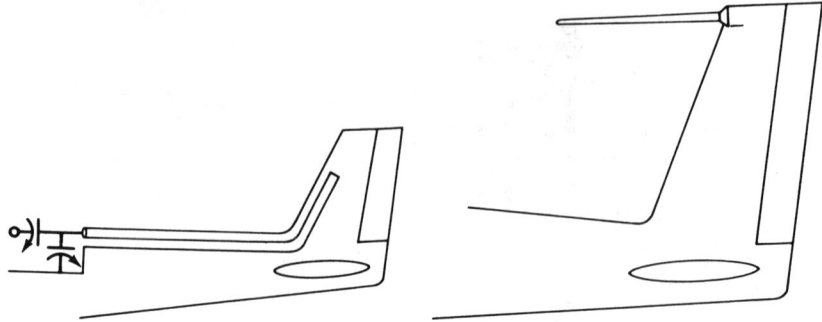

Fig. 18.14 (a) The HF flush slot antenna, and (b) a Boeing 707 style "stinger" antenna.

view of a craft as massive as a 747, it seems hard to believe that any slight extra weight could be a serious consideration, but airlines can quote a cost per hour of flying time *per pound*. Every single pound must be accounted for and justified.

Vibration and shock are also somewhat more of a concern with aircraft than with shipboard equipment, although perhaps no more than with land mobile equipment. Since the aircraft environment is usually sheltered, aircraft couplers do not have to be weatherproof. This is a decided advantage in designing for shock, vibration, and weight requirements.

Clearly, mechanical design requirements differ from land mobile to shipboard to aircraft equipment. The talent for producing suitable designs usually requires experience in the field of concern.

19
Reflections in Space and Antennas above Earth

One of the jobs for which the power of a computer can be put to good use is the calculation of reflections. Such calculations have always been tedious, and the availability of a well thought out program makes it possible to solve problems that might otherwise be glossed over or overlooked. We shall begin with some simpler special cases.

TRANSMISSION LINE REFLECTION ANALOGY

For normal incidence, a complete duality exists in space and transmission-line phenomena. As a result, we will be able to use some of the relationships of Chap. 12.

Whenever a wave traveling in one medium encounters an interface with a second medium of different characteristic impedance, it is natural to expect that a portion of incident wave will be reflected while another portion is transmitted. Using a transmission-line analogue for *normal* (right-angle) incidence, Fig. 19.1 shows the quantities involved. From the development of Eq. 11.27, we obtain for the first medium, Z_{M1},

$$Z_{M1} = \sqrt{\frac{\mu_1}{\varepsilon_1}} \quad \text{or} \quad \sqrt{\frac{\rho_1 + j\mu_1\omega}{\sigma_1 + j\varepsilon_1\omega}}$$

and similarly for Z_{M2},

$$Z_{M2} = \sqrt{\frac{\mu_2}{\varepsilon_2}} \quad \text{or} \quad \sqrt{\frac{\rho_2 + j\mu_2\omega}{\sigma_2 + j\varepsilon_2\omega}}$$

where

ρ = resistivity, in ohm-meters
σ = conductivity, in mhos per meter

The first formulation is for the lossless or dielectric case and the second

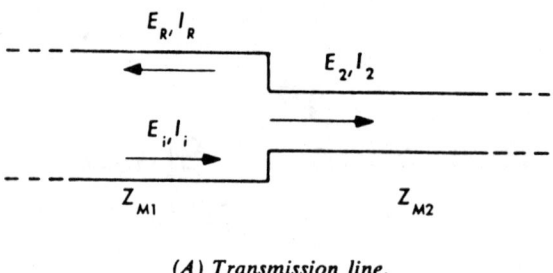

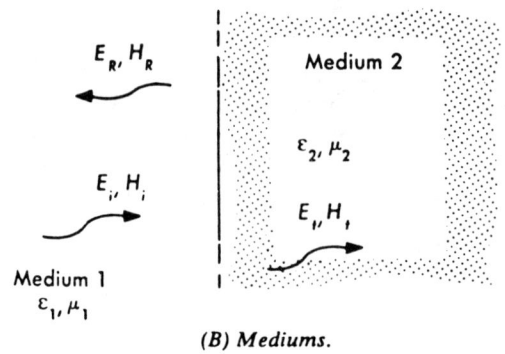

Fig. 19.1 Transmission-line reflection analogy.

for the more general case. As noted in Chap. 12, the characteristic impedances may be purely real if

$$\rho = \sigma = 0$$

$$\mu = \varepsilon = 0$$

or

$$\rho\varepsilon = \sigma\mu \tag{19.1}$$

The case of Eq. 19.1 is for attenuation but nevertheless purely real.

Of course, in each medium, the relation between the electric field and the magnetic field for the forward and reverse waves is, by definition,

$$Z_M = \frac{E}{H}$$

Now, from Eq. 12.7 we may write

$$E_R = E_i \left(\frac{Z_{M2} - Z_{M1}}{Z_{M2} + Z_{M1}} \right) \tag{19.2}$$

and the driving force for the transmitted wave is simply the electric field at the interface, i.e., the vector addition,

$$E_t = E_i + E_R$$

Let us consider a few elementary cases. For an air or vacuum and perfect conductor interface, $Z_{M1} = 120\pi$-ohms, but $\sigma_2 \gg \rho_2$; μ_2; ε_2. Therefore,

$$Z_{M2} = 0$$

and

$$E_R = E_i \left(\frac{0 - 120\pi}{0 + 120\pi} \right) = -E_i$$

so that

$$E_t = E_i + (-E_i) = 0$$

Thus the reflection is complete and antiphased, and the transmitted wave is nonexistent.

For an air or vacuum and plastic interface, $Z_{M1} = 120\pi$-ohms and

$$\rho_2 = 0 \qquad \sigma_2 = 0, \qquad \mu_2 = \mu_0 \qquad \varepsilon_2 \cong 2.5\varepsilon_0$$

(for plexiglass, lucite, etc.). Then,

$$Z_{M2} \cong 120\pi \sqrt{\frac{1}{2.5}} = 76\pi$$

Thus,

$$E_R = E_i \left(\frac{76\pi - 120\pi}{76\pi + 120\pi} \right)$$
$$= -0.225 E_i$$

and

$$E_t = E_i + E_R = 0.775 E_i$$

The incident power is

$$\frac{E_i^2}{Z_{M1}} = P_i$$
$$P_i = \frac{E_i^2}{120\pi}$$

and

$$P_R = \frac{(-0.225 E_i)^2}{120\pi}$$
$$= \frac{0.0507}{120\pi} E_i^2$$

and

$$P_t = \frac{(0.775 E_i)^2}{76\pi}$$
$$P_t = \frac{0.600 E_i^2}{76\pi} = \frac{0.95 E_i^2}{120\pi}$$

The addition of P_R and P_t shows that power is conserved. The 95-percent transmission, 5-percent reflection figure is reasonably good for clear plastics at radio frequencies and plastics and glass at optical frequencies. Telescopes for viewing the sun often use nonsilvered primary or diagonal mirrors to obtain the 13-dB reflection loss.

FINITE THICKNESS

If medium $M2$ is of finite thickness (see Fig. 19.2), the impedance at the interface, including the effect of a backward wave in $M2$, must be used in Equation 19.2. This interface impedance is a function of the second interface impedance and the thickness of $M2$. This problem may be solved in sequence: finding the reflection at the second interface first and then translating to a first interface impedance to find a first interface reflection. From Eq. 12.16 or Program 12.2 we may write the second interface impedance

REFLECTIONS IN SPACE AND ANTENNAS ABOVE EARTH 295

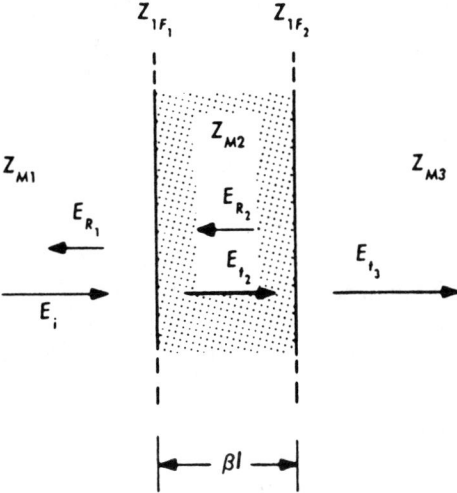

Fig. 19.2 Two-interface problem.

$$Z_{1F_1} = Z_{M2}\left(\frac{Z_{M3}\cos\beta l + jZ_{M2}\sin\beta l}{Z_{M2}\cos\beta l + jZ_{M3}\sin\beta l}\right)$$

Some special cases are of interest here. For instance, for $\beta l = n\pi$ (n = any integer), then

$$Z_{1F_1} = Z_{M2}\left(\frac{Z_{M3}}{Z_{M2}}\right) = Z_{M3}$$

and the second medium is invisible!

For $\beta l = m\pi/2$ (m = an odd integer),

$$Z_{1F_1} = Z_{M2}\left(\frac{Z_{M2}}{Z_{M3}}\right) = \frac{Z_{M2}^2}{Z_{M3}}$$

Then, if $Z_{1F_1} = Z_{M1}$ is desired,

$$Z_{M2} = \sqrt{Z_{M1}Z_{M3}}$$

The layer $M2$ acts as a quarter-wave transformer matching Z_{M3} to Z_{M1} without loss, though a standing wave exists in $M2$. This is the principle on which antireflection optical coatings work. The odd purplish reflections noted in these devices (lenses and prisms) attest that the extreme red and violet are reflected whereas the center of the visual spectrum is passed.

For $\beta l \ll \pi/2$,

$$Z_{1F_1} = Z_{M2}\left(\frac{Z_{M3}}{Z_{M2}}\right) = Z_{M3}$$

and $M2$ is again invisible! This case is used in the design of thin-skin radomes at radio frequencies. Reverting to transmission-line tricks, two thin-skins spaced a quarter-wave apart will cancel the small residuals resulting from the nonzero thickness if the intervening space is Z_{M3}.

OBLIQUE INCIDENCE

The transmission-line analogy breaks down for incidence angles that depart appreciably from the normal. The reflected wave is no longer constrained to follow backward along the path of the incident wave and the transmitted wave is bent or refracted toward the normal if the velocity in $M2$ is slower than the velocity in M_1 (see Fig. 19.3).

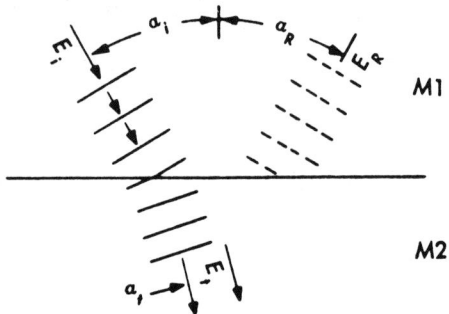

Fig. 19.3 Oblique incidence problem.

Without a lengthy derivation, it should be relatively apparent that $\alpha_R = \alpha_i$ since any phase shift occurring on reflection will take place equally on all portions of the phase (or wave) front, and the first portion to strike will be the first to reflect. A standing-wave structure will exist in $M1$. However, the VSWR will be lowered by the power in the transmitted wave.

For the transmitted wave, a somewhat different phenomenon takes place. We have seen that the velocity of propagation in any medium is

$$v_M = 1/\sqrt{\mu_M \varepsilon_M}$$

Consider the wavefronts in $M1$ and $M2$ in Fig. 19.4; here, v_A must match on both sides of the interface so that

$$v_A = \frac{v_1}{\sin\alpha_i} = \frac{v_2}{\sin\alpha_t}$$

and from the relation $v_M = 1/\sqrt{\mu_M \varepsilon_M}$,

$$\frac{1}{\sqrt{\mu_1 \varepsilon_1} \sin\alpha_i} = \frac{1}{\sqrt{\mu_2 \varepsilon_2} \sin\alpha_t}$$

This is called *Snell's law of refraction*; it was discovered by Willebrod Snell in 1621. It may be transposed as follows:

$$\sin\alpha_i = \frac{\sqrt{\mu_2 \varepsilon_2} \sin\alpha_t}{\sqrt{\mu_1 \varepsilon_1}} \tag{19.3}$$

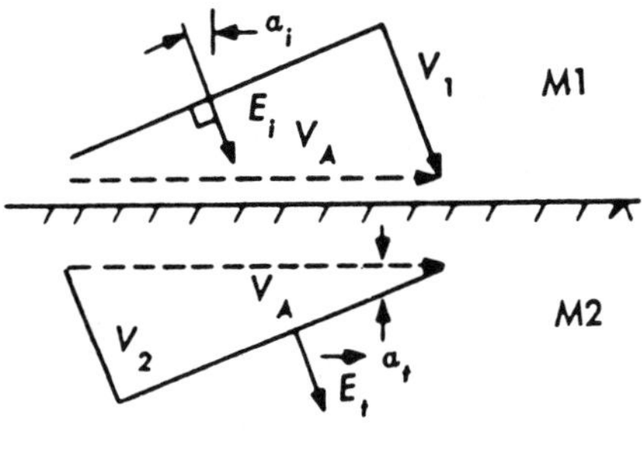

Fig. 19.4 Snell's Law.

Because of the limited range of $\sin\alpha$ (0 to +1; the negative values are not physically significant), an interesting special case exists. If $\mu_1 \varepsilon_1 \geq \mu_2 \varepsilon_2$, transmission cannot exist beyond

$$\sin\alpha_i \leq \frac{\sqrt{\mu_2 \varepsilon_2}}{\sqrt{\mu_1 \varepsilon_1}}$$

since $\sin\alpha_t \leq 1$.

For larger values of α_i, complete reflection exists. This is the mechanism by which a 45°/90° prism bends a light beam 90°. To accomplish

298 EXPLORING ANTENNAS AND TRANSMISSION LINES

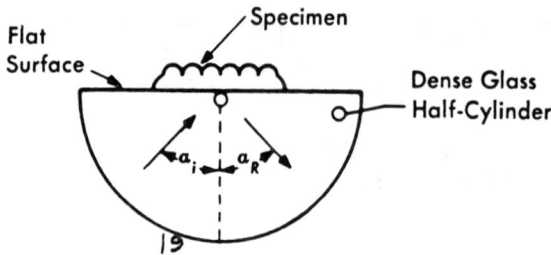

Fig. 19.5 Extinction refractometer.

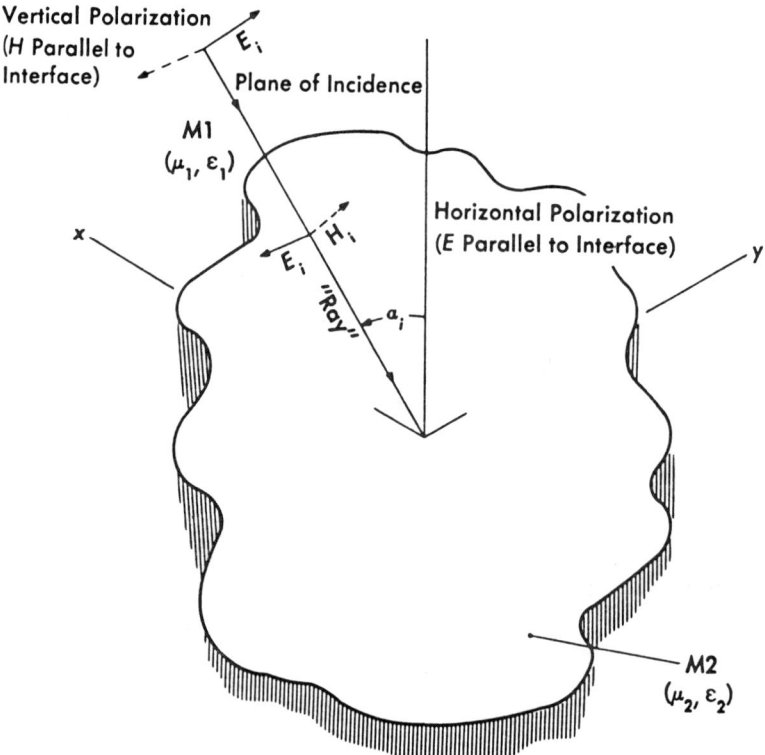

Fig. 19.6 Oblique-incidence geometry.

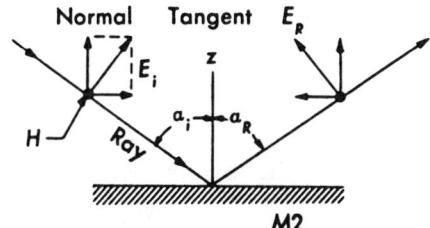

Fig. 19.7 Resolution of components normal and tangent to M2.

this, the velocity ratio (index of refraction ratio) must be greater than

$$\sin 45° = 0.707 = \frac{\sqrt{\mu_2 \varepsilon_2}}{\sqrt{\mu_1 \varepsilon_1}}$$

This "angle of extinction" is frequently used in refractometers (Fig. 19.5) to measure liquids, salves, etc., since they may be poured or smeared on the optically flat surface of a dense glass cylinder and the extinction angle measured without optical surface preparation. The instrument is self-calibrating against air.

BREWSTER'S ANGLE

To this point, we have not considered the effect of wave polarization. Indeed, in the normal-incidence case, a TEM wave is always polarized parallel to the surface. However, in the oblique-incidence case, polarization must be taken into account. From Fig. 19.6, we see that the incident wave may be resolved into components parallel to and perpendicular to the plane containing the incident "ray" and the normal to the interface.

Figure 19.7 illustrates the resolution of E_i (or H_i) into components normal and tangent to the $M1 - M2$ interface. In the Snell's law analysis, we matched the phase velocity of the waves on both sides of the interface. Here, we are concerned with the wave impedance parallel to the interface and must therefore reduce the field component parallel to the plane of incidence to its component parallel to the interface.

For H parallel to the interface,

$$Z_{M1\tan} = \frac{E_{i\tan}}{H_i} = \frac{E_i \cos \alpha_i}{H_i} \qquad (19.4)$$

and

$$Z_{M2\tan} = \frac{E_{t\tan}}{H_t} = \frac{E_t \cos\alpha_t}{H_t} \qquad (19.5)$$

In each medium, however, the relation between E and H is

$$\frac{E}{H} = \sqrt{\frac{\mu}{\varepsilon}} \quad \text{and} \quad E = H\sqrt{\frac{\mu}{\varepsilon}} \qquad (19.6)$$

and from

$$E_R = E_i\left(\frac{Z_{M2\tan} - Z_{M1\tan}}{Z_{M2\tan} + Z_{M1\tan}}\right) \qquad (19.7)$$

this equation follows:

$$E_R = E_i\left[\frac{\sqrt{\mu_1/\varepsilon_1}(H_i/H_i)\cos\alpha_i - \sqrt{\mu_2/\varepsilon_2}(H_t/H_t)\cos\alpha_t}{\sqrt{\mu_1/\varepsilon_1}(H_i/H_i)\cos\alpha_i + \sqrt{\mu_2/\varepsilon_2}(H_t/H_t)\cos\alpha_t}\right] \qquad (19.8)$$

Simplifying,

$$E_R = E_i\left[\frac{\sqrt{\mu_1/\varepsilon_1}\cos\alpha_i - \sqrt{\mu_2/\varepsilon_2}\cos\alpha_t}{\sqrt{\mu_1/\varepsilon_1}\cos\alpha_i + \sqrt{\mu_2/\varepsilon_2}\cos\alpha_t}\right] \qquad (19.9)$$

But since α_i and α_t are related by Snell's law (Eq. 19.3),

$$\sin\alpha_t = \frac{\sqrt{\mu_1\varepsilon_1}}{\sqrt{\mu_2\varepsilon_2}}\sin\alpha_i \qquad (19.10)$$

By trigonometry,

$$\cos x = \sqrt{1 - \sin^2 x} \qquad (19.11)$$

Now let us consider the case in which $\mu_2 = \mu_1$. Substituting α_t in Eq. 19.8, we obtain

$$\cos\alpha_t = \sqrt{1 - \sin^2\alpha_t} \qquad (19.12)$$
$$= \sqrt{\frac{\varepsilon_1}{\varepsilon_2}}\sqrt{\frac{\varepsilon_2}{\varepsilon_1} - \sin^2\alpha_i}$$

REFLECTIONS IN SPACE AND ANTENNAS ABOVE EARTH 301

Substituting Eq. 19.9 into Eq. 19.6, we obtain

$$E_R = E_i \left[\frac{\sqrt{\frac{\mu_1}{\epsilon_1}}\cos\alpha_i - \sqrt{\frac{\mu_2}{\epsilon_2}}\sqrt{\frac{\epsilon_2}{\epsilon_2}}\sqrt{\frac{\epsilon_2}{\epsilon_1} - \sin^2\alpha_i}}{\sqrt{\frac{\mu_1}{\epsilon_1}}\cos\alpha_i + \sqrt{\frac{\mu_2}{\epsilon_2}}\sqrt{\frac{\epsilon_2}{\epsilon_2}}\sqrt{\frac{\epsilon_2}{\epsilon_1} - \sin^2\alpha_i}} \right] \quad (19.13)$$

and, by manipulation,

$$E_R = E_i \left[\frac{(\epsilon_2/\epsilon_1)\cos\alpha_i - \sqrt{(\epsilon_2/\epsilon_1) - \sin^2\alpha_i}}{(\epsilon_2/\epsilon_1)\cos\alpha_i + \sqrt{(\epsilon_2/\epsilon_1) - \sin^2\alpha_i}} \right] \quad (19.14)$$

for H parallel to the interface and $\mu_1 = \mu_2$ (vertical polarization when $\alpha_i = 90°$). Equation 19.10 obviously reduces to Equation 19.2 for normal incidence ($\alpha = 0°$).

A very special solution exists for this equation when the numerator disappears so that

$$\frac{\epsilon_2}{\epsilon_1}\cos\alpha_i = \sqrt{\frac{\epsilon_2}{\epsilon_1} - \sin^2\alpha_i}$$

This equation may be solved for

$$\tan\alpha = \sqrt{\frac{\epsilon_2}{\epsilon_1}} \quad (19.15)$$

The angle, $\alpha = \tan^{-1}\sqrt{\epsilon_2/\epsilon_1}$, is called *Brewster's angle*.

At Brewster's angle, the impedance at the interface is perfectly matched, and all the energy is transmitted from a wave with H parallel to the interface (vertical polarization).

ELECTRIC FIELD PARALLEL TO INTERFACE

For the electric field vector parallel to the interface, we must match the H_{itan} components. Equations 19.4 and 19.5 become

$$Z_{M1\tan} = \frac{E_i}{H_{itan}} = \frac{E_i}{H_i\cos\alpha_i} \quad (19.16)$$

and

$$Z_{M2\tan} = \frac{E_t}{H_{ttan}} = \frac{E_t}{H_t\cos\alpha_t} \quad (19.17)$$

By a similar procedure, these may be solved to yield

$$E_R = E_i \left[\frac{\cos\alpha_i - \sqrt{(\varepsilon_2/\varepsilon_1) - \sin^2\alpha_i}}{\cos\alpha_i + \sqrt{(\varepsilon_2/\varepsilon_1) - \sin^2\alpha_i}} \right] \qquad (19.18)$$

If $\varepsilon_2/\varepsilon_1 > 1$, this function increases monotonically with α_i and exhibits no zero-reflection angle.

The sign difference in E_R/E_i shown at $\alpha_i = 0°$ (where the cases are identical) is due to the algebraic convention of the wave equations and the fact that a forward-going reflected wave in the horizontally polarized case travels in the backward direction for the incident wave.

The curves of Fig. 19.8 show that the reflected wave at Brewster's angle will be 100-percent horizontally polarized. This was an experimentally observed fact by Sir David Brewster about 1816 and served as one of the stepping stones in the redevelopment of the wave theory of light. Brewster noted that the total light reflected from a glass (or other transparent) surface went through a well-defined minimum and that the reflected beam would then be 100-percent transmitted or partially reflected by a second glass surface at the "polarization angle," depending on its orientation (as shown in Fig. 19.9). Faraday's discovery of the rotation of the plane of polarization of light on reflection from a magnet face gave the first clear evidence of the magnetic nature of light. And the demonstration by Weber and Kohlrausch in 1856 that the relation between the cgs electrostatic and electromagnetic units is 3×10^{10}, (the velocity of light in centimeters per second, as measured by the occultation of the moons of Jupiter by Roemer in 1675) added further fuel. Under this bombardment and the development of the sinusoidal-wave interference theory by Thomas Young in 1801, the corpuscular theory of Newton gave way to the earlier wave theory of Huygens and Hooke (of 1665). The inability of Newton's corpuscular theory to explain either polarization or interference phenomena finally crumbled it. (The theory was supported for some 150 years by Newton's reputation alone.)

The publication of Maxwell's equations in 1873 tied up this work in a neat package, and, following the demonstration of electromagnetic waves by Hertz in 1888, physicists rushed back into the breach, demonstrating Brewster's phenomenon with microwaves in a great burst of activity between 1888 and 1900. Thus Brewster, who never personally accepted the wave theory, was a major factor in the revival of it.

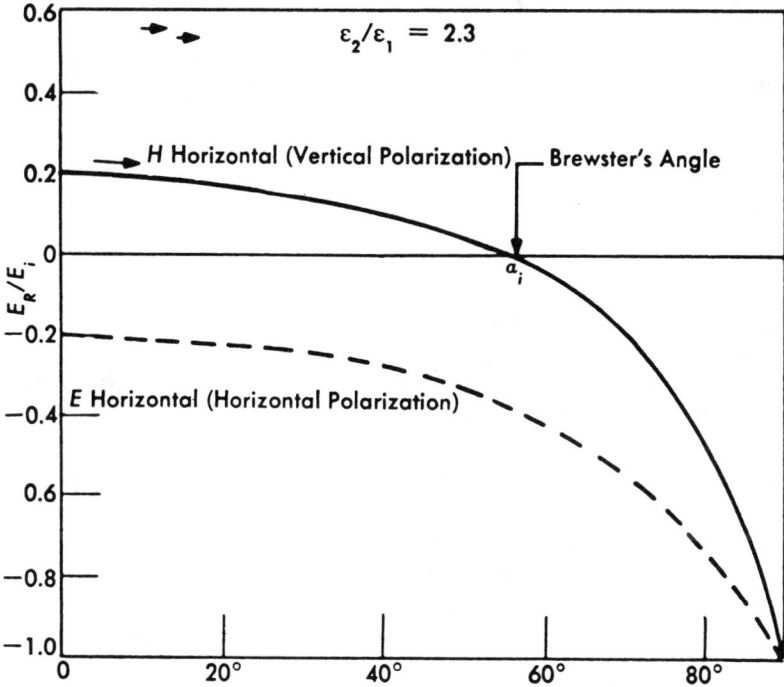

Fig. 19.8 Plot of Eqs. 19.7 and 19.18 for typical sample case.

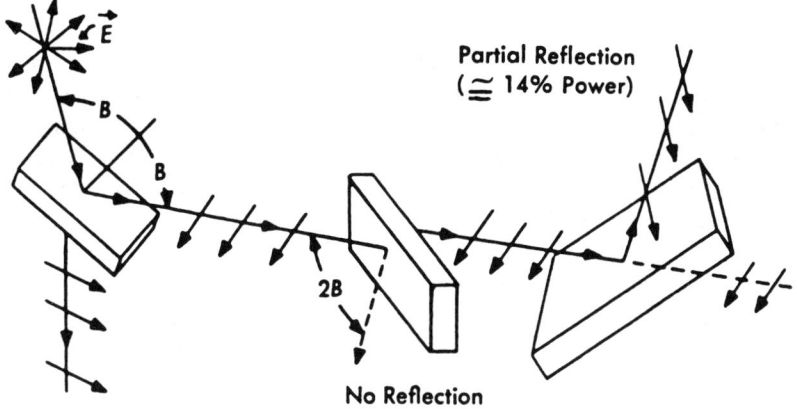

Fig. 19.9 Brewster's angle apparatus.

IMPERFECT DIELECTRICS

Equations 19.14, 19.15, and 19.18 may be extended by substitution to include the effects of imperfect dielectrics. The term $\varepsilon_2/\varepsilon_1$ is simply replaced by

$$\eta^2 = \frac{\varepsilon_2 - j\sigma_2/\omega\varepsilon_0\varepsilon_2}{\varepsilon_1 - j\sigma_1/\omega\varepsilon_0\varepsilon_1} \tag{19.19}$$

where

η = reflection coefficient (complex)
σ = conductivity in mhos per meter
ε_0 = dielectric constant of free space (8.85×10^{-12} farad per meter)
ε_n = relative dielectric constant of the medium

In a lossy dielectric, the reflection coefficients become complex, and the vertically polarized coefficient may pass from positive to negative without passing through zero; thus, only a pseudo-Brewster angle exists. Typical values for the above parameters at 1-MHz are shown in Table 19.1. At higher frequencies, Brewster's angle falls rapidly for poor soil.

Reflection coefficients calculated by means of Eq. 19.19, once substituted into Eqs. 19.14 and 19.18, can be used to assign an amplitude and a phase to a virtual or *image* antenna that is hypothetically located beneath the surface of the earth at the same depth as the height of an actual antenna above its surface (see Fig. 19.10). The two antennas then become a two-element array similar to those discussed in Fig. 4.1, with the important exception that both the amplitude and phase of the image element must be changed as a function of angle. In the principal planes of the antenna, that is, those parallel and perpendicular to the E vector, it is sufficient to use only the one reflection coefficient formula. In an oblique plane where the E vector is neither parallel nor perpendicular to the plane of incidence, however, both the parallel and perpendicular components must be calculated and then combined with proper attention given to phasing. This computational technique was first expounded by C. C. B. Feldman of Bell Labs.[1]

Program 19.1 solves Eqs. 19.14 and 19.18 for various angles as a function of frequency, dielectric constant, and soil conductivity. After the prompting preamble for the data entry, lines 265 through 324 calculate and print the results of Eq. 19.19. Note that the equation has been solved

---

[1] "The Optical Behavior of the Ground for Short Radio Waves," *Proc. IRE*, June 1933, pp. 764–801.

Table 19.1 Conductivity, Relative Dielectric Constant, and Brewster's Angle at 1-MHz

SUBSTANCE	σ (MHOS/METER)	ε_n	BREWSTER'S ANGLE, α_i
Seawater	4.0	80	89.4°
Good soil	0.02	30	88.5°
Poor soil	0.001	4	85.5°

Medium 1 is air: $\sigma_1 = 0$ and $\varepsilon_1 = 1$

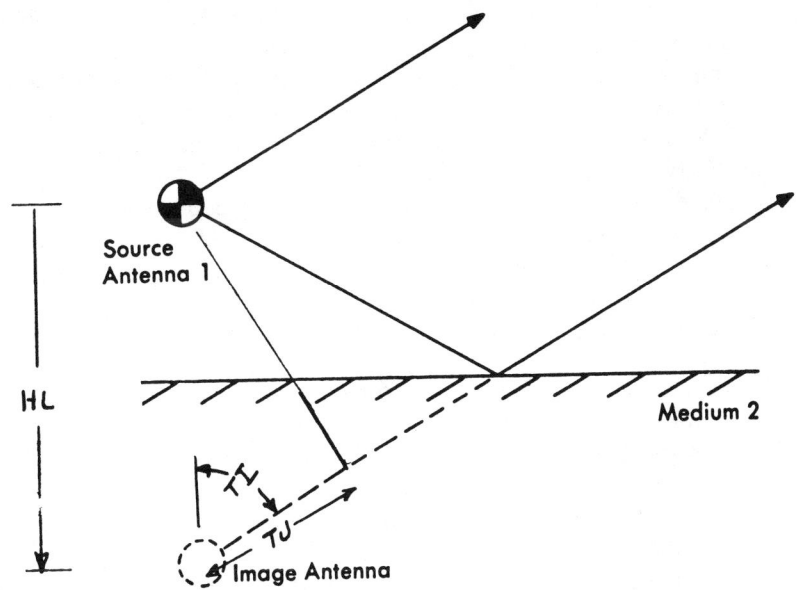

Fig. 19.10 Horizontally polarized image antenna.

with the assumption that the conductivity of medium 1 is zero and that the dielectric constant is equal to that of free space.

Lines 350 through 421 solve Eq. 19.14; lines 500 through 590 solve Eq. 19.18. One can imagine the amount of effort that must have gone into the manipulations performed by Feldman and associates in grinding out antenna patterns by slide-rule and log-table methods.

If the coefficients near the zenith are not required, a starting angle can be inserted somewhere, for example, in line 242. Similarly, smaller increments may be inserted in line 630 and a stop limit in line 640. The 10-degree increments used here are not sufficiently fine to locate Brewster's angle precisely.

306 EXPLORING ANTENNAS AND TRANSMISSION LINES

The program will operate with conductivities of zero and can be used to duplicate the curves of Fig. 19.8.

Table 19.2 shows the output of the program for "good soil," which is defined as moist, loamy Midwestern farm soil. In contrast, "poor soil" is defined as very dry sandy soil of low conductivity and high silica content.

HORIZONTAL DIPOLE PATTERNS

One could, of course, enter the calculated reflection coefficients into the array pattern calculation, but to do so would be tedious. Instead, writing the pattern calculation into the routine will allow one to enter only the specifications and let the computer do the rest.

Program 19.2 uses the coordinates and geometry of Fig. 19.10, with the pattern calculation appended starting at line 600. If the E-plane pattern is desired, it is necessary to use vertical polarization because the E vector lies in the plane of incidence.

Sample patterns for three surfaces—seawater, good earth, and poor earth—are to be found in Fig. 19.11. It may be seen that the principal effect of even very poor soil is one of attenuation; the pattern shape is affected only slightly. The height selected for the patterns is 3/8-wavelength. This height is about optimum for skywave F2 propagation from 0 to 800-miles.

The field strengths, which are relative to one another, are representative of the relative signal provided the antennas have equal current. Because of the differences in mutual impedance from surface to surface, this conditional equality of current does not necessarily imply that the antennas enjoy equal input power as well, although the differences are probably not great.

For single-hop F2-layer propagation, the launch angles are determined by the height of the F2-layer, which is approximately 380-km in the daytime and 300-km at night. For ranges of less than 800-miles, the launch angles always exceed 20-degrees. The dipole should be broadside to targets at ranges greater than 200-miles. For ranges less than 200-miles, the antenna is essentially omnidirectional at the steeper angles (above 50-degrees).

As the antenna is raised, the dip in the pattern in the vertical direction will deepen, becoming a complete null for the seawater condition at heights of an odd number of half wavelengths. Although the launch angle will lower slightly for heights above $\lambda/2$, the horizontal antenna is not a good choice for propagation at angles less than 20-degrees and ranges greater than 800-miles.

```
125 PI = 3.14159
127  HOME
150  PRINT "ENTER FREQUENCY IN MHZ"
160  INPUT F
170 W = 2 * 3.14159E6 * F
180  PRINT "ENTER RELATIVE DIELECTRIC CONSTANT"
190  PRINT "OF THE REFLECTING MEDIUM"
200  INPUT E6
210 ER = 8.84195E - 12 * E6
220  PRINT "ENTER CONDUCTIVITY OF MEDIUM"
230  PRINT "IN MHOS PER METER"
240  INPUT CD
258  REM >>>>>>>>>>>>>>>>>>>>>>>>>>>>>>>>>>>>>>>>>
260  REM  THE CALCULATION FOR THE COEFFICIENT
264  REM ----------------------------------------
265 D2 = W * ER
270 JN =  - CD / D2: REM  THE IMAG PART OF NUMERATOR.
280  REM   JD=0 FOR AIR
290 NM =  SQR ((E6 * E6) + (JN * JN))
300 PH =  ATN (JN / E6)
301  HOME : PR# 1
302  PRINT "AT  ";F;"  MHZ
303  PRINT "FOR DIELECTRIC CONSTANT = ";E6
304  PRINT " AND CONDUCTIVITY = ";CD;" MHOS PER METER"
310  PRINT "MAGNITUDE = ";NM;" AT ANGLE ";180 * PH / PI;" DEGREES"
320  PRINT "REAL =  ";E6;" AND IMAGINARY=";JN
322  PRINT ""
324  PR# 0
330  REM ----------------------------------------
340  REM   VERTICAL POLARIZATION
345  REM ----------------------------------------
350 C1 =  COS (TI):C2 =  SIN (TI)
351  PR# 1
352  PRINT "FOR INCIDENCE ANGLE=  ";180 * TI / PI;"  DEGREES"
360 C3 = C2 * C2
362 C4 = E6 - C3
364 C5 =  SQR ((C4 * C4) + (JN * JN))
365 C5 =  SQR (C5)
366 T3 = .5 *  ATN (JN / C4)
370 C6 = C5 *  COS (T3)
380 C7 = C5 *  SIN (T3)
385  REM  THE VALUE OF THE RADICAL
386 C8 = C1 * NM
387 C9 = C8 *  COS (PH)
388 CA = C8 *  SIN (PH)
390 CR = C9 - C6
392 CQ = CA - C7
394 CP =  SQR ((CR * CR) + (CQ * CQ))
396 T4 =  ATN (CQ / CR)
400 CO = C9 + C6
402 CN = CA + C7
404 CM =  SQR ((CO * CO) + (CN * CN))
406 T5 =  ATN (CN / CO)
408  REM  THE DENOMINATOR
410 CL = CP / CM
412   IF CR < 0 THEN CL = CL * ( - 1)
415 T6 = T4 - T5
417  PRINT ""
418  PRINT "FOR VERTICAL POLARIZATION"
420  PRINT "MAGNITUDE=  ";CL
421  PRINT "PHASE ANGLE=  ";180 * T6 / PI;"  DEGREES"
```

Program 19.1

```
490  REM  ********************************************
485  REM  HORIZONTAL POLARIZATION
498  REM  ////////////////////////////////////////////
500  CW = C6 - C1
510  CV =   SQR ((CW * CW) + (C7 * C7))
520  T7 =  ATN (C7 / CW)
530  REM  THE NUMERATOR
540  CU = C6 + C1
550  CT =   SQR ((CU * CU) + (C7 * C7))
560  T8 =  ATN (C7 / CU)
570  REM  THE DENOMINATOR
580  CS =  - CV / CT
590  T9 = T7 - T8
600  PRINT ""
602  PRINT "FOR HORIZONTAL POLARIZATION"
604  PRINT "MAGNITUDE =  ";CS
606  PRINT "PHASE ANGLE= ";180 * T9 / PI
620  PRINT ""
630  TI = TI + (PI / 18)
640   IF TI < (PI / 2) + .01 GOTO 350
650    PR# 0
2000  END
```

Program 19.1 (*continued*)

For greater ranges, vertical polarization is preferred since it allows radiation to be obtained at lower angles. The fundamental limitation of horizontal polarization is its phase reversal of the image for high-conductivity earths. If one considers it, this phase reversal is demanded by the boundary condition requiring the E vector to be zero in an infinitely conducting plane. When the conductivity is finite, it is possible to have some radiation on the horizon, but, as a practical matter, the dipole will not radiate much below 20-degrees unless it is elevated by many wavelengths. The horizontal polarization of TV and FM is practical only becuase the antennas involved are always mounted many tens of wavelengths above the earth. The horizontal antenna is easier to construct, particularly in directive arrays.

The fact that the horizontal dipole will not operate single-hop beyond a range of 800-miles does not mean that it cannot be used for communications beyond this range because it can also operate in a multi-hop mode. Any additional hops come at some cost in signal strength since the signal must penetrate the absorbing layers at least two additional times and must reflect from the ionosphere and the earth. A two-hop signal is usually 20 to 30-dB weaker than a single-hop signal from the same station. The multi-hop path has a lower OWF (optimum working frequency) than the single-hop path.

It should be noted that polarization is important only during the transmission of skywave propagation since the downcoming wave is

Table 19.2 Output of Program 19.1 for "Good Soil"

```
AT 2 MHZ
FOR DIELECTRIC CONSTANT = 30
 AND CONDUCTIVITY = .02 MHOS PER METER
MAGNITUDE = 30.5941169 AT ANGLE -11.3099405 DEGREES
REAL =   30 AND IMAGINARY=-5.99999915

FOR INCIDENCE ANGLE=  0  DEGREES

FOR VERTICAL POLARIZATION
MAGNITUDE=  .695125114
PHASE ANGLE= -2.10946479  DEGREES

FOR HORIZONTAL POLARIZATION
MAGNITUDE =  -.695125114
PHASE ANGLE= -2.10946479

FOR INCIDENCE ANGLE=  10  DEGREES

FOR VERTICAL POLARIZATION
MAGNITUDE=  .691274458
PHASE ANGLE= -2.14102771  DEGREES

FOR HORIZONTAL POLARIZATION
MAGNITUDE =  -.698936366
PHASE ANGLE= -2.07842811

FOR INCIDENCE ANGLE=  20  DEGREES

FOR VERTICAL POLARIZATION
MAGNITUDE=  .679209361
PHASE ANGLE= -2.24169993  DEGREES

FOR HORIZONTAL POLARIZATION
MAGNITUDE =  -.710388601
PHASE ANGLE= -1.98599807

FOR INCIDENCE ANGLE=  30  DEGREES

FOR VERTICAL POLARIZATION
MAGNITUDE=  .657222062
PHASE ANGLE= -2.4323922  DEGREES

FOR HORIZONTAL POLARIZATION
MAGNITUDE =  -.729527147
PHASE ANGLE= -1.83425898

FOR INCIDENCE ANGLE=  40  DEGREES

FOR VERTICAL POLARIZATION
MAGNITUDE=  .621793847
PHASE ANGLE= -2.76105798  DEGREES

FOR HORIZONTAL POLARIZATION
MAGNITUDE =  -.756393722
PHASE ANGLE= -1.62681763

FOR INCIDENCE ANGLE=  50  DEGREES

FOR VERTICAL POLARIZATION
MAGNITUDE=  .566065917
PHASE ANGLE= -3.34083011  DEGREES

FOR HORIZONTAL POLARIZATION
MAGNITUDE =  -.790975601
PHASE ANGLE= -1.36895163

FOR INCIDENCE ANGLE=  60  DEGREES

FOR VERTICAL POLARIZATION
MAGNITUDE=  .475924208
PHASE ANGLE= -4.49919929  DEGREES

FOR HORIZONTAL POLARIZATION
MAGNITUDE =  -.833135513
PHASE ANGLE= -1.06772209

FOR INCIDENCE ANGLE=  70  DEGREES

FOR VERTICAL POLARIZATION
MAGNITUDE=  .318453237
PHASE ANGLE= -7.79069347  DEGREES

FOR HORIZONTAL POLARIZATION
MAGNITUDE =  -.882525913
PHASE ANGLE= -.731975688

FOR INCIDENCE ANGLE=  80  DEGREES

FOR VERTICAL POLARIZATION
MAGNITUDE=  -.0493305795
PHASE ANGLE=  75.5680574  DEGREES

FOR HORIZONTAL POLARIZATION
MAGNITUDE =  -.938495335
PHASE ANGLE= -.372171589

FOR INCIDENCE ANGLE=  90  DEGREES

FOR VERTICAL POLARIZATION
MAGNITUDE=  -.999985157
PHASE ANGLE=  8.13686993E-05  DEGREES

FOR HORIZONTAL POLARIZATION
MAGNITUDE =  -.999999515
PHASE ANGLE= -2.84480287E-06
```

completely depolarized by Faraday rotation. No net benefit derives from using the same polarization during reception as was used during transmission.

```
125 PI = 3.14159
127  HOME
130 X0 = 135:Y0 = 165
150  PRINT "ENTER FREQUENCY IN MHZ"
160  INPUT F
162  PRINT "ENTER HEIGHT IN METERS"
164  INPUT H
165 HL = 4 * PI * H * F * 1E6 / 3E8
170 W = 2 * 3.14159E6 * F
180  PRINT "ENTER RELATIVE DIELECTRIC CONSTANT"
190  PRINT "OF THE REFLECTING MEDIUM"
200  INPUT EG
210 ER = 8.84195E - 12 * EG
220  PRINT "ENTER CONDUCTIVITY OF MEDIUM"
230  PRINT "IN MHOS PER METER"
240  INPUT CD
258  REM ))))))))))))))))))))))))))))))))))))))
260  REM  THE CALCULATION FOR THE COEFFICIENT
264  REM ------------------------------------------
265 D2 = W * ER
270 JN =  - CD / D2: REM  THE IMAG PART OF NUMERATOR.
280  REM  JD=0 FOR AIR
290 NM =  SQR ((EG * EG) + (JN * JN))
300 PH =  ATN (JN / EG)
302  HGR2
304  HCOLOR= 3
306  POKE  - 12524,0
308  POKE  - 12525,64
310  POKE  - 12529,255
350 C1 =  COS (TI):C2 =  SIN (TI)
490  REM ****************************************
495  REM  HORIZONTAL POLARIZATION
498  REM /////////////////////////////////////////
500 CW = C6 - C1
510 CV =  SQR ((CW * CW) + (C7 * C7))
520 T7 =  ATN (C7 / CW)
530  REM  THE NUMERATOR
540 CU = C6 + C1
550 CT =  SQR ((CU * CU) + (C7 * C7))
560 T8 =  ATN (C7 / CU)
570  REM  THE DENOMINATOR
580 CS = CV / CT
590 T9 = T7 - T8
592  REM ))))))))))))))))))))))))))))))))))))))
594  REM  THE H PLANE PATTERN
596  REM %%%%%%%%%%%%%%%%%%%%%%%%%%%%%%%%%%%%%%
600 TJ = HL *  COS (TI)
605  REM  SPACE PHASE DELAY
610 TK = TJ + T9
620  REM  SPACE PLUS REFLECTION PHASE
630 CK = 1 - (CS *  COS (TK))
640 CJ = CS *  SIN (TK)
650 CI =  SQR ((CK * CK) + (CJ * CJ))
```

Program 19.2

```
652 CI = CI * .5
660 REM   MAGNITUDE OF VOLTAGE IN (TI) DIRECTION
700 XP = X0 + (135 * CI *   SIN (TI))
710 XQ = X0 - (135 * CI *   SIN (TI))
720 YP = Y0 - (135 * CI *   COS (TI))
730  IF TI = 0 THEN M = XP:Q = XQ:N = YP
740 HPLOT M,N TO XP,YP
750 HPLOT Q,N TO XQ,YP
760 M = XP:Q = XQ:N = YP
770 TI = TI + PI / 36
780  IF TI < 1.6 THEN   GOTO 350
790 HPLOT 0,165 TO 279,165
800 HPLOT 133,165 TO 133,163 TO 137,163 TO 137,165
810 TI = 0
820 R = 135
830 XP = X0 + R *   SIN (TI)
840 XQ = X0 - R *   SIN (TI)
850 YP = Y0 - R *   COS (TI)
860  IF TI = 0 THEN M = XP:Q = XQ:N = YP
862  IF KK <  > 0 THEN KK = 0: GOTO 890
870 HPLOT M,N TO XP,YP
880 HPLOT Q,N TO XQ,YP
888 KK = 1
890 M = XP:Q = XQ:N = YP
900 TI = TI + PI / 18
910  IF TI < 1.6 GOTO 830
915 TI = 0
920  IF R = 135 THEN R = 95.5: GOTO 830
2000  END
```

Program 19.2 (*continued*)

THE VERTICAL DIPOLE

For single-hop propagation in the range of 800 to 2000-miles, the advantage belongs to vertical polarization. Since the reflection coefficient is not negative at angles exceeding Brewster's angle, vertically polarized antennas reach a maximum at the horizon. Two thousand miles is usually considered to be about as far as one can transmit in a single hop since it is virtually impossible to make a shortwave antenna radiate at an angle much below 5-degrees. Although a vertically polarized antenna radiates well down to that level, below it, the groundwave enters into the act and attenuates the signal much more rapidly than on a skywave path.

The chief disadvantage of the vertical antenna is that it usually requires a counterpoise if it is fed from the base. This requirement is minimized by the vertical half-wave antenna, which feeds with a very high impedance at the base. It requires a coupling network, but that is generally cheaper than the counterpoise required by a quarter-wave antenna. For maximum

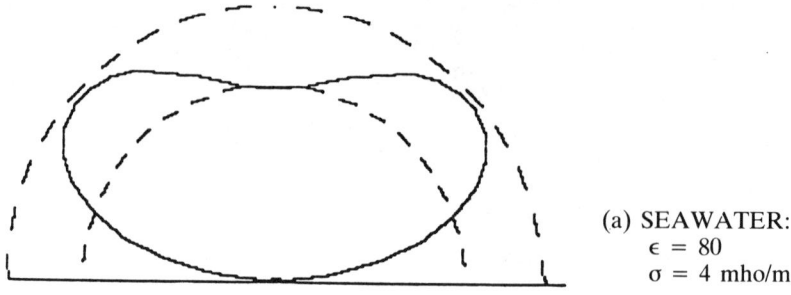

(a) SEAWATER:
$\epsilon = 80$
$\sigma = 4$ mho/m

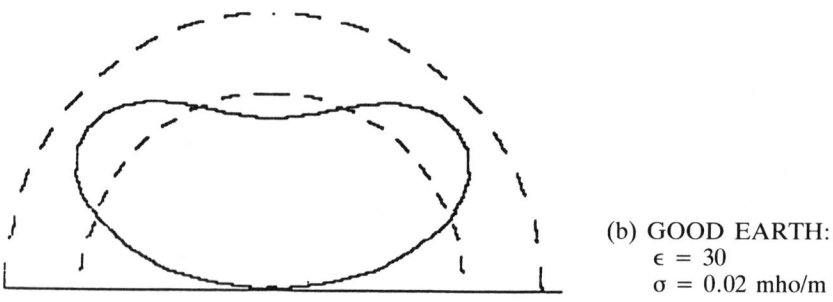

(b) GOOD EARTH:
$\epsilon = 30$
$\sigma = 0.02$ mho/m

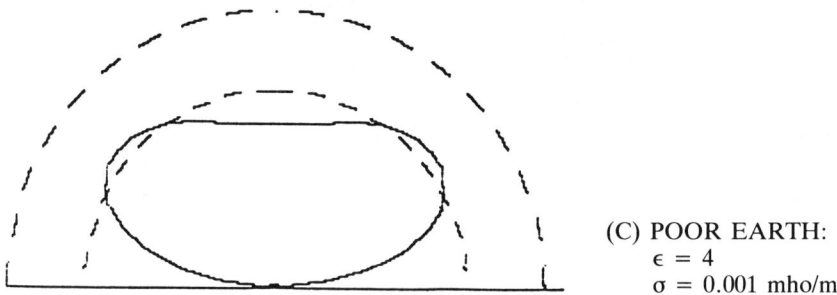

(C) POOR EARTH:
$\epsilon = 4$
$\sigma = 0.001$ mho/m

Fig. 19.11 Computer-calculated (H-plane) radiation pattern (at 2 MHz) for horizontal half-wave dipole over various surfaces at a height of $3/8\lambda$.

```
125 PI = 3.14159
127  HOME
150  PRINT "ENTER FREQUENCY IN MHZ"
160  INPUT F
170 W = 2 * 3.14159E6 * F
180  PRINT "ENTER RELATIVE DIELECTRIC CONSTANT"
190  PRINT "OF THE REFLECTING MEDIUM"
200  INPUT EG
210 ER = 8.84195E - 12 * EG
220  PRINT "ENTER CONDUCTIVITY OF MEDIUM"
230  PRINT "IN MHOS PER METER"
240  INPUT CD
245  DIM X(90,2)
258  REM ))))))))))))))))))))))))))))))))))))))
260  REM  THE CALCULATION FOR THE COEFFICIENT
264  REM ----------------------------------------
265 D2 = W * ER
270 JN =  - CD / D2: REM  THE IMAG PART OF NUMERATOR.
280  REM   JD=0 FOR AIR
290 NM =  SQR ((EG * EG) + (JN * JN))
300 PH =  ATN (JN / EG)
301  HOME : PR# 1
302  PRINT "AT  ";F;"  MHZ
303  PRINT "FOR DIELECTRIC CONSTANT = ";EG
304  PRINT "AND CONDUCTIVITY ";CD;" MHOS PER METER";
322  PRINT ""
324  PR# 0
330  REM ----------------------------------------
340  REM  VERTICAL POLARIZATION
345  REM ----------------------------------------
346 TI = PI / 180
347  FOR I = 1 TO 90
350 C1 =  COS (TI):C2 =  SIN (TI)
360 C3 = C2 * C2
362 C4 = EG - C3
364 C5 =  SQR ((C4 * C4) + (JN * JN))
365 C5 =  SQR (C5)
366 T3 = .5 *  ATN (JN / C4)
370 C6 = C5 *  COS (T3)
380 C7 = C5 *  SIN (T3)
385  REM  THE VALUE OF THE RADICAL
386 C8 = C1 * NM
387 C9 = C8 *  COS (PH)
388 CA = C8 *  SIN (PH)
390 CR = C9 - C6
392 CQ = CA - C7
394 CP =  SQR ((CR * CR) + (CQ * CQ))
396 T4 =  ATN (CQ / CR)
398  IF CQ < 0 AND CR < 0 THEN T4 = T4 - PI
400 CO = C9 + C6
402 CN = CA + C7
404 CM =  SQR ((CO * CO) + (CN * CN))
406 T5 =  ATN (CN / CO)
407  IF CN < 0 AND CO < 0 THEN T5 = T5 - PI
408  REM  THE DENOMINATOR
410 CL = CP / CM
415 T6 = T4 - T5
```

Program 19.3

```
500  REM   ::::::::::::::::::::::::::::::::::::::::::::::
510  REM       THE HALF-WAVE DIPOLE PATTERN
520  REM   ::::::::::::::::::::::::::::::::::::::::::::::
525 T7 = TI
530  DN =  SIN (T7)
540   IF DN = 0 THEN DN = .0001
550 E0 = ( COS ((PI / 2) *  COS (T7))) / DN
560  REM   ))))))))))))))))))))))))))))))))))))
570  REM       THE SPACE PHASE ANGLE
580  REM   ((((((((((((((((((((((((((((((((((((
600 T8 = PI *  SIN (T7)
610 E1 = E0 * CL
620 T9 = T6 + T8
630 ER =  - (E1 *  COS (T9)) + E0
640 EJ = (E1 *  SIN (T9))
650 ET =  SQR ((ER * ER) + (EJ * EJ))
655   IF ET > EM THEN EM = ET
660 X(I,1) = TI
665 X(I,2) = ET
667  PRINT ET,TI * 180 / PI
670 TI = TI + PI / 180
690  NEXT I
700  REM   ::::::::::::::::::::::::::::::::::::::::::::::
710  REM       THE GRAPHICS ROUTINE
720  REM   ::::::::::::::::::::::::::::::::::::::::::::::
740 X0 = 0:Y0 = 190
750  HGR2
760  HCOLOR= 3
770  POKE  - 12524,0
780  POKE  - 12525,64
790  POKE  - 12529,255
800  FOR I = 1 TO 90
810 X = 190 * (X(I,2) / EM) *  SIN (X(I,1))
820 Y = 190 - (X(I,2) / EM) *  COS (X(I,1)) * 190
830  IF I = 1 THEN X0 = X:Y0 = Y
840  HPLOT X0,Y0 TO X,Y
850 X0 = X:Y0 = Y
860  NEXT I
870  HPLOT 0,0 TO 0,190
880  HPLOT 0,190 TO 190,190
890 A = PI / 18
900 X1 = 190 *  COS (A)
910 X2 = 185 *  COS (A)
920 Y1 = 190 - 190 *  SIN (A)
930 Y2 = 190 - 185 *  SIN (A)
940  HPLOT X1,Y1 TO X2,Y2
950 A = A + PI / 18
960  IF A < PI / 2 GOTO 900
2000  END
```

Program 19.3 (*continued*)

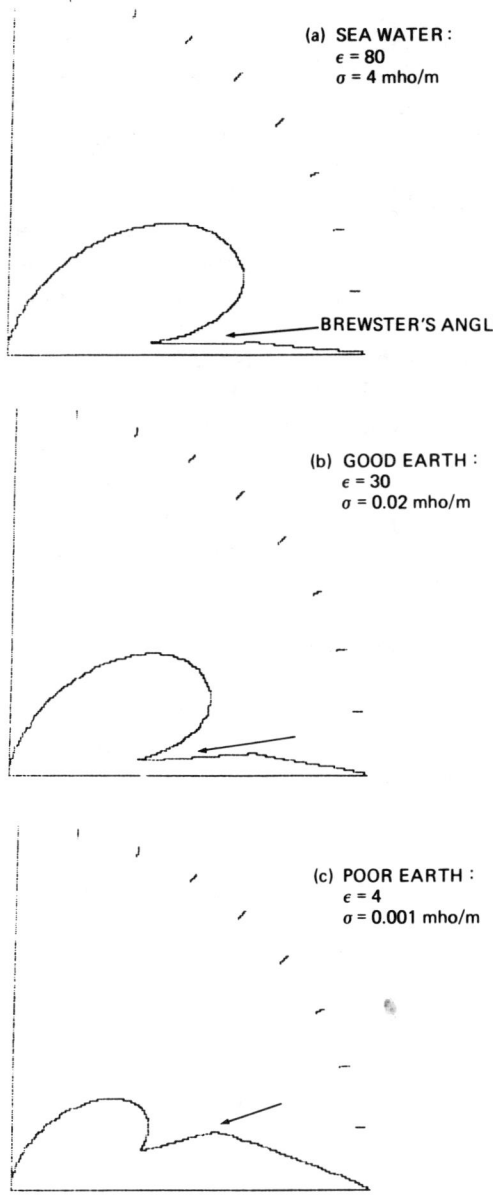

Fig. 19.12 E-plane radiation patterns at a height of $\lambda/2$ over various surfaces.

effectiveness, the counterpoise should consist of 120 or more radials, each a half wave long. In 1934, Feldman, Friis, and Sharpless showed that a vertical half-wave antenna demonstrates a consistent gain advantage of 4-dB over a vertical quarter-wave antenna despite the fact that the calculated directivity advantage is only 1.6-dB.[2] They attributed the difference to losses in the counterpoise.

Program 19.3 computes the pattern for a vertical half-wave antenna above earth surfaces of various properties. In operation, it differs from the horizontal antenna program in that it completes all of its calculations before starting to plot. This variation is not necessary, but it allows us to show a different way to use a FOR-NEXT routine.

The patterns of Fig. 19.12 are relative to one another. The frequency selected is higher than that for the horizontal antenna because long-haul operations usually require a higher frequency. It may be seen that the ground reflection significantly reshapes the low angles, most noticeably at the Brewster's angle notch. The data should be taken with a grain of salt at angles below about 5-degrees. The groundwave propagation effect in that region will considerably alter the shape of the pattern. It may be seen that the "poor earth" case is particularly hurt by poor soil conductivity. Furthermore, even seawater has a significant effect; the seawater pattern shown is noticeably different from the half circle usually exhibited by an antenna above a perfectly conducting groundplane.

Both this program and Program 19.2 can serve as models for programs for other antenna arrangements.

---

[2]"The Determination of the Angle of Arrival of Short Radio Waves," *Proc. IRE*, Jan. 1934, pp. 47, 48.

Index

Aether, 5, 48
Airey rings, 99
Amplitude monopulse, 113
Annular slot (flush) antenna, 288
Antenna coupler, 250, 275
Aperture taper, 52
Armstrong, Edwin, 10
Arrays; even, 48
    odd, 47
AS-1729, 274
ATN (Basic), 17

Base spring, 271
Baseband video, 176
Base-loaded antenna, 254
BASIC, 17
Bose, J. C., 7
Branley, E., 7
Brewster, David, 6
Brewster's Angle, 299
Brueckmann, H., 274

Capacitance-whip antenna, 265
Cardioid pattern, 126
Cartesian coordinates, 14
Cassagranian telescope, 79
Choke dipole, 262
Circuits; distributed, 179
    lumped, 179
Cloverleaf antenna, 262
Coherer, 7
Collimator, 70
Coma, 99
Complex notation, 11
Conical scan, 109
Conical transmission line, 220
Coordinate transformation, 20
Correlation interferometer, 157
Cosecant squared pattern, 130
Cosine integral, 241
Coulomb, C. A., 4
Coulomb's law, 4

De Forest, Lee, 8
Descartes, René, 15
DFT (Discrete Fourier Transform), 133
Dielectrics, 2
    imperfect, 302
Diffraction, 296
Dipole, 209, 223
    elementary, 213
Dipole antenna, 223
Dipole field, 211
Dipole patterns, 232
Distributed circuits, 179
Downlead, 248

Edison, T. A., 8
Edison effect, 8
Ellipse, 76
Endfire antenna, 61
Eulers equation, 15, 63

Fan antenna, 287
Far-field zone, 35
Faraday, Michael, 4
Feldman, C. B., 304
FFT/IFT, 135
Field, Cyrus, 4, 161
Flush antenna, 289
Focus, 89, 99
Fourier, J. B. J., 132
Franklin, Benjamin, 2, 211
Fraunhofer region, 35
Fresnel region, 35
Friis transmission formula, 25

Gain, 25
    rule of thumb for, 28
Great circle calculations, 22
Gregorian telescope, 5, 78
Ground independent antennas, 274

Hairpin antenna, 263
Heaviside, Oliver, 175
Helacyl II antenna, 256

Helmholtz, Heinrich, 7
Henry, Joseph, 4
Hertz, Rudolph Heinrich, 7
Hooke, Robert, 5
Hula hoop antenna, 255
Huygens, Christian, 30
Huygens wavelets, 30
Hyperbola, 76

Imaginary components, 14
Integration by computer, 18
Interference, wave, 48, 147
Interferometer, 48, 147
Isotropic radiator, 26

j (the imaginary operator), 13
Joule, James Prescott, 3

Kelvin, Lord (William Thompson), 5
Kohlrauch, 6, 302

Line length, measurement of, 206
Light; velocity of, 5
  nature of, 5
Loop antenna, 258

Marconi, Guglielmo, 7
Maxwell, James Clerk, 3
Monopulse, 112
Morse, Samuel F. B., 4

Near-field zone, 35
Newton, Sir Isaac, 6
Newton's rings, 48
Newtonian telescope, 73

Oak Beam Test, 270
Objective mirror, 73

Paraboloid, 73
Pascal, Blaise, 5
Pattern formation, 32, 38
Pattern integration, 27
Phase angle, 39
Phase errors, 119
Phase monopulse, 119
Phase steering, 58
Plunging, 101

Quadratic error, 90
Quasi-static case, 216

Radar range, 130
Radiation resistance, 217
Rectangular-to-polar transformation, 16
Rhombic antenna, 234
Ring array, 65
Roemer, Ole, 5

Shorted line, 181
Sidelobes, 52
Silentype (Apple), 46
Smith, P. H., 187
Smith, Chart construction, 201
Snell's law of refraction, 297
Spherical aberration, 71
Squint, 92
Stacked-beam antenna, 106
Standing wave, 176, 224
Steinmetz, Charles Proteus, 11
Surge impedance, 179

Telegraphers equation, 162
Thompson, William (Lord Kelvin), 5, 161
Traveling wave, 161
Traveling-wave antenna, 231
Tribble, 27

Unfilled arrays, 147

Vectors, 11, 13, 43
Vector mathematics, 13, 15
Vee antenna, 234
Volta, Allesandro, 4
Voltage standing-wave ratio (VSWR), 180
VOR (Very High-frequency Omnirange), 127

Weber, Wilhelm, 6, 302
Wheeler, Harold, 247
Whip antenna impedance table, 278
Wullenweber array, 65

Young, Thomas, 6, 48
Young's interferometer, 48, 147